Complete PCB Design Using OrCAD® Capture and PCB Editor

Scan the below QR codes to access the book companion site and
to download the OrCAD Trial version.

Complete PCB Design Using OrCAD® Capture and PCB Editor

Second Edition

KRAIG MITZNER

Independent Consultant, Washington, United States

BOB DOE

Parallel Systems Ltd, Bracknell, United Kingdom

ALEXANDER AKULIN

Co-founder of PCB Soft Ltd., PCB Technology, Moscow, Russia

ANTON SUPONIN

PCB Soft, Moscow, Russia

DIRK MÜLLER

FlowCAD, Feldkirchen, Germany

ELSEVIER

ACADEMIC PRESS

An imprint of Elsevier

Academic Press is an imprint of Elsevier
125 London Wall, London EC2Y 5AS, United Kingdom
525 B Street, Suite 1650, San Diego, CA 92101, United States
50 Hampshire Street, 5th Floor, Cambridge, MA 02139, United States
The Boulevard, Langford Lane, Kidlington, Oxford OX5 1GB, United Kingdom

Notices
Knowledge and best practice in this field are constantly changing. As new research and experience broaden our understanding, changes in research methods, professional practices, or medical treatment may become necessary.

Practitioners and researchers must always rely on their own experience and knowledge in evaluating and using any information, methods, compounds, or experiments described herein. In using such information or methods they should be mindful of their own safety and the safety of others, including parties for whom they have a professional responsibility.

To the fullest extent of the law, neither the Publisher nor the authors, contributors, or editors, assume any liability for any injury and/or damage to persons or property as a matter of products liability, negligence or otherwise, or from any use or operation of any methods, products, instructions, or ideas contained in the material herein.

British Library Cataloguing-in-Publication Data
A catalogue record for this book is available from the British Library

Library of Congress Cataloging-in-Publication Data
A catalog record for this book is available from the Library of Congress

ISBN: 978-0-12-817684-9

For Information on all Academic Press publications
visit our website at https://www.elsevier.com/books-and-journals

Publisher: Mara Conner
Acquisition Editor: Tim Pitts
Editorial Project Manager: Ana Claudia A. Garcia
Production Project Manager: Nirmala Arumugam
Cover Designer: Alan Studholme

Typeset by MPS Limited, Chennai, India

Working together
to grow libraries in
developing countries

www.elsevier.com • www.bookaid.org

Contents

Acknowledgments

The authors would like to thank Cadence for the use of the software and the technical assistance during the development of this book.

The authors would also like to thank their families and friends for their support and encouragement without which this book would not have been completed.

And last, but not the least, a hearty thank you goes to the readers of the first book. Your enthusiasm and feedback provided the motivation to create the second edition of this book.

Introduction

The OrCAD PCB design solutions provide a comprehensive suite of tools for design entry, using OrCAD Capture, mixed-signal circuit simulation using OrCAD PSpice, and physical layout using OrCAD PCB Editor for mainstream electronic circuit designs.

OrCAD PCB Editor is based on Allegro PCB Editor, so this book will be useful to new Allegro PCB Editor users as well. Allegro PCB Editor is a powerful, full-featured design tool. While OrCAD PCB Editor has inherited many of those features, including a common file format, it does not possess all of the capabilities available to the Allegro PCB tiers, such as Allegro High-Speed Option, Analog/RF Option, FPGA System Planner, Design Planning, and Miniaturization Option. Consequently most of the basic tools and features are described here, but only a few of the more-advanced tools are covered, as outlined later.

Chapter 1, Introduction to printed circuit board design and computer-aided design, introduces the reader to the basics of printed circuit board (PCB) design. The chapter begins by introducing the concepts of computer-aided engineering, computer-aided design (CAD), and computer-aided manufacturing. The chapter then explains how these tools are used to design and manufacture multilayer PCBs. Many 3D pictures are used to show the construction of PCBs. Topics such as PCB cores and layer stack-up, apertures, D-codes, photolithography, layer registration, plated through holes, and Gerber files are explained.

Chapter 2, Introduction to the printed circuit board design flow by example, leads new users of the software through a very simple design example. The purpose of the example is to paint a "big picture" of the design flow process. The example begins with a blank schematic page and ends with the Gerber files. The circuit is ridiculously simple, so that it is not a distraction to understanding the process itself. Along the way, some of PCB Editor's routing tools are briefly introduced along with some of the other tools, which sets the stage for Chapter 3, Project structures and the PCB Editor tool set.

Chapter 3, Project structures and the PCB Editor tool set, provides an overview of the OrCAD project files and structure and explains PCB Editor's tool set in detail. The chapter revisits and explains some of the actions performed and tools used during the example in Chapter 2, Introduction to the printed circuit board design flow by example. Gerber files are also explained in detail.

Chapter 4, Introduction to industry standards, introduces some of the industry standards organizations related to the design and fabrication of PCBs (e.g., IPC and

JEDEC). PCB performance classes and producibility levels are also described, along with the basic ideas behind standard fabrication allowances. These concepts are described here to help the reader realize some of the fabrication issues up front to help minimize board failures and identify some of the guides and standards resources that are available for PCB design.

Chapter 5, Introduction to design for manufacturing, addresses the mechanical aspect of PCB design—design for manufacturability. The chapter explains where parts should be placed on the board, how far apart, and in what orientation from a manufacturing perspective. OrCAD PCB Editor's design rule checker is then considered relative to the manufacturing concepts and IPC's courtyard concepts. To aid in understanding the design issues, manufacturing processes such as reflow and wave soldering, pick-and-place assembly, and thermal management are discussed. The information is then used as a guide in designing plated through holes, surface-mount lands, and PCB Editor footprints in general. Tables summarize the information and serve as a design guide during footprint design and PCB layout.

Chapter 6, Printed circuit board design for signal integrity, addresses the electrical aspect of PCB design. Several good references are available on signal integrity, electromagnetic interference, and electromagnetic compatibility. This chapter provides an overview of those topics and applies them directly to PCB design. Topics such as loop inductance, ground bounce, ground planes, characteristic impedance, reflections, and ringing are discussed. The idea of "the unseen schematic" (the PCB layout) and its role in circuit operation on the PCB is introduced. Look-up tables and equations are provided to determine required trace widths for current handling and impedance as well as required trace spacing for high-voltage designs and high-frequency designs. Various layer stack-up topographies for analog, digital, and mixed-signal applications are also described. The design examples in Chapter 9, Printed circuit board design examples, demonstrate how to apply the layer stack-ups described in this chapter. A demonstration on how to use PSpice to simulate transmission lines to aid in circuit design and PCB layout is also provided.

Chapter 7, Making and editing capture parts, explains how to construct Capture parts using the Capture Library Manager and Part Editor and the PSpice Model Editor. Heterogeneous and homogeneous parts are developed in examples using four methods. Different methods are used depending on whether a part will be used for simple schematic entry, design projects intended for PCB layout, PSpice simulations, or all of these. The chapter also demonstrates how to attach PSpice models to Capture's schematic parts using PSpice models downloaded from the Internet and basic PSpice models developed from functional Capture projects. The Capture parts can then be used for both PSpice simulations and PCB layout as demonstrated in Chapter 9, Printed circuit board design examples.

Detailed coverage of padstacks and footprints is covered in Chapter 8, Making and editing footprints. The chapter begins with an overview of PCB Editor's symbols library, describes the various types of symbols, and explains the anatomy of a footprint. Then a detailed description of the padstack (as it relates to PCB manufacturing described in Chapters 1, Introduction to printed circuit board design and computer-aided design, and 5, Introduction to design for manufacturing) is given, as it is the foundation of both footprint design and PCB routing. Design examples are provided to demonstrate how to design discrete through-hole and surface-mount devices and how to use the footprint design wizard. The OrCAD Library Builder is also introduced in this chapter.

Chapter 9, Printed circuit board design examples, provides four PCB design examples that use the material covered in the previous chapters. The first example is a simple analog design using a single op amp. The design shows how to set up multiple plane layers for positive and negative power supplies and ground. The design also demonstrates several key concepts in Capture, such as how to connect global nets, how to assign footprints, how to perform design rule checks, how to use the Capture part libraries, how to generate a bill of materials (BOM), and how to use the BOM as an aid in the design process in Capture and PCB Editor. The design also shows how to perform important tasks in PCB Editor, such as how to set up a board outline, place parts, and modify padstacks. Intertool communication (such as annotation and back annotation) between Capture and PCB Editor is also demonstrated. The second design is a mixed digital/analog circuit. In addition to the tasks demonstrated in the first example, the design also demonstrates how to set up and use split planes to isolate analog and digital power supplies and grounds. Other tasks include using copper areas on routing layers to make partial ground planes, setting up split power and ground planes, and defining anticopper areas on plane and routing layers. The third example uses the same mixed digital/analog circuit from the second example but demonstrates how to use multiple-page schematics and off-page connectors to add PSpice simulations to a Capture project used for PCB layout, all within a single project design. It also demonstrates how to construct multiple, separated power and ground planes and a shield plane to completely isolate analog from digital circuitry. The use of guard rings and guard traces is also demonstrated.

The fourth example is a high-speed digital design, which demonstrates how to design transmission lines, stitch multilayer ground planes, perform pin/gate swapping, place moated ground areas for clock circuitry, and design a heat spreader.

The last part of Chapter 9, Printed circuit board design examples, includes a short discussion about the differences between using negative and positive planes in PCB design.

Chapter 10, Artwork development and board fabrication, describes taking the PCB design from the CAD stage through fabrication. A simple design example shows how to produce the artwork (Gerber) files for a PCB design. PCB Editor is then used to review the artwork files before they are sent to a manufacturer. PCB Editor is also used to generate a drawing (DXF - Drawing eXchange Format) file that can be opened and edited with many drawing applications, so they can be 3D modeled to review form, fit, and function. The chapter also describes how to create a custom report that can be used for pick-and-place machines during the assembly process.

Chapter 11, Component information system, introduces OrCAD Capture component information system (CIS) as a solution to manage part properties (including part information required at each step in the PCB design process, from implementation through manufacturing) within your schematic designs. It explains what CIS means, how it should be configured using a database, an open database connectivity driver and a database configuration file for the transfer of the different properties. This chapter also describes which PSpice properties must be defined and how to use the same schematic design for PCB layout and simulation in the same flow. Finally it shows how to create variants and CIS BOM.

Chapter 12, Signal integrity simulation with OrCAD, gives a brief description of OrCAD Signal Explorer, OrCAD PCB SI, and Sigrity ERC tools which are dedicated to help locating and solving the signal integrity problems, described deeply in Chapter 6, Printed circuit board design for signal integrity.

Ancillaries for this book are available at book companion site that can be found on www.elsevier.com/books/complete-pcb-design-using-orcad-capture-and-pcb-editor/mitzner/978-0-12-817684-9. At the website for this book, you should find supplementary content including the design files and component libraries used in writing this book. The design files include the design flow example from Chapter 2, Introduction to the printed circuit board design flow by example, the Capture parts from Chapter 7, Making and editing capture parts, the footprint and padstack symbols from Chapter 8, Making and editing footprints, the board layout files from Chapter 9, Printed circuit board design examples, and the manufacturing files from Chapter 10, Artwork development and board fabrication. The newest version of the OrCAD demo can be obtained from the cadence www.orcad.com website, look for the FREE TRIAL link to request and download it. You will need the 64-bit Windows™ operating system to run OrCAD 17.2. Be sure to make your computer and domain name in latin alphabet to allow the smooth OrCAD Trial installation.

During the development of this text the authors received technical assistance from Cadence, so many of the procedures and concepts were written from lessons learned and feedback from Cadence. While every effort has been made to ensure technical accuracy of the material, it is possible that some of the steps outlined in the design examples could be done more efficiently or in other ways. Please work closely with your board manufacturer during the procurement and fabrication process to ensure that the end product will be what you intended.

CHAPTER 1

Introduction to printed circuit board design and computer-aided design

Contents

Computer-aided design and the OrCAD design suite

Before digging into the details of PCB Editor, we take a moment to discuss computer–aided engineering (CAE) tools in general. CAE tools cover all aspects of engineering design from drawings to analysis to manufacturing. Computer-aided design (CAD) is a category of CAE related to the physical layout and drawing development of a system design. CAD programs specific to the electronics industry are known as electronic CAD or electronic design automation (EDA). EDA tools reduce development time and cost because they allow designs to be simulated and analyzed prior to purchasing and manufacturing hardware. Once a design has been proven through drawings, simulations, and analysis, the system can be manufactured. Applications used in manufacturing are known as computer-aided manufacturing (CAM) tools. CAM tools use software programs and design data (generated by the CAE tools) to control automated manufacturing machinery to turn a design concept into reality.

So how does OrCAD®/Cadence fit into all of this? Cadence owns and manages many types of CAD/CAM products related to the electronics industry, including the OrCAD design suite. The OrCAD design suite can be purchased through resellers

Complete PCB Design Using OrCAD® Capture and PCB Editor
DOI: https://doi.org/10.1016/B978-0-12-817684-9.00001-1

(a list of resellers can be found on the OrCAD.com website), which packages different combinations of CAD/CAM applications, including Capture, PSpice, and PCB Editor, to suit customers' needs. Although these applications can operate individually, bundling the individual tools into one suite allows for intertool communication.

Capture is the centerpiece of the package and acts as the prime EDA tool. Capture contains extensive parts libraries that may be used to generate schematics that stand alone or interact with PSpice, PCB Editor, or both simultaneously. A representation of a Capture part is shown in Fig. 1.1.

The pins on a Capture part can be mapped into the pins of a PSpice model or the pins of a physical package in PCB Editor. PSpice is a CAE tool that contains the mathematical models for performing simulations, and PCB Editor is a CAD tool that converts a symbolic schematic diagram into a physical representation of the design. Netlists are used to interconnect parts within a design and connect each of the parts with its model and footprint. In addition to being a CAD tool, PCB Editor also functions as a front-end CAM tool by generating the data on which other CAM tools operate when manufacturing the PCB. Combining all three applications into one package produces a powerful set of tools to efficiently design, test, and build electronic circuits. The key to successful project design and production is in understanding the PCB itself and knowing how to use the tools that build the PCB.

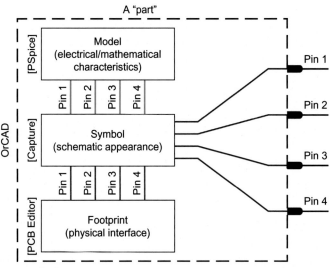

Figure 1.1 The pieces of a "part."

Printed circuit board fabrication

We now look at how PCBs are manufactured for a better understanding of what we are trying to accomplish with PCB Editor and why. A PCB consists of two basic parts: a substrate (the board) and printed wires (the copper traces). The substrate provides a structure that physically holds the circuit components and printed wires in place and provides electrical insulation between conductive parts. A common type of substrate is FR4, which is a fiberglass/epoxy laminate. It is similar to older types of fiberglass boards but is flame resistant. Substrates are also made from Teflon, ceramics, and special polymers.

Printed circuit board cores and layer stack-up

During manufacturing the PCB starts out as a copper clad substrate as shown in Fig. 1.2.

A rigid substrate is a C-stage laminate (fully cured epoxy). The copper cladding may be copper-plated onto the substrate or copper foil glued to the substrate. The thickness of the copper is measured in ounces of copper per square foot, where 1.0 oz/ft^2 of copper is approximately 1.2−1.4 mil (0.0012−0.0014 in. / 0.03−0.035 mm) thick. It is common to drop "/ft^2" and refer to the thickness only in oz. For example, you can order 1 oz copper on a 0.125-in.-thick FR4 substrate.

A substrate can have copper on one or both sides. Multilayer boards are made up of one or more single- or double-sided substrates called *cores*. A core is a copper–plated epoxy laminate. The cores are glued together with one or more sheets of a partially cured epoxy, as shown in Fig. 1.3.

The sheets are also referred to as *prepreg* or *B-stage laminate*. Once all of the cores are patterned (described next) and aligned, the entire assembly is fully cured in a heated press.

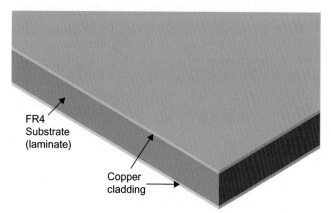

Figure 1.2 A double-sided copper clad FR4 substrate.

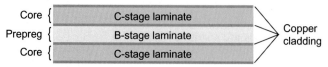

Figure 1.3 Cores and prepreg.

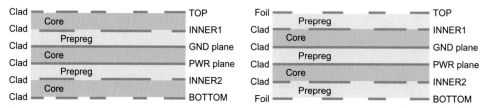

Figure 1.4 Two stack-up methods for a six-layer board: left, multicore, outer clad; right, multicore, outer foil.

There are three methods of assembling the cores when making a multilayer board. Fig. 1.4 shows the first two methods in an example with four routing layers and two plane layers. Fig. 1.4 (left) shows three (double-sided) cores bonded together by two prepreg layers, while Fig. 1.4 (right) shows the same six layers made of two cores, which make up the four inner layers, bonded together by one prepreg layer. The outer layers in this panel are copper foil sheets bonded to the assembly with prepreg.

The routing layers in Fig. 1.4 are shown as patterned copper segments, and the plane layers are shown as solid lines. The inner layers are patterned prior to bonding the cores together. The outer layers are patterned later in the process, after the cores have been bonded and cured and most of the holes have been drilled. Because the outer layers are etched later and copper foil is typically less expensive than copper cladding, the stack-up shown in Fig. 1.4 (right) is more widely used.

The third method uses several fabrication techniques by which highly complex boards can be fabricated, as illustrated in Fig. 1.5. This circuit board may have a typical four-layer core stack-up at its center, but additional layers are built up layer by layer on the top and the bottom, using sequential lamination techniques. The techniques can be used to produce blind and buried vias as well as typical plated through–hole vias, nonplated holes, and back-drilled plated through holes. Resistors and capacitors can also be embedded into the substrate. More about blind vias is discussed in later chapters (Chapter 8: Making and editing footprints and Chapter 9: Printed circuit board design examples).

Printed circuit board fabrication process

The copper traces and pads seen on a PCB are produced by selectively removing the copper cladding and foil. Two methods are commonly used for removing the

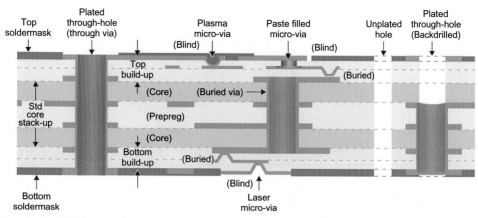

Figure 1.5 A built-up, multitechnology, PCB stack-up. *PCB*, Printed circuit board.

unwanted copper: wet acid etching and mechanical milling. Acid etching is more common when manufacturing large quantities of boards because many boards can be made simultaneously. One drawback to wet etching is that the chemicals are hazardous and must be replenished occasionally, and the depleted chemicals must be recycled or discarded. Milling is usually used for smaller production runs and prototype boards. During milling the traces and pads are formed by a rotating bit that grinds the unwanted copper from the substrate. With either method a digital map is made of the copper patterns. The purpose of CAD software like OrCAD PCB Editor is to generate the digital maps.

Note: Only one layer is considered in the following explanation of the fabrication process.

Photolithography and chemical etching

Selectively removing the copper with etching processes requires etching the unwanted copper while protecting the wanted copper from the etchant. This protection is provided by a polymer coating (called *photoresist*) deposited onto the surface of the copper cladding, as shown in Fig. 1.6.

The photoresist is patterned into the shape of the desired printed circuit through a process called *photolithography*. The patterned resist protects selected areas of the copper from the etchant and exposes the copper to be etched.

The two steps to photolithography are patterning the photoresist and developing it. Patterning is accomplished by exposing the resist to light [typically ultraviolet (UV)], and developing it is accomplished by washing it in a chemical bath. The two types of photoresist are positive resist and negative resist. When positive resist is exposed to UV light, the polymer breaks down and can be removed from the copper. Conversely, negative resist shielded from UV light is removed.

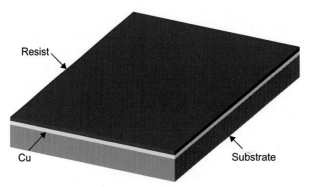

Figure 1.6 A copper clad board coated with photoresist.

A mask is used to expose the desired part of the photoresist. A mask is a specialized black-and-white photographic film or glass photoplate on which a picture of the traces and pads is printed with a laser photoplotter. Two types of masks are shown in Fig. 1.7.

The masks are examples of a trace connected to a pad. Fig. 1.7 (left) shows a positive mask used to expose positive photoresist, and Fig. 1.7 (right) shows a negative mask used to expose negative photoresist. Masks that will be used repeatedly are sometimes produced on glass photoplates instead of film.

The mask is placed on top of the photoresist, as shown in Fig. 1.8, and the assembly is exposed to the UV light. The dark areas block UV light and the white (transparent) areas allow the UV light to hit the photoresist, which imprints the circuit image into the photoresist. A separate mask is used for each layer of a circuit board. OrCAD PCB Editor generates the data that the photoplotter uses to make these masks.

Another way of exposing the photoresist is by using a programmable laser to "draw" the pattern directly onto the photoresist. This is a newer technique, called *laser direct imaging* (LDI). A benefit of the LDI process is that it uses the same data as the photoplotters but no masks are required.

After the photoresist has been exposed (either with the mask and UV or with the laser), it is washed in a chemical called the *developer*. In the case of positive resist the resist breaks down during exposure and is removed by the developer. In the case of negative resist the UV light cures the resist, and only the unexposed resist is removed by the developer. Common developers are sodium hydroxide (NaOH) for positive resist and sodium carbonate (Na$_2$CO$_3$) for negative resist. Once the resist has been exposed and developed, a circuit image made of the photoresist is left on the copper, as shown in Fig. 1.9.

Next, the board is etched in a corrosive solution, such as alkaline ammonia or cupric chloride. The etching solution does not significantly affect the photoresist but

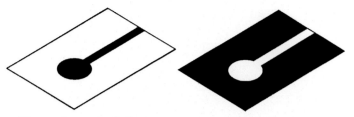

Figure 1.7 Photolithography masks: (left) positive mask and (right) negative mask.

Figure 1.8 Positive photomask on photoresist-coated board.

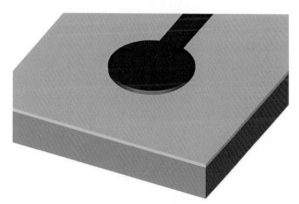

Figure 1.9 Developed photoresist on copper.

attacks the bare copper and removes it from the substrate, leaving behind the resist–coated copper, as shown in Fig. 1.10.

Some processes use a plated tin alloy as the etch resist. The tin alloy plating is more resistant to etchants and preprepares the copper surface for solder processes.

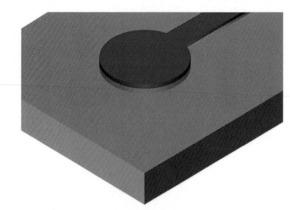

Figure 1.10 Unwanted copper removed by etching.

In this case, photolithography processes are used to selectively plate the circuit pattern onto the copper surfaces prior to etching.

When polymer etch resists are used, the photoresist is cleaned from the copper with a resist stripper, leaving behind the copper traces. Fig. 1.11 shows the final patterned copper. When metal etch resists are used, the plating is typically left in place. Holes for the leads and so forth are not etched into the pads because they are drilled after all of the cores have been glued together (later in the process) to ensure proper alignment of the holes between board layers.

Mechanical milling

As mentioned previously, milling is an alternative to etching. To mill the board a computer numerical control (CNC) machine is programmed with the digital map of the board and grinds away the unwanted copper. The unwanted copper can be completely removed (like that in Fig. 1.11), or just enough copper may be removed to isolate the pads and traces from the bulk copper, as shown in Fig. 1.12. Removing only enough copper to isolate the traces from the bulk copper reduces milling time but can affect the impedance of the traces.

Layer registration

After the inner layers have been patterned, the cores are aligned (called *registration*) and glued together. Registration is critical because the pads on each layer need to be properly aligned when the holes are drilled. Registration is accomplished using alignment patterns (called *fiducials*) and tooling holes in the board, which slide onto guide pins. With the cores in place and properly aligned a heated press cures the assembly.

After the assembly is cured, holes are drilled for through-hole component leads and vias. The drilling process inevitably heats the laminate due to friction between the

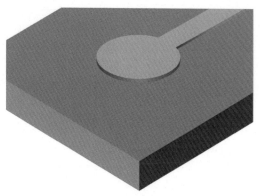

Figure 1.11 Copper pad and trace after etching and resist stripping.

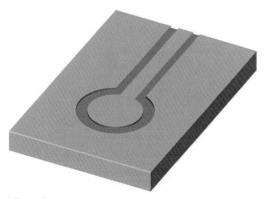

Figure 1.12 A mechanically milled trace.

laminate and the high-speed drill bit. This tends to soften the laminate and smear it across the walls of the drilled copper. After the drilling processes are finished the assembly is placed into a bath to etch back the laminate slightly and clean the faces of the copper pad walls. This is called *laminate etchback* or *desmear*.

Once the holes have been drilled and desmeared, a physical path exists between pads on different layers, but as Fig. 1.13 (left) shows, there is no electrical connection between them. To make electrical connections between pads on different layers the board is placed into a plating bath that coats the insides of the holes with copper, which electrically connects the pads, hence the term *plated* through holes. The plating thickness varies but is typically about 1 mil (0.001 in.) thick. The cutaway view of Fig. 1.13 (right) shows a plated through hole on an internal layer of a PCB. The top and bottom copper is actually patterned *after* the plating process is finished because the plating process would replate the areas where copper had been removed.

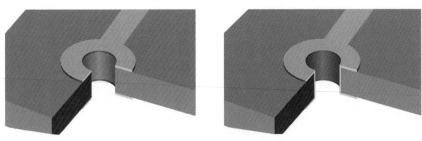

Figure 1.13 Holes are drilled into the board and copper plated: (left) a nonplated through hole and (right) a plated through hole.

Not all layers have traces. Some layers are planes (Fig. 1.14). Plane layers are typically used to provide low-impedance (resistance and inductance) connections to power and ground and to provide easy access to power and ground at any location on the board. Plane layers and impedance issues are discussed in Chapter 6, Printed circuit board design for signal integrity. The leads of components are connected to ground or power by soldering them into plated through holes. Since copper conducts heat well, soldering to a plane layer could require an excessive amount of heat, which could damage the components or the plating in the hole (called the *barrel*). Thermal reliefs are used, as shown in Fig. 1.14, to reduce the path for heat conduction but maintain electrical continuity with the plane.

Since ground and power planes are often inner layers and signal layers will likely be above and below them, in some instances, a via will run through a plane layer but must not touch it. In this case a "clearance" area (shown in Fig. 1.15) is etched into the plane layer around the via to prevent a connection to the plane. The clearance is larger than the normal pad size to ensure that the plane stays isolated from the plated hole.

After the through holes are plated the top and bottom layers are patterned using the photolithography process as described for the inner layers. After the outer copper has been patterned the exposed traces and plated through holes can be tinned (although tinning is sometimes deferred until later). Nonplated holes (such as for mounting holes) may be drilled at this time.

Next, a thin polymer layer is usually applied to the top and bottom of the board. This layer [shown in *dark green* (dark gray in print version) in Fig. 1.16] is called the *soldermask* or *solder resist*. Holes are opened into the polymer using photolithography to expose the pads and holes where components will be soldered to the board. The soldermask protects the top and bottom copper from oxidation and helps prevent solder bridges from forming between closely spaced pads. Sometimes openings in the soldermask are not made over small or densely placed vias (called *tenting a via*). Tented vias are protected from having chemicals such as flux from becoming trapped inside

Figure 1.14 A connection to a plane layer through a thermal relief.

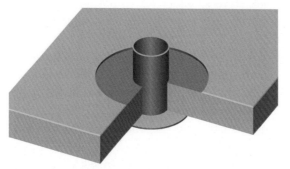

Figure 1.15 A clearance area provides isolation between a plated hole and a plane.

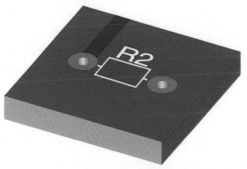

Figure 1.16 Final layers are the soldermask [*dark green* (dark gray in print version)] and silk screen (*white*).

the hole. Tenting also prevents solder migration into the hole, which could lead to poor solder joints on small components that are close to and connected to the via.

Finally, markings (called the *silk screen*) are placed on the board to identify where the components are to be placed. The silk screen is shown in *white* in Fig. 1.16.

Function of OrCAD PCB Editor in the printed circuit board design process

PCB Editor is used to design the PCB by generating a digital description of the board layers for photoplotters and CNC machines, which are used to manufacture the boards. Separate layers are used for routing copper traces on the top, bottom, and all inner layers; drill hole sizes and locations; soldermasks; silk screens; solder paste; part placement; and board dimensions. These layers are not all portrayed identically in PCB Editor. Some of the layers are shown from a positive perspective, meaning what you see with the software is what is *placed onto* the board, while other layers are shown from a negative perspective, meaning what you see with the software is what is *removed from* the board. The layers represented in the positive view are the board outline, routed copper, silk screens, solder paste, and assembly information. The layers represented in the negative view are drill holes and soldermasks. Copper plane layers are handled in a special way, as described next.

Fig. 1.17 shows routed layers (top and bottom and an inner, for example) that PCB Editor shows in the positive perspective. The background is black and the traces and pads on each layer are a different color to make it easier to keep track of visually. The drill holes are not shown because, as mentioned already, the drilling process is a distinct step performed at a specific time during the manufacturing process.

Fig. 1.18 shows examples of drill symbols and silk screen representations used by PCB Editor (traces not shown). The *red shapes* (gray in print version) are examples of drill symbols that indicate locations of the drill holes and are used in conjunction with a drill chart (Fig. 1.19), which produces ID numbers for the different drill tools. The white print is the silk screen discussed previously.

Fig. 1.20A shows the soldermask with the patterned holes that allow access to pads, and Fig. 1.20B shows the negative representation used by PCB Editor. Here, the black background is actually the polymer film and the *green circles* (gray in print version) are the holes in the soldermask.

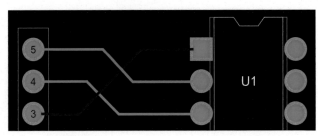

Figure 1.17 Copper in routed layers.

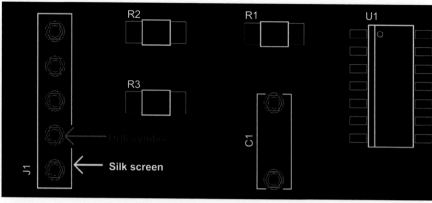

Figure 1.18 Circuit layout with drill symbols and silk screen on a PCB.

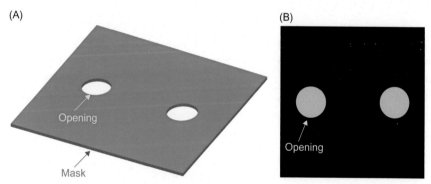

Figure 1.19 Drill chart with drill symbols and specifications.

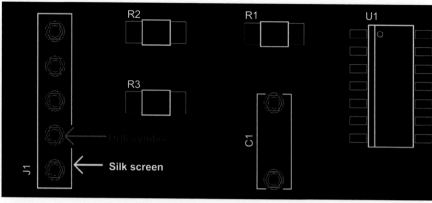

Figure 1.20 Soldermask layer and negative view: (A) soldermask and (B) negative view in PCB Editor.

Plane layers may be processed as negative layers in PCB Editor and are used to generate negative image Gerber files. However, PCB Editor displays plane layers in the positive view. Fig. 1.21A shows a physical copper plane layer with a thermal relief for a pin, Fig. 1.21B shows the negative view, and Fig. 1.21C shows the representation displayed by PCB Editor in "what you see is what you get" mode.

(A)   (B)   (C)

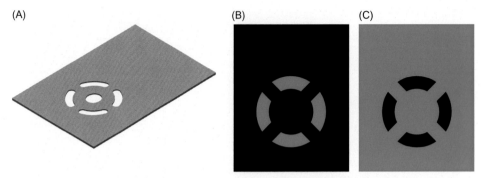

Figure 1.21 Copper in a plane layer: (A) copper plane with thermal relief, (B) negative view as processed, and (C) displayed in WYSIWYG mode. *WYSIWYG*, What you see is what you get.

Design files created by PCB Editor

PCB Editor format files

When you are designing a board, a PCB Editor works with and saves the design in a format that is efficient for the computer. The PCB Editor designed file has a .brd extension. When you are ready to fabricate the board, PCB Editor processes the design and converts it into a format that the photoplotters and CNC machines can use. These files are called *Gerber and drill files*.

Artwork (Gerber) files

Artwork (Gerber) files are created for each of the etch and mask layers discussed previously, and drill files are created for both plated and nonplated holes. PCB Editor generates as many as 30 or so layer files to describe various manufacturing aspects of the PCB. Some examples of these files, their extensions, and their functions are listed in Table 1.1.

These and other files (such as combined format IPC2581) that PCB Editor generates are discussed in greater detail in the next two chapters and in the PCB design examples.

Printed circuit board assembly layers and files

Several layer files generated by PCB Editor are not part of the actual fabrication process. These files are used for automated assembly of a finished board and are mentioned only briefly here. The first layer is the solder-paste layer. It is used to make a contact mask (or *stencil*) for selectively applying solder paste onto the PCB's pads so that components can be reflow soldered to the board. There may be one solder-paste layer for the top side of the board and one for the bottom side, as indicated in Table 1.1. The second layer file is the assembly layer, which contains information for board assemblers as to the part type, its position, and its orientation on the board. As

Table 1.1 PCB project, Gerber and Drill files, their extensions, and their functions.

File name and extension	Function
ProjectName.brd	Main board project file
Assembly_Top.art	Top side assembly
Solderpaste_Top.art	Top side solder paste
Silkscreen_Top.art	Top side silk screen
Soldermask_Top.art	Top side soldermask
TOP.art	Top side copper (usually routing)
INNER1.art	Inner layer 1 (usually routing)
INNER2.art	Inner layer 2 (usually routing)
INNERx.art	Inner layer x (usually routing)
PWR.art	Power layer (a plane layer)
GND.art	Ground layer (a plane layer)
BOTTOM.art	Bottom side copper (usually routing)
Soldermask_Bottom.art	Bottom side soldermask
Silkscreen_Bottom.art	Bottom side silk screen
Solderpaste_Bottom.art	Bottom side solder paste
Assembly_Bottom.art	Bottom side assembly
ProjectName.rou	Board outline cutting path
ProjectName.drl	Drill hole data

with the solder–paste layer, there may be one assembly layer for the top side of the board and one for the bottom side. PCB design for the various soldering and assembly processes is discussed in Chapter 5, Introduction to design for manufacturing.

The purpose of this chapter has been to introduce you to the process by which PCBs are manufactured. The purpose of the next chapter is to show you how to use OrCAD PCB Editor to design a board and generate the files needed to manufacture the PCB.

CHAPTER 2

Introduction to the printed circuit board design flow by example

Contents

Now that we have covered the construction of a PCB and know PCB Editor's role in it, we go through a simple design example so that you get a feel for the overall design process. This simple example sets the stage for Chapter 3, Project structures and the PCB Editor tool set, in which we dig deeper into the details of the process and learn more about PCB Editor itself and the PCB design examples.

Overview of the design flow

This section illustrates the basic procedure for generating a schematic in Capture and converting the schematic to a board design in PCB Editor. The basic procedure is as follows:

1. Start Capture and set up a PCB project using the PC board wizard.
2. Make a circuit schematic using OrCAD® Capture.
3. Use Capture to generate a PCB Editor netlist and automatically start PCB Editor and open the project as a `.brd` file.
4. Make a board outline.
5. Position the parts within the board outline.
6. Route the board.
7. Generate manufacturing data files.

Complete PCB Design Using OrCAD® Capture and PCB Editor
DOI: https://doi.org/10.1016/B978-0-12-817684-9.00002-3
17

Creating a circuit design with Capture

Snap To Grid button

If you do not have a full version of the OrCAD portfolio installed, you can get the OrCAD Trial installed from this website: http://www.orcad.com/free-trial. Before installing check if you are running the 64-bit Microsoft Windows operating system and your computer and domain name are set in latin alphabet. If you are using an older version of PCB Editor, most of the following information in this book still apply, but some of the dialog boxes and menu items may be different.

Starting a new project

Before you make a PCB layout, you need to have a circuit to lay out. You use Capture to make the schematic, so the first step is to start the Capture application by clicking the Windows Start button on your task bar and navigate to All Programs → Cadence Release 17.2-2016 → OrCAD Products → Capture CIS or look up the capture.exe in your Cadence installation folder. When you run it you can choose the license from a list by selecting File → Change Product. In the dialog window, you will see all available licenses. Once Capture is running, you should have a blank Capture session frame and a session log. Go to the File dropdown menu and navigate to File → New and click Project as shown in Fig. 2.1.

The New Project dialog box in Fig. 2.2 will pop up. Type a name for your project, then select the PC Board Wizard radio button. If you feel comfortable selecting your own location to save the project, you can do that (use the Browse... button), or you can use the default location for now (just remember where it is). Click OK.

After you click OK, the PCB Project Wizard dialog box shown in Fig. 2.3 (left) will pop up. For now, circuit simulation will not be performed, so leave the Enable project simulation box unchecked (we look at circuit simulation in the PCB design examples in Chapter 9). Click Next.

After you click Next, the PCB Project Wizard dialog box shown in Fig. 2.3 (right) will pop up. This box allows you to add specific libraries to your project. Scroll down until you find the Discrete.olb library, highlight it by clicking on it, then click the Add >> button; then click Finish. This completes the project setup.

You should have a Project Manager window in the left side of the Capture session frame, as shown in Fig. 2.4.

You may also have a Schematic window in the work space. If the schematic is not open, expand the *projectname*.dsn directory by clicking the " + " box to the left of the *projectname*.dsn icon (where *projectname* is the name you gave your project while using the project setup wizard). Click the " + " box next to the Schematics folder, then double click the file called Page1. The Schematic page should open. If you do

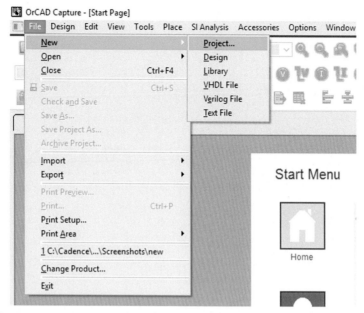

Figure 2.1 Starting a new project in Capture.

not see the dots, it means your grid is turned off. *The grid must be turned on, and the snapping to grid button must be turned on, to properly place and connect parts. To do it,* select View → Grid. If the grid is on, the grid dots are visible. To set the snapping to grid, click the Snap to Grid button, ⊞. If the snapping to grid is on, the Snap to Grid button is gray, ⊞, instead of red. You can use Options → Preferences to set a "pale blue" color, for example, for Grid and set the Grid Style to Lines.

Note: The menu bar at the top of the Capture session frame changes depending on whether you are working with the Project Manager window or the Schematic Page window. If you need to access options or tools for the Project Manager (or Schematic page), you must have that window active. To make the desired window active, click on its title bar or select the desired window from the Window menu. When inactive, it is gray; and when active, it is blue (or whatever colors you set up in Windows). Also, for projects that have PSpice simulation capabilities, an additional toolbar is displayed, which is not shown in Fig. 2.4.

Placing parts

Place Part tool button

To add parts to your schematic, make the schematic page active and select Part from the Place dropdown menu, or press the Place Part tool button ⊞, or press P on your keyboard. The Place Part pane shown in Fig. 2.5 will appear. In the Libraries

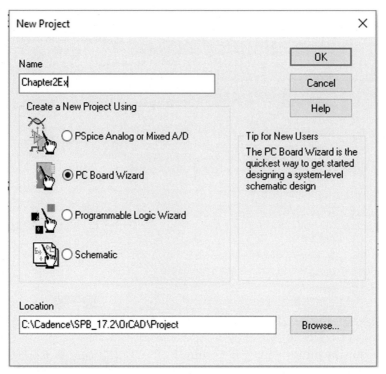

Figure 2.2 New Project dialog box.

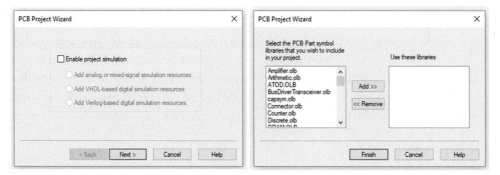

Figure 2.3 PCB Project Wizard dialog boxes: (left) simulation selection, (right) parts library selection.

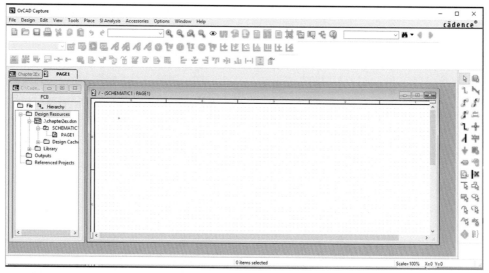

Figure 2.4 Example view of a new project.

selection box in the bottom of the pane, click over the DISCRETE library that was included by PC Board Wizard. Then, in the Part List box, click R (for resistor). You should see its symbol in the preview window. In order to place it to the schematic page, double click its name in the Part List window, or hit Enter. After that you will immediately return to the Schematic page and have a resistor tagging along with your mouse pointer, as shown in Fig. 2.6. Place one resistor on the page (left click the mouse or press Space), then hit the ESC key or right click the mouse and select End Mode from the pop-up menu.

In the Libraries window, you may have libraries different from that shown in Fig. 2.5. At the very least, you have included the DISCRETE library recently into the Libraries window using the wizard. If, for some reason, you do not see any parts or the DISCRETE library is not there, you can follow along for now to get an overview of the process or you can find and add another library to your project.

To add a part library to the project, select Part from the Place dropdown menu as described previously. In the Place Part dialog box shown in Fig. 2.5, press the Add Library... button   to bring up the Browse File dialog box shown in Fig. 2.7.

Find and select the desired library and click Open. All symbol libraries in OrCAD Capture have the file extension *.OLB. You can look for such files in the installation folder, or download them from the component supplier websites. You can also find a resistor ready for PSpice simulation in the PSpice library folder. To add it, double click

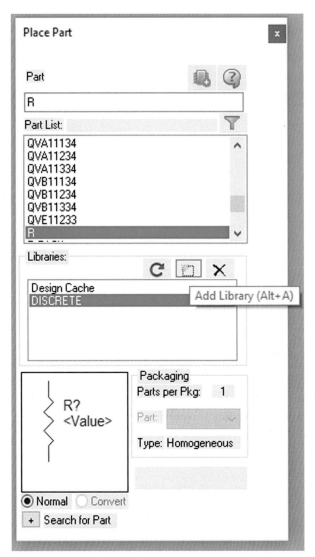

Figure 2.5 Place Part dialog box.

the PSpice folder, select the analog.olb library, and click Open. You should now be back to the Place Part dialog box, and the library you just added should be shown in the Libraries list box. Find and select R from the Part List selection box and click OK.

Before placing more parts, we assign a footprint to the existing part and make copies of it. *To assign a footprint to a part*, double click the part on schematic sheet to display the Property Editor spreadsheet (see Fig. 2.8). The cells may be shown vertically

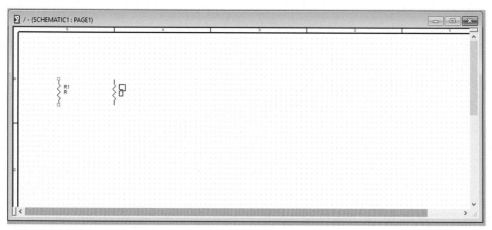

Figure 2.6 Placing the parts.

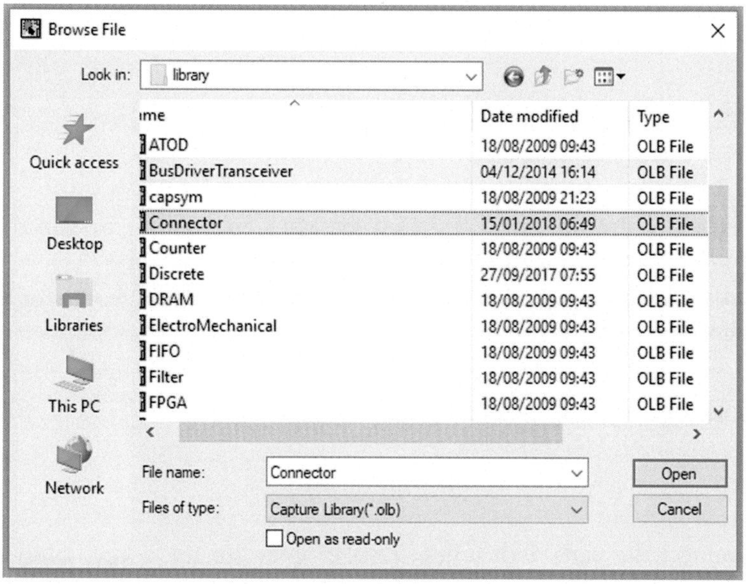

Figure 2.7 Add a library using the Browse File dialog box.

or horizontally. Find the PCB Footprint cell and enter res400. This is a basic through-hole, axial lead resistor footprint. Finding and selecting footprints is described in greater detail in Chapter 8, Making and editing footprints. If the PCB Footprint variable is not shown, make sure that the <Current properties> option is selected in the Filter by: list. Once the footprint has been assigned, close the spreadsheet.

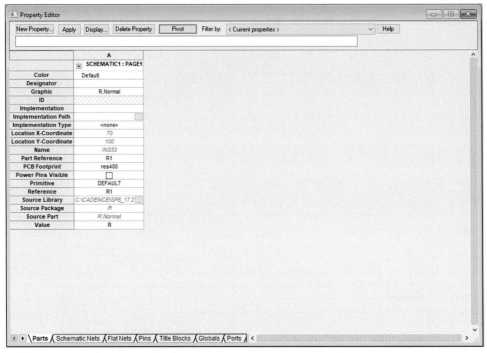

Figure 2.8 Assigning a footprint to a part with the Property Editor.

Next, duplicate the resistor. To do so, left click the resistor to select it and use standard Windows copy and paste procedures (Ctrl + C and then Ctrl + V, or Edit → Copy and then Edit → Paste from the menu) to place several copies of the resistor onto the schematic page.

Wiring (connecting) the parts

Place Wire tool button

Next, connect the parts with wires. *To place wires*, hit the W key, select Wire from the Place dropdown menu, or press the Place wire tool button, . The cursor will turn into a crosshair. Place the cursor on a box at the end of one of the resistor's leads and left click to start a wire (see Fig. 2.9).

Click on the end of another resistor lead to complete that wire. The crosshair will persist, so you can continue placing wires. Finish connecting wires to the resistors however you wish. Once you have finished connecting the circuit, press the ESC key or right click and select End Wire to stop the place wire cursor and get the pointer back. You can also connect two pins by dropping the pins of two parts together and dragging one part to draw the wire. If you inadvertently click near a lead but not on it, the wire may appear to be connected but may not be (that is why it is important to

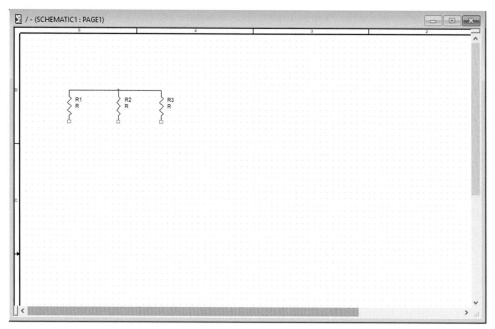

Figure 2.9 Connecting the parts with wires.

have the snapping to grid enabled). If the connections are not made properly, you will have problems when attempting to generate a netlist. You will be able to tell if a connection you made to a component was completed properly because the box at the end of the lead will disappear. At this point do not worry about power supplies or ground connections; this is just a "big picture" exercise to demonstrate the design flow process.

It is possible to connect the pins automatically. Choose Place → Auto wire and select:

- Two points—it will connect two points selected by user. The program tries not to cross the existing symbols or connections. If it's impossible to avoid the crossing, the program will make the connection anyway.
- Multiple points—it will connect all points selected by user. After you choose this variant you should sequentially click on each pin that should be connected to this net, then right click and choose Connect.
- Connect to bus—it will connect the pins selected by user to the existing bus. The bus is an object visually drawn as a wide line that represents two or more nets with the identical text name and the sequential numerical index, for example, A0, A1, and A2. In order to connect several pins to the bus you should sequentially click on each pin, then click right mouse button, choose Connect to bus, and then click on the desired bus line. The connections between selected pins and the bus will be created automatically. The left upper pin will be connected to the first net of a bus, and other pins, that is, top-to-bottom and left-to-right will be connected to

the next nets in the bus sequentially. The bus line should be drawn in advance opposite to all selected pins to allow the proper connection.

Creating the PCB Editor netlist in Capture

Once all the connections are complete, the next step is to create a netlist (a set of files that describes the circuit). Several types of netlists are possible, but you want *to generate a PCB Editor netlist.* Begin by making the `Project Manager` window active (instead of the `Schematic Page` window) and select the `.dsn` icon by left clicking it once. If the Schematic page is active, the `Tools` menu will not be available. Minimize the Schematic page if necessary to get to the Project Manager. Select `Tools` → `Create Netlist` from the `Tools` menu. The `Create Netlist` dialog box will pop up as shown in Fig. 2.10.

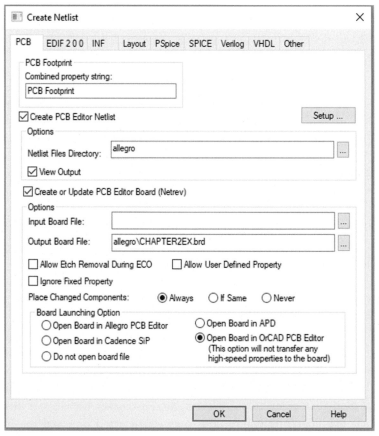

Figure 2.10 Creating a netlist for PCB Editor.

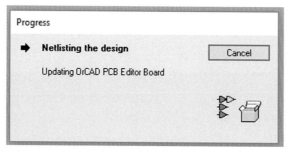

Figure 2.11 The netlist Progress dialog box.

From the `Create Netlist` dialog box, select the `PCB` tab. Select the `Create or Update PCB Editor Board (Netrev)` option. For the time being, use the default output board filename and path. Select the `Open Board in OrCAD PCB Editor` option to automatically launch the PCB Editor application. Click `OK` to start the process.

Capture will display a warning text box stating "`Directory 'Design Path\project-name\allegro'`" `specified as Netlist directory does not exist. Do you want to create it?` Click `Yes`. Capture will display a warning text box stating `Design Path\project-name.dsn will be saved prior to netlisting`. Click `OK`.

You should then see the progress box shown in Fig. 2.11 as the netlist is generated. Capture will then generate the netlist files, report the results in the `Session log`, and launch PCB Editor. If you had the `View Output` option checked, the three netlist files will also be displayed in Capture. Leave Capture open so that it and PCB Editor can communicate with each other if necessary. This will allow you to go back and review the circuit if you need to when you are working in PCB Editor.

Designing the printed circuit board with PCB Editor
The PCB Editor Window

Once the PCB Editor application is running, you should end up in the board layout environment shown in Fig. 2.12.

Fig. 2.12 shows six areas of the work environment. These areas are discussed in detail in Chapter 3, Project structures and the PCB Editor tool set, and only briefly introduced here. The design window is where you work on your board. With the design window and the various tools controlled by the control panel, menus, and toolbars, you can control various aspects of components, traces, and planes. Examples of controllable aspects include whether items are visible or not, whether they are fixed or movable, component location and orientation, and trace and board characteristics, to name just a few. The command window displays messages and prompts and allows you to enter commands. The worldview window gives you a bird's-eye view of the

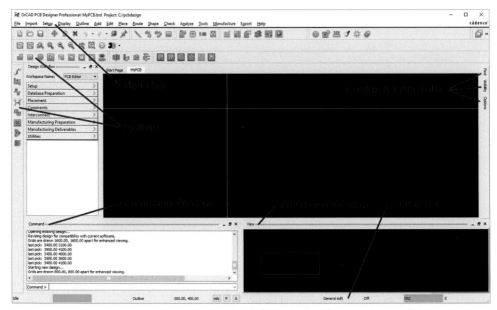

Figure 2.12 The PCB Editor environment and initial view.

location and size of the design widow relative to the actual PCB. The status window lets you know if the program is busy or idle and shows the coordinates of the cross-hairs. The P and A buttons are related to the display and coordinates but are described later.

PCB editor begins with nothing in the design window and parts are not visible until they are placed, so we need to draw a board outline and place the parts.

Controlling the view

Zoom tools

Zoom to Points button

The initial size of the design window is about 33 in. wide by 22 in. tall. To make a realistic board size for this simple example you need to zoom in. There are a couple of ways to do that. The zoom tools on the tool palette,

, are Zoom to Points, Fit, In,

Out, Previous, Selection, and Redraw, respectively. Click the Zoom to Points button, 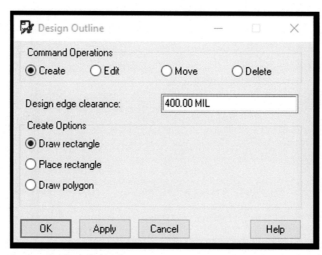, place the cursor at about coordinate position 4600, 3600 (units in mils, i.e., thousandths of an inch) then left click and release. Move the cursor down and to the left until it is less than or close to coordinates 0, 0. As the mouse is moved, a rectangular box with crossbars will become visible. Left click and release again to finish the Zoom to Points command. The design window covers a work area that is a little more than 4 in. by 3 in., which is plenty large for the few components in this example.

Drawing the board outline

To draw the board outline, select Outline → Design... from the menu. The Design Outline dialog box will be displayed, as shown in Fig. 2.13. With the dialog box still displayed, move your mouse over to the design area and left mouse click at point 0, 0 (or close to it). While watching the cursor coordinates on the status bar (lower right side of the design window), move the mouse up about 4 in. (4000 mil) and to the right about 3 in. (3000 mil) and left click. A dashed outline will be displayed, and the Design Outline dialog box changes to Edit mode. The Design Edge Clearance: determines how far from the edges of the board the place keep-in and route keep-in boundaries will be. The default value is 400 mil. For this design example, enter 200 mil in the text box and click OK.

When you click OK, another rectangle is automatically drawn inside the board outline. Actually there are two rectangles, one directly on top of the other. The two

Figure 2.13 New Design Outline dialog box.

outlines are the route keep-in and package keep-in outlines. The route keep-in outline defines the edge of any plane layer and the boundary within which traces must stay. The package outline defines the area within which all components must reside.

If you hover your mouse over the inner rectangle, the data tips box will tell you which outline it is. You can turn that layer off to view the other rectangle. If the data tip says Package Keepin (or Route Keepin), move your mouse over to the Options tab on the right of design window (see Fig. 2.14) to display the pane, and select Package Keepin class from the list. Left click the colored box next to the All subclass. The box will turn black and the place keep-in outline will disappear. You should then see the other outline, and the data tip should tell you it is the Route Keepin/All outline (or the Package Keepin). You can turn this one off by selecting Route Keepin class from the Options tab as described previously.

You can select and view the different boundaries by selecting them from the class list in the Options pane or the Visibility pane in the control panel area at the right of the design window (see Fig. 2.12). The class and subclass for each outline so far is

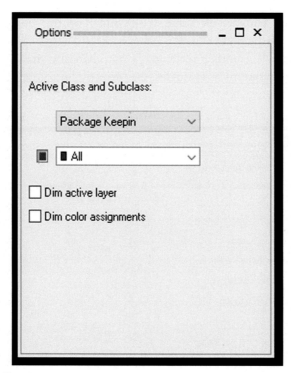

Figure 2.14 The Options pane.

`Board Geometry & Design_Outline`, `Route Keepin & All`, and `Package Keepin & All`, respectively. Classes and subclasses are discussed in greater detail in Chapter 3, Project structures and the PCB Editor tool set and Chapter 8, Making and editing footprints, and the PCB design examples in Chapter 9.

If some objects are overlapped, and it's hard to select the right one, you can use the right-click and choose `Reject` to change the selection to another object (this method works only if some command was activated before selecting the object).

Placing parts

`Manual Place button`

Parts can be placed manually or automatically. Multiple automatic placement modes are available, but we wait until the PCB design examples to use them. *To place parts manually*, select `Place` → `Components Manually...` from the menu; or click the `Manual Place` button, , on the toolbar. The manual placement mode is activated. If you don't see the `Placement` dialog box (Fig. 2.15), right click the mouse and choose `Show`. Click the `Components by refdes` box; this will automatically select all the components. Leave the `Placement` dialog box displayed (i.e., do not dismiss it by clicking the `Close` or `Cancel` button). Move the mouse cursor over to the design window; a component will be attached to it. Left click on the design window to place the part. The next part in the list will automatically be attached to the mouse cursor. The box next to the part that was just placed will be unchecked (it is no longer in the queue). Once all the parts are placed, dismiss the `Placement` dialog box by clicking `Close` button.

Moving and rotating parts

`Move button`

To move parts, select the `Move` button, , click and release the left mouse button to select a part. Move the part to the desired location then click and release the left mouse button to place the part. You can continue moving other parts or right click and select `Done` from the pop-up menu to deactivate the move tool. *To rotate a part*, use the move tool as just described to select the part, then right click and select `Rotate` from the pop-up menu. It's also possible first to select the part and then to choose the function, for example, to press the `Move` button. This way you can use the preselect mode of operation with almost any command.

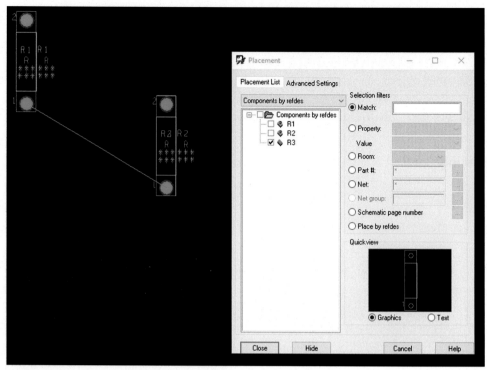

Figure 2.15 Manually placing components.

Fig. 2.16 shows the component footprint as well as the silkscreen and assembly details. The parts of the footprint are as follows (subclass and class are listed):

Text:

1. Assembly top (Comp Value)
2. Silkscreen top (Comp Value)
3. Assembly top (Ref Des)
4. Silkscreen top (Ref Des)
5. Assembly top (Tolerance)
6. Silkscreen top (Tolerance)
7. Assembly top (User Part)
8. Silkscreen top (User Part)

Objects:

1. Place_Bound_Top (Package Geometry)
2. Pin_Number (Package Geometry)
3. Assembly_Top (Package Geometry)
4. Silkscreen_Top (Package Geometry)

Color button

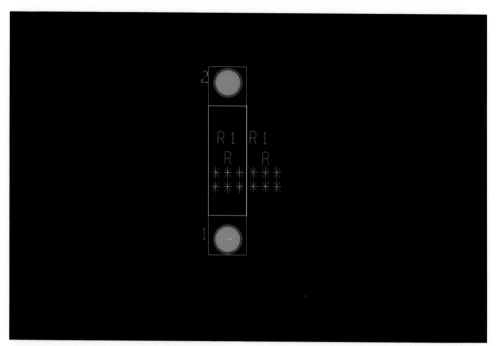

Figure 2.16 Parts of a component footprint.

You can change the visibility of the various parts of the footprint (and PCB as a whole) using the Color dialog box. Click the Color button, , or select Setup → Colors... from the menu. *To control the visibility of text objects*, select Components from the Group list, as shown in Fig. 2.17. *To control the visibility of package objects*, select Geometry / Package Geometry from the Group list as shown in Fig. 2.18. *To control the visibility of padstacks*, select Stack-Up from the Group list. You control the visibility of the various parts by checking or unchecking the boxes then clicking Apply. You can customize the color of any text, etch, or detail object by selecting a color from the palette of available colors then clicking the color square of the item you want to customize. More will be said about footprint composition and construction in Chapter 8, Making and editing footprints.

Routing the board
Using the autorouter
The next step is to route the board. *To route the board using the autorouter*, select Route → PCB Router → Route Automatic... from the menu; the Automatic Router dialog box shown in Fig. 2.19 will be displayed. The autorouter can be run in three modes. The different options will be demonstrated in the PCB design examples in Chapter 9. For the time being, leave everything as it is and click the Route button.

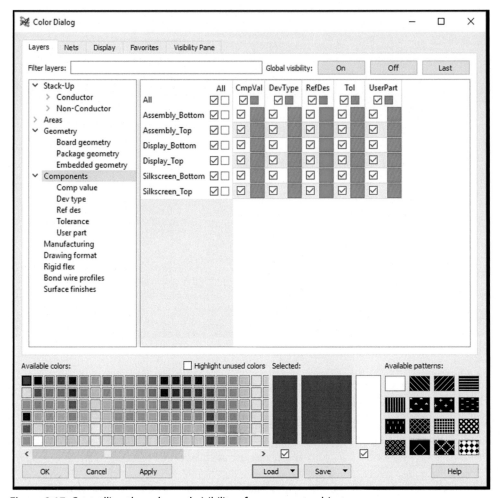

Figure 2.17 Controlling the color and visibility of components objects.

The autorouter will begin routing the board and a status box will be displayed (Fig. 2.20). When the routing is complete the status box will be dismissed. Fig. 2.21 shows how the board might look after routing.

In complex designs, you need to know how to use the manual routing tools to preroute critical traces and clean up after the autorouter. The manual routing tools are described next.

Manual routing

Delete button

Add Connect button

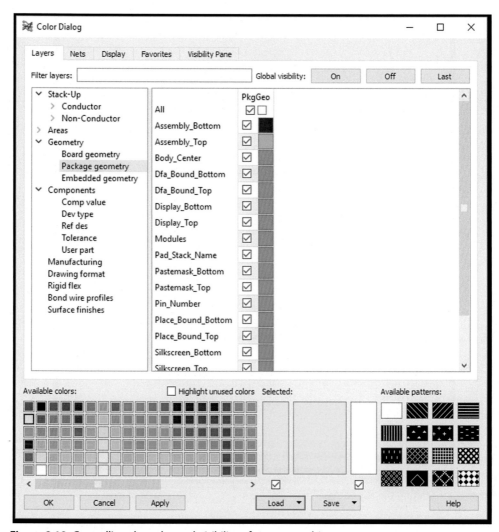

Figure 2.18 Controlling the color and visibility of geometry objects.

In Chapter 3, Project structures and the PCB Editor tool set, and the PCB design examples in Chapter 9, we take an extensive look at manually routing traces, so the manual routing is only briefly introduced here. First, unroute (rip up) the traces routed by the autorouter. *To rip up a trace*, select the `Delete` button, , then choose the `Clines` and `Vias` objects in the `Find` pane. The types of objects, which are not selected will not be available for the operation. Then left click the trace you want to rip up. The trace will be highlighted. To complete the rip-up, click the trace again, select another trace, or right click and select `Done` from the pop-up menu. Ripping up multiple (or all) traces from a list is demonstrated in the PCB design examples. If you prefer

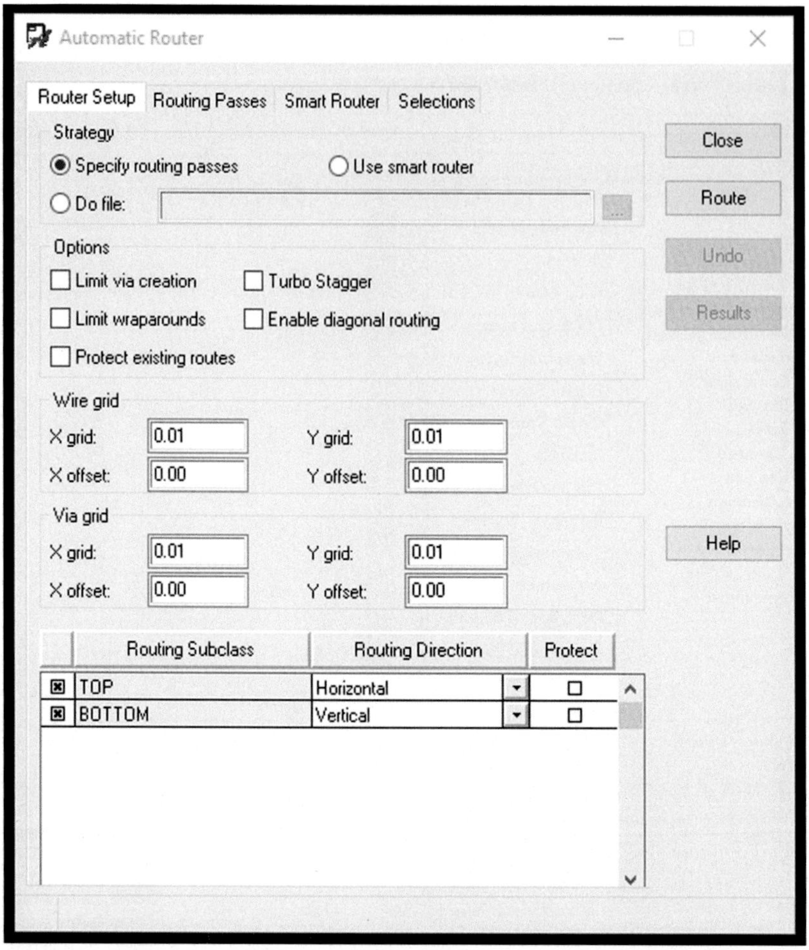

Figure 2.19 Starting the autorouter.

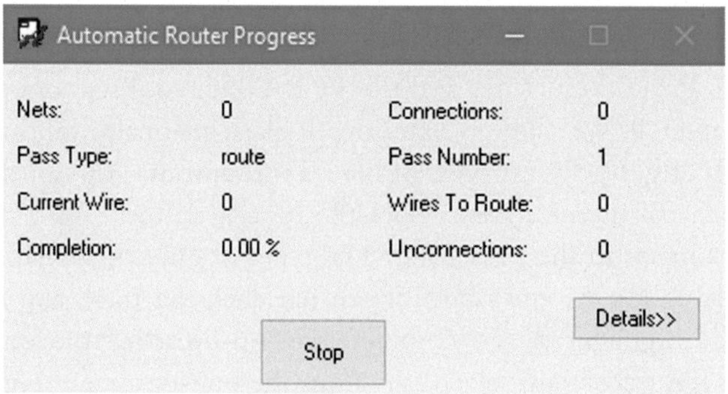

Figure 2.20 Automatic Router Progress dialog box.

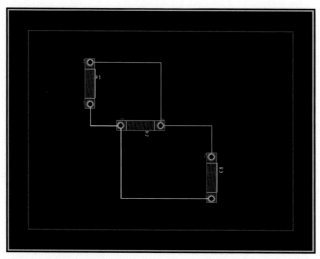

Figure 2.21 Example of a routed board.

to use the preselect mode you should first choose Clines and Vias in the Find panel, then select the needed traces using mouse, and press Delete button.

To begin manual routing, click the Add Connect button, , on the toolbar. Left click a rat's nest to begin routing. Left click again to place a vertex. Clicking on a pin ends the route, or you can right click and select Finish to end the route. To stop routing, right click and select Done from the pop-up menu. You can right click and select Oops to roll back one step.

Note: If the route tool does not seem to be working, display the Find pane and make sure that Ratsnest box is checked.

To add a via, select a net with the Add Connect tool, place a vertex, right click, and select Add Via from the pop-up menu. A via will be placed at the vertex and routing will continue on the alternate layer listed in the Options pane.

Slide button

If you want *to push (move) a trace* without rerouting it, use the Slide button,. Left click the trace, move your mouse pointer to the desired location, and left click again to place the trace to a new point. You can use the preselect mode as well.

By using the control panel's Options pane, you can change routing properties, such as the active layer (Act), alternate layer (Alt), the trace width, and line type. *To select a different routing layer*, use the Act dropdown list to select the desired layer. When you are finished routing, right click and select Done from the pop-up menu.

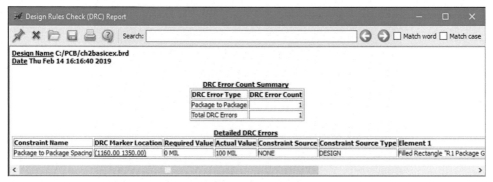

Figure 2.22 An example DRC report. *DRC,* Design rule checker.

These are just a few of the basic tools you need to know about. In the PCB design examples, you will see how to use more of the tools and control routing properties, such as setting minimum and maximum trace widths and setting up plane layers.

Performing a design rule check

After you have completed routing your board, you should check for errors. *To check for errors,* run the design rule checker (DRC) by selecting `Check → DRC Update` from the menu. A message will be displayed in the console window (at the bottom of the screen) as to whether there were errors. If there are errors, the message in the console window does not go into detail about what they are, just that they exist. If you want to see a description of the errors, run a DRC report. *To run a DRC report,* select `Export → Quick Reports → Design Rules Check Report` from the menu. An example of a DRC report is shown in Fig. 2.22. The report shows what would be reported if R2 and R3 were too close to each other (not the circuit in Fig. 2.21).

Also you can use `Tools → DRC Browser` to see all DRC errors as a table grouped by DRC type.

Creating artwork for manufacturing

`Artwork button`

At this stage, PCB Editor has generated a design file that fully describes your board. This file is optimized for viewing, editing, and saving on your computer, but it is not in the format that many PCB manufacturers use for fabricating boards. The most common type of file system used in PCB manufacturing is the Gerber file system. PCB Editor has the capability of translating its `.brd` file to Gerber files. In PCB Editor, this is called *manufacturing* the design. PCB Editor allows considerable control over the manufacturing. As a result, several steps are required to generate all of the manufacturing files, especially with more complicated designs. Setting up the manufacturing process and generating the Gerber files is described in detail in Chapter 10, Artwork development and board fabrication. A quick overview is given here.

To create artwork for your design, select Export → Gerber... from the menu or click the Artwork button, 📷 , on the toolbar. The Artwork Control Form shown in Fig. 2.23 is used to create the artwork. Artwork files and NC drill and router file creation are described in detail in Chapter 10, Artwork development and board fabrication.

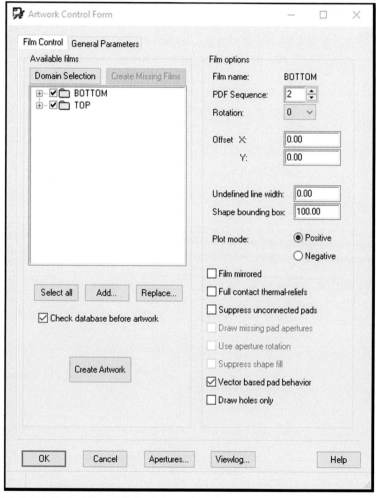

Figure 2.23 Artwork Control Form dialog box.

Congratulations, you have routed your first PCB using OrCAD PCB Editor! The objectives of this chapter were to demonstrate the basic steps of designing a circuit schematic and using PCB Editor to design a PCB. The following is a summary of the process:

1. Start Capture and set up a PCB project using the PCB wizard.
2. Make a circuit schematic using OrCAD Capture.
3. Use Capture to generate a PCB Editor netlist and automatically start PCB Editor.
4. Make a board outline.
5. Position the parts within the board outline.
6. Route the board.
7. Generate manufacturing artwork.

In the next chapter, we cover the design flow in greater detail and learn more about the PCB Editor tool set; and in Chapter 8, Making and editing footprints, and Chapter 9, Printed circuit board design examples, we learn more about footprint design, PCB layer design, and the steps to manufacturing the board design.

CHAPTER 3

Project structures and the PCB Editor tool set

Contents

This chapter explains what you did when making the simple design in Chapter 2, Introduction to the printed circuit board design flow by example, and why. It also introduces and describes the PCB Editor tool set in greater detail, so that you will be well equipped to lay out more complicated boards in the printed circuit board design examples in Chapter 9, Printed circuit board design examples.

Project setup and schematic entry details

Capture projects explained

When you set up your project by following the File → New → Project menu path, you had several options from which to choose: a project, a design, a library, a VHDL file, a Verilog file, or a Text file. The options we are most interested in are projects and libraries, and we look at those in great detail throughout the book. VHDL and

Verilog files are used in field programmable gate array projects and are not discussed here. A Text file is simply a text file (for making project notes, for example).

After you selected the New Project option to begin setting up your project, four more options were available to you in the New Project dialog box: PSpice Analog or Mixed-Signal A/D, PC Board Wizard, Programmable Logic Wizard, or Schematic. A flow diagram of these options and suboptions is shown in Fig. 3.1 (see also Fig. 2.2). PSpice Analog or Mixed-Signal A/D is used to simulate analog and/or digital circuits using PSpice. PSpice is used to develop and test models (Chapter 7: Making and editing Capture parts), perform circuit simulations [Chapter 7: Making and editing Capture parts, and Chapter 9, PCB design examples], and simulate transmission lines

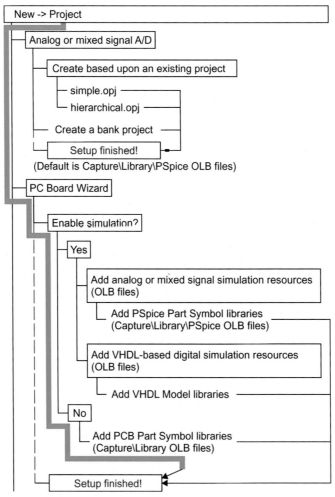

Figure 3.1 New project design flow and options.

(PCB design examples). For now, we work mostly with the second option, the `PC Board Wizard`, while we focus on designing PCBs. The next option, `Programmable Logic Wizard`, is for working with programmable devices and is not discussed in this text. A `Schematic` is basically just a design file with only a schematic and a parts cache. The thick green line in Fig. 3.1 shows the path you followed in Chapter 2, Introduction to the printed circuit board design flow by example, that is, `File → New → Project → PC Board Wizard → No Simulation`.

There are four types of OrCAD® libraries. As shown in Fig. 3.2, these four libraries have three file extensions. There are two types of `OLB` libraries, a `LIB` library and a symbols library, which contains `.dra` files as well as some other files that are explained in detail in Chapter 8, Making and editing footprints. Inside the *capture* folder is a *library* folder, which contains one of the types of `OLB` files, and a folder called *PSpice*, which contains the other types of `OLB` files. The `OLB` files located directly in the Capture *library* folder contain simple schematic part symbols and are the ones we used in Chapter 2, Introduction to the printed circuit board design flow by example. The libraries located in the *PSpice* folder contain parts with schematic symbols, too, but the parts also contain PSpice templates, which are links to specific PSpice models. The PSpice models to which templates point are located in the main OrCAD *PSpice* folder (shown at the bottom of Fig. 3.2). Individual models are grouped into various PSpice

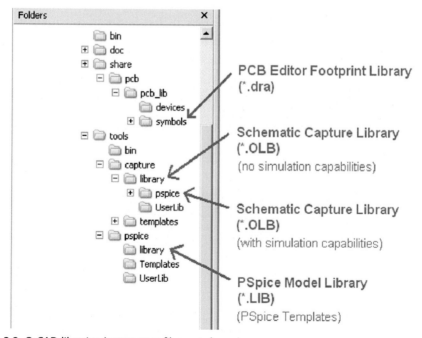

Figure 3.2 OrCAD libraries (extraneous files not shown).

library files, which contain the .LIB extension. The *share* folder contains the footprint models (called *symbols*), which are files with the .dra extension. Most parts in Capture are capable of having a PSpice template or a footprint assigned to them, but only some parts have them preassigned. It may seem backward, but only parts from the PSpice part library (located in the *capture/library/PSpice* folder) have preassigned templates and footprints. So, in Chapter 2, Introduction to the printed circuit board design flow by example, when you worked with the PCB project wizard, you selected libraries from the *capture/library* folder, which had parts with schematic symbols but had no PSpice simulation capabilities or footprints assigned to them.

Once you clicked Finish in the PCB Project Wizard box, Capture opened up the Project Manager window shown in Fig. 3.3. Behind the scenes, Capture generated two files: an OrCAD project file (name.OPJ) and a design file (name.DSN). These files will be in the directory you chose when you set up the project.

As you look at the Project Manager window, you can see that a project contains three folders labeled Design Resources, Outputs, and Referenced Projects. Initially, the Outputs and Referenced Projects folders are empty. When a netlist is created, netlist files are placed in the Outputs folder. The Design Resources folder contains one design (represented by the icon) and a Library folder. A project can have only one design, but a design can have several subfolders, which in turn may contain several different items. The Library folder contains links to the libraries used by your design. We discuss library management in Chapter 7, Making and editing Capture parts. The design contains at least one schematic folder (the root folder) and a Design Cache folder. A design can contain multiple schematic folders, and each schematic folder can contain multiple schematic pages. The Design Cache folder contains a record of each part you

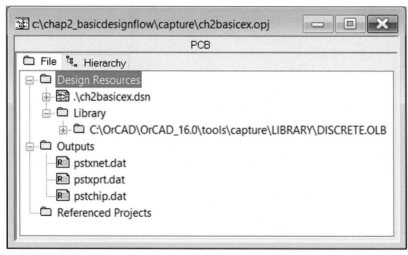

Figure 3.3 Project Manager window.

used in your design. If you modify one of the parts on a schematic page, Capture makes a copy of it (leaving the original part in the library unchanged) and adds a record of the modified part to the design cache. A design with one schematic folder and one or more schematic pages connected together by off page connectors is called a *flat design*. A design with more than one schematic folder and one or more schematic pages per folder or that containing hierarchical blocks is called a *hierarchical design*. Hierarchical designs are not discussed here. For more information on project details, see Chapter 2, Introduction to the printed circuit board design flow by example, or see the Capture User's Guide, under "Starting a New Project."

Capture part libraries explained

Once the project is set up, the next step is to open the schematic page (if it is not already open) and begin placing parts from the `Place` dropdown menu. When the `Place Part` dialog box opens (see Fig. 3.4), it shows a list of libraries (in the `Libraries`

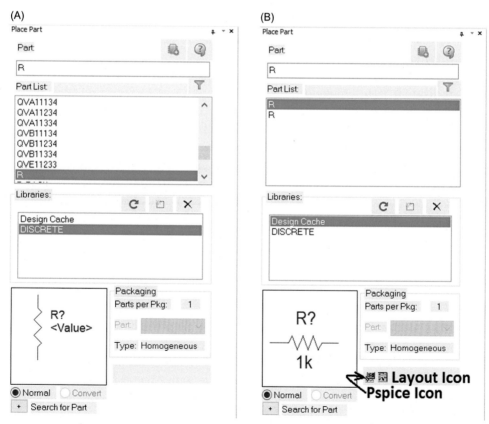

Figure 3.4 Place Part from Capture or Capture/PSpice library: (A) Capture parts, (B) PSpice parts.

window) and a list of the parts (in the Part List window) within a library. If you select a part within one of the libraries, you can see what the part looks like in the preview window, as pointed out in Fig. 3.4A. The libraries listed in the Libraries window are ones that were added by the PC Board Wizard when you set up the current project and libraries that were added to the list during previous projects. You can add other libraries to the list by clicking the Add Library... button. Likewise, you can remove a library from the list by selecting it and clicking on the Remove Library button. The libraries that are listed may be from the Capture or the PSpice library. It is not obvious which library it is from just by looking at the name (compare Fig. 3.4A and B). However, if you have Windows ToolTips turned on and hold your mouse over the library name or the part name, the ToolTips text box will show the path of the library. From the path, you can tell which type of library it is. Another important note is that, if a PSpice library (not a PCB Editor library) is associated with the Capture part, you will see one or both of the icons near the part preview window, as shown in Fig. 3.4B. If no icons are present, it means that it is just a Capture schematic part that has no footprint or PSpice template (models) assigned to it.

As you place parts in your design, Capture keeps track of the parts and stores the information in a database file generated the moment you place your first part. This file has .DSN extension, and it is written to a hard drive every time you press the Save icon. It is recommended to set the Autobackup interval and path in the Options menu, which will store the additional copy of your job, to allow the easy recovery. After you finish your design and tell Capture to make the PCB Editor netlist, Capture generates three more files that describe which parts were used and how they were connected. The files are pstxnet.dat (the netlist file), pstxprt.dat (the reference designator and device type file), and pstchip.dat (a device definition file also used for pin swapping, etc.); and they are located in the Outputs folder. These files are also used when back-annotating information from PCB Editor to Capture. More on back-annotation is discussed in the PCB design examples in Chapter 9. For the most part, these files are used behind the scenes, and you need not know much about them for what we cover in this text. If you want to know more about them, please see *Allegro® User Guide: Transferring Logic Design Data*.

Another file that gets created is the netrev.1st file, which is a netlisting report generated when the PCB Editor netlist files are created. If there is a problem during the netlisting process, this is where the errors and warnings are documented. If you have trouble trying to create a netlist, open this file with Microsoft Word or another text application and search for the word *error* or *warning*. An example of a netlist error is given in the PCB design examples in Chapter 9.

Understanding the PCB Editor environment and tool set
Terminology

Before we start talking about the tool set, we examine a couple of terms that should be defined up front, so the descriptions of the tools make more sense. These important terms are as follows:

- Classes and subclasses—another way of thinking about layers and objects on layers. From the PCB perspective, there are the following layers: routing layers, plane layers, soldermask layers, and the like. But being able to group or separate things into classes gives you more control over the design. Generally speaking, a class is a broad grouping of things and a subclass is a more specific thing that can be grouped into a class. For example, `Etch` (defined later) is a class, and `Top` and `Bottom` are subclasses of `Etch`. Some subclasses are distinct but have the same name. They are distinct because they belong to different classes. An example of this situation is the `Silkscreen` subclass, one of which belongs to the `Board Geometry` class (lines or text belonging to the board), another belongs to the `RefDes` (text only, belonging to components only) class. By having separate subclasses of silk screen (or soldermask, say), you can control the color and visibility of each type of silk-screen object. As you go through the design examples, this will become very clear.

- Clines—where *cline* is short for *connection line*. A cline is a routed copper trace that makes electrical connections to pins. Clines are also called *etch*. Unrouted connection lines are called *net*s or *rat's nest*. Lines that make up graphic objects are just lines.

- Etch—objects on routing and plane layers that become copper patterns during the manufacturing process. Etch objects are on subclasses that belong to the Etch class and can be traces (clines), areas (rectangles or other shapes), or text. Etch objects can be connected to nets or may stand alone.

- Elements—a general term for things that you put into your board design. You can have text elements, graphic elements (e.g., silk-screen outlines), and components (footprint symbols).

- Frectangle—a filled rectangle. A frectangle is drawn using one of the Shape tools as described later. Frectangles can have special properties, and not all rectangles are frectangles. Some of the differences between plain rectangles and frectangles are described later. There are also dynamic copper shapes, which can be rectangles but are different from frectangles. Frectangles and dynamic copper objects are demonstrated in the design examples.

- Symbols—elements you can place in your design that can be made up of one or more other elements. For example, a component footprint is called a *symbol* (specifically a footprint symbol), which can comprise padstacks, outline details, and text elements. Other types of symbols include flash symbols, mechanical symbols, and drawing symbols, to name a few. Symbols are used extensively in Chapter 8, Making and editing footprints, and Chapter 9, the PCB design examples.

- Functions—a part of an integrated circuit (e.g., a logic gate).

PCB Editor windows and tools

At this point in Chapter 2, Introduction to the printed circuit board design flow by example, PCB Editor was launched, the board design (`name.brd`) was opened with all default settings, and a blank work area was presented to you. From this point you made a board outline, placed the parts on your board, autorouted the board, and produced the artwork for it. To be fluent at these tasks, you need to know how to use the PCB Editor tools. We now take a tour of the PCB Editor environment and tool set.

Note: From the time you start your design in Capture to the time the artwork is produced in PCB Editor, nearly 30 files can be generated, which together fully describe your design. If you save more than one project in the same folder, it can become very cluttered. It is a good idea to set up a "`MyProjects`" folder that contains subfolders for each project.

The design window

The `Design` window is the working environment for a board design (see Fig. 2.12). From the `Design` window, you have access to the tools you need to handle parts, route traces, and perform back annotations (design updates from PCB Editor to Capture). Hundreds of tasks can be executed from the `Design` window menus. Since they are covered in detail in the *Allegro® User's Guide*, a detailed discussion of the menus is not given here, but the key menu options are discussed in the examples as the need arises. The toolbar is discussed next.

The toolbar groups

By default, several most important toolbar groups are displayed. You can move them wherever you want or turn them off. *To add or change toolbar groups*, select `Setup` → `More` → `Customize Toolbar...` from the menu. This section does not describe the tools in great detail but serves as an introduction only. The use of the tools are demonstrated in the design examples and described further in *Allegro® PCB Editor User Guide: Getting Started with Physical Design*.

Application mode group

PCB Editor works in several application modes: (a) General edit, (b) Placement edit, (c) Etch Edit, (d) Signal Integrity, (e) Shape edit. The differences between modes are subtle, and most of the time you need not consciously choose between one or the other. You can execute most commands in either mode, but you will find that some commands are easier to perform in one mode than the other and you may have to wrestle with the Find and Options panes settings to do so. You can determine which application mode you are in by looking at the application box in the status bar (described later). See *Allegro® PCB Editor User Guide: Getting Started with Physical Design* for further information. All of the documentation is available via Help menu of the tool, as well as in PDF format from the Cadence support site (a login is required to access it).

Edit group

The edit group is made up of the following tools: (a) Move, (b) Copy, (c) Delete, (d) Undo/Redo, and (e) Fix/Unfix. These tools can be used on graphical objects, text, and etch. The Delete tool deletes (erases) objects, but it is used also to rip up traces (it does not delete nets) and unplace parts. The Fix/Unfix tool is used to prevent objects (e.g., shapes and text) from being moved or deleted, and it prevents routed traces (clines) from being ripped up.

View group

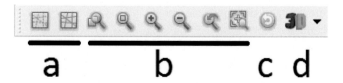

The view group consists of the following tools:

(a) Unrats All/Rats All—controls the visibility of all unrouted nets (also known as *rat's nest*). You can also control the visibility of individual nets using the `Rats Off` property with the Constraint Manager (described later) or by selecting `Unrats All` then selecting `Display` → `Show Rats` → `Net` in concert with the `Find` filter tab to then display specific nets.

(b) Zoom—allows you to zoom to a specific area (Zoom points), the entire design (Zoom fit), in or out, to a previous view, or to a selection (by drawing a box across a part and some nets, you use Zoom selection to view everything that was highlighted).

(c) Redraw—refreshes the display.

(d) 3D viewer—allows you to visualize the whole project with attached STEP-models of the parts in realistic 3D, almost as a real PCB. You can also view in 3D just the selected objects, such as components, nets, traces, vias, or shapes. For the flexible cables it's possible to simulate the bending. Also there is a powerful capability of checking the intersection between the components and other objects.

Setup group

The following tools constitute the setup group:

(a) Grid Toggle—allows you to turn the grids on or off. The displayed grid resolution depends on the settings in the `Define Grid` dialog box (from `Setup` menu) and the type of layer (class) that is active. Classes related to etch layers use the `Etch` grid resolution and nonetch-related classes use the `Non Etch` resolution. If the grid resolution is too dense for a given Zoom level, PCB Editor automatically adjusts the grid density view but not the functionality.

(b) Color Palette—allows you to control the color and visibility of classes and subclasses. The `Color` dialog box is described in greater detail later.

(c) Shadow Toggle—controls the intensity of visible layers. Shadow mode makes highlighted items stand out more clearly.

(d) Stack-up—launches the `Cross-section Editor` dialog box, where you define your board's stack-up as described later.

(e) Constraint Manager—launches the Constraint Manager, where you set autorouter specifications (e.g., trace width and spacing) and design rule constraints. The Constraint Manager is discussed further later.

(f) Setup parameters—controls the design parameters for this design.

Shapes group

These are drawing tools and are different from the Add (line and rectangle) group objects described later. They are used to make shapes that interact with the design rule checker (DRC) tool and are used to create dynamic shapes that adjust automatically (e.g., plane layers that heal and you can plow through) when interacting with other Etch objects. You can also make static shapes with them, which are used for keep-in/-out areas and the like. Shapes can be filled or unfilled, static or dynamic. Some classes, such as the Package Keep-in class, automatically know what type of shape it needs (unfilled), so even if you unknowingly choose a static filled rectangle, it will automatically become unfilled. Many other classes do not. For example, you can place a filled rectangle on the GND plane, but it may not connect to pins properly. Copper areas on negative plane layers should be dynamic shapes. Many of the icons are hidden after the first installation and can be restored through Setup → More → Customize Toolbar.

(a) Shape Add tools—The following tools are used to make the shapes: Shape Add (polygon), Shape Add Rect(tangle), and Shape Add Circle.

(b) Shape Select—used to edit existing shapes (stretch shapes, add or move vertices, etc.).

(c) Edit Shape Boundary—is used to add or remove areas to a shape at the shape's boundaries. It is similar to using the Merge and Void commands but is easier to use and can be used only on the edge of a shape.

(d) Shape Void tools (available from Shape → Manual Isolation/Cavity menu)—are used to remove areas of shapes. For example, if you want to create an opening in the plane layers because a mounting hole is inserted in the design and has a route keep-out area, then you can use a void to make the opening.

(e) Delete Unconnected Copper tool (available from Shape → Delete Unconnected Copper menu)—deletes pieces of planes or shapes that become detached from the main shape due to spacing requirements around groups of pins. These floating pieces can become EMI issues and should be removed.

Note: When making changes that affect dynamic shapes some changes may not take effect immediately. To rebuild the shapes, select Shape → Global dynamic parameters, choose Shape fill tab and press Update to Smooth button.

Manufacture group

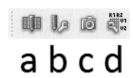

(a) NC Drill Legend—is used to set up then display the drill legend (drill chart). Drill symbols in parts on your design then also become visible.

(b) NC Drill Param(eters)—opens the `NC (numerical control) Parameters` dialog box, where you can set up drill format (Enhanced Excellon format, etc.).

(c) Artwork—opens the `Artwork Control Form` dialog box, where you set up the classes and subclasses for which you want manufacturing data generated.

(d) Silkscreen Param—automatically generates the silk-screen drawing, to cut the lines over soldermask openings, move the texts to a free space, and rotate the text to proper position. The new subclass with generated silk-screen drawing is called `Manufacturing/Autosilk TOP/BOTTOM`.

Display group

(a) Show Element—displays a text box that provides information and properties about the object you select (nets, shapes, components, etc.). Use it in concert with the `Find` filter to locate objects or nets without actually having to click on them.

(b) Show Constraints—is similar to Show Element but reports only constraints.

(c) Show Measure—measures the distance between two pick points and displays the measurement in a pop-up text box.

(d) Assign color—allows one to assign the user-selected color to any object. It's also possible to select the pattern while assigning the color.

(e) Highlight/Dehighlight—marks objects to make them easier to see. Also used in conjunction with intertool communication between Capture and PCB Editor.

Miscellaneous group

a b

(a) Report—opens the Reports dialog box so that you can pick a report to view (e.g., DRC, bill of materials). Selecting this button is the same as selecting Export → Reports from the menu.

(b) DRC update—updates the DRC status. It does the same thing as selecting Check → DRC Update from the menu. You can then use Check → Design Status to see how many errors there are, or you can use the Report button to show the list of errors if there are any.

Place group

a b c

(a) Place Manual—displays the Placement dialog box so that you can place parts or other objects from the list.

(b) Place Manual—H—does the same thing as Place Manual but does not automatically show the Placement dialog box. To show the dialog box, right click and select Show from the pop-up menu.

(c) Swap pins/components—allows one to swap two selected pins or two selected components.

Route group

a b c d e f g h

The route group after installation contains just three tools: add connect, slide, and delay tune. But there are more tools available from the Route menu, and some of them can be added to the toolbar via Setup → More → Customize toolbar, similar to shown at the above picture.

(a) Add Connect—is used to manually route traces (produce clines).

(b) Slide—is used to move or stretch a routed trace. It automatically uses autorouter algorithms to prevent most (but not all) DRCs.

(c) Delay Tune—is used to elongate the traces. It is not enabled in basic OrCAD PCB Editor Standard.

(d) Spread between voids—moves the group of clines so that the spacing between all of them becomes equal.

(e) Create Fanout—is used with `Setup` → `Design Parameters` → `Route` → `Create Fanout Parameters` to define how fanouts should be done. Select this tool then click on a pin or component to make it happen. All `Fanout` settings are also available in the `Options` pane.

(f) Custom Smooth—is used to reroute routed traces to make the routing more efficient and remove unnecessary corners and route distance. It performs the same basic function as `Gloss` but is used to work on one trace at a time. Select this tool then the trace you want smoothed.

(g) Vertex—is used to add a vertex to a line or cline (use Shape Select to add or move vertices on shapes).

(h) Auto route—runs the autorouter (the same as `Route` → `PCB Router` → `Route Automatic`).

Add group

These objects are graphical in nature only. Use these tools for making graphics and text on silk-screen and assembly layers. The tools are

(a) Add Line.

(b) Add Text.

(c) Edit Text.

Note: Edit Text is used to change the text characters. If you need to change the text properties (size and spacing etc.), you need to use the `Options` pane and `Setup` → `Design Parameters` → `Text` → `Setup Text Sizes`.

Control panel with foldable window panes

The Control Panel is an area containing several tabs that show or hide collapsible window panes. The panes are shown in Fig. 3.5 in their default condition (no tools or commands active). The panes are dynamic, so what is displayed depends on what tool is active or what type of object you selected at that moment. From these panes, you

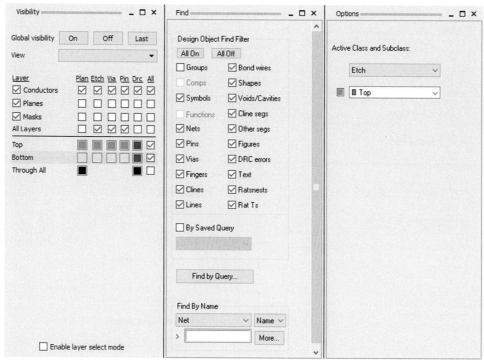

Figure 3.5 Control panel window panes.

can control visibility and selectability of objects and what particular tools can do (extent of effect, etc.) when you use them.

Visibility pane

The Visibility pane is a shortened version of the Color dialog box (described later). It provides a handy way to turn on and off routing and plane layers (or specific elements on those layers). The colored boxes and check boxes are switches you can use to toggle on and off specific items or entire rows or columns. The colored box in the Options pane (described later) works the same way, and with it, you can control the visibility of more elements using the class and subclass lists, or predefined views. It's possible to customize the set of subclasses available with this pane.

Find filter pane

The Find pane acts as a selection filter for tools and commands. By selectively checking object boxes, you can restrict which objects will be selected when you perform mouse picks in your design. The order of the object types indicates a level of hierarchy, so if all objects are enabled and you attempt to perform a mouse pick in a

(A)

(B)

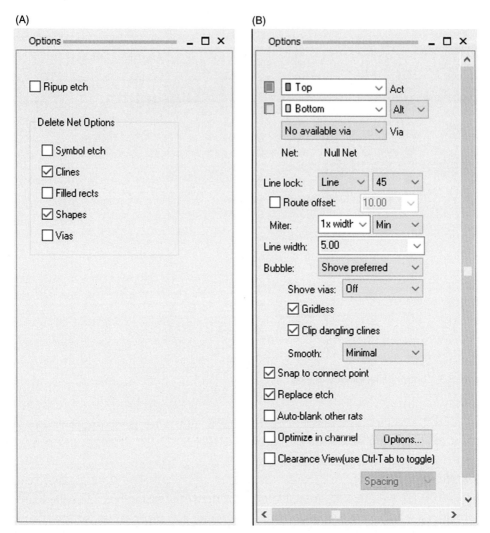

Figure 3.6 Options pane with (A) Delete tool and (B) Add Connect tool active.

congested area of your board, the top-down order indicates which object type will likely be picked. Unchecking an object prevents that type of object from being selected. *Note:* If you attempt to perform a command on an object and PCB Editor does not let you do it, check the Find pane; chances are, the box for that type of object has been unchecked. It's important to remember that for each command its own state of checked/unchecked boxes is stored. So when you have activated a certain PCB Editor command (while no object is selected), the state of check boxes in the Find pane is switched to another state which you have setup earlier exactly for this command. This is very useful feature though it can be a little bit confusing for beginners.

You can further narrow searches and selections by using the Find By Name area. Select the type of thing you are looking for from the dropdown list then click the More... button. A dialog box will pop up, which will allow you to pick specific objects from a list by name or property.

The Find By Query tool is another, even more powerful, function to find the needed objects.

Options pane

The Options pane is very dynamic and you will use it often. Its appearance is determined by what tool you have selected or what command you are running. Fig. 3.6 gives an example, where the left part shows what the Options pane looks like when the Delete tool is active and the right part shows what the Options pane looks like when the Add Connect tool is active. Along with the Visibility pane and the Color dialog box, you can turn on or off layers (classes) or parts of layers (subclasses) by toggling the colored squares. When you first start out learning PCB Editor, it is easy to forget about the Options pane, but you want to keep it in mind, as it gives you significant control over your tools. You might want to pin this one up until you get used to relying on it.

Command window pane

The Command window (Fig. 3.7) provides you with information, gives instructions, and allows you to enter commands at the Command prompt. Most of the commands you often use are also located on the toolbar and in the menus, but the Command window can give you greater control of the tools (if you know the commands). The Cadence Help can be used to search for commands, sorted alphabetically. So, for example, if you are looking for instructions on how to use the Command window to perform a

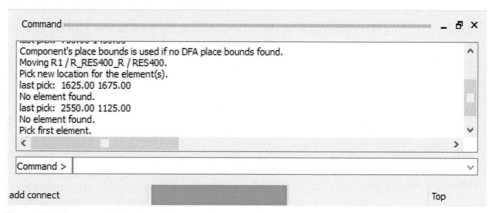

Figure 3.7 Command window pane.

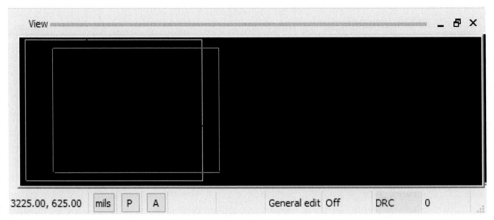

Figure 3.8 View window pane.

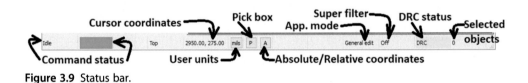

Figure 3.9 Status bar.

manual routing task (add connect) use the Help → Search... menu and search for Add Connect in the "A Commands" document.

World view window pane

The World View pane shown in Fig. 3.8 is located in the lower right-hand corner of the design window and gives you a bird's-eye overview of large boards. The white square represents your monitor, and the green outline represents the PCB's outline. View pane is interactive. If you use the highlight tool and select a part or net (whether by clicking on it or by selecting it from a list using the Find pane), the object will be displayed in the View. If you right click inside of it, a pop-up menu is displayed that lets you change the size and location of the view in the design area.

Status bar

The status bar (shown in Fig. 3.9) is located along the bottom of the design window, below the Command window and View window. At the far left of the bar the Command Status section lets you know if a command is active and running. If a tool has been selected (e.g., Move), the name or function of the tool is listed at the far left; and when a command is running, the colored box turns red. When the command has finished executing, the box turns green again.

The x, y coordinates indicate the location of your cursor. The accuracy of the display is based on the design accuracy, which you can change in the Design Parameters dialog box (select Setup → Design Parameters → Design from the menu). The coordinates are absolute (based on the origin of the design) or relative (based on the last mouse pick). You can toggle between the two at any time using the A (or R) button, as described later.

The P button is interactive. Clicking it produces a Pick dialog box, which you can use to enter coordinate points in the work area using the keyboard rather than selecting a point with the mouse. This is useful for drawing outlines and so forth on large designs, so that you need not pan around the design trying to find and select a particular point. The x and y coordinates in the Pick box are entered with space between them, not a comma.

The A (or R) button determines whether the coordinates are absolute or relative. Absolute coordinates show the cursor position relative to the design origin. Relative coordinates show the cursor x and y distance relative to the last pick point you made (whether a tool was active or not).

The Application Mode text box informs you of whether PCB Editor is in Etch Edit, General Edit or any other mode (described previously). The Super Filter allows you to set quickly only one of object types to be selectable. Don't forget to turn it off when you finished working with that type of objects. The DRC Status box lets you know at a glance if the design rule check is up to date and if any errors exist by the color of the box. If the box is any color other than green (see Fig. 3.8), then you need to update the DRC and use a DRC report to locate any errors if they exist.

Color and visibility dialog box

Color button

The Color dialog box (Fig. 3.10) is used to define custom colors for classes and subclasses and allows you to control the visibility of specific objects belonging to those classes (checked boxes indicate the object is visible, open boxes indicate they are invisible). The Color dialog box can be displayed by selecting the Color button, , on the toolbar. As mentioned previously, the Color dialog box performs some of the same functions as the colored and check boxes in the Visibility and Options panes, but as the figure shows, you have much greater control of objects and layers.

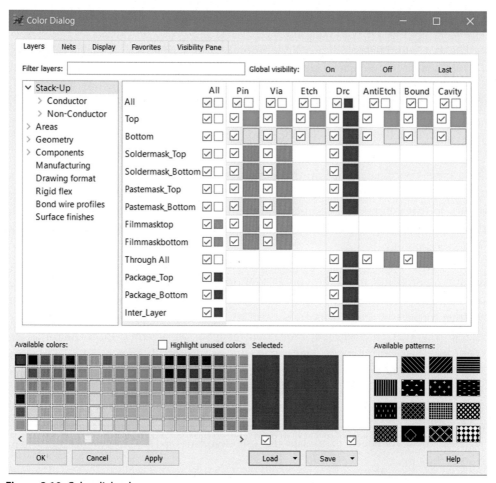

Figure 3.10 Color dialog box.

Cross-section Editor (layer stack-up) dialog box

Xsection button

The Cross-section Editor dialog box, shown in Fig. 3.11, is where you define your board's layer stack-up. From this dialog box, you can add or delete layers, define their physical and electrical properties (if you want to), and define positive and negative properties to routing and plane layers, respectively. The Cross-section Editor dialog box is displayed using the Xsection button, . Examples of setting up layer stack-ups are given in the PCB design examples in Chapter 9.

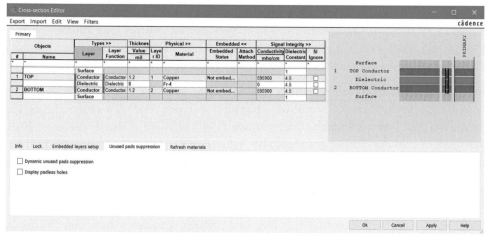

Figure 3.11 Cross-section Editor dialog box.

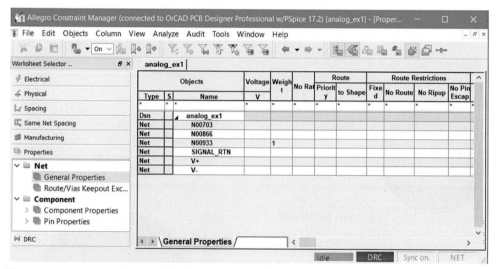

Figure 3.12 Constraint Manager dialog box.

Constraint manager

`Cmgr button`

The Constraint Manager, shown in Fig. 3.12, is where you set routing and component placement rules for your board design. The Constraint Manager can be launched using the `Cmgr` button, , by selecting `Setup → Constraints` from the menu, or by typing `cmgr` at the prompt in the `Command` window. The Constraint

Manager has several tabs and each tab has several views (indicated by folders and icons). From these tabs and views, you have very precise control over routing characteristics for every net (e.g., trace width and spacing, what vias are used, routing enabled, locking traces, etc.), every component, and shapes.

Note: the Constraint Manager takes a moment to load, but if it is taking a long time to load, check the Command window. This tool will not load if a command is still active, and the Command window (and status bar) will tell you if that is the case. If so, you need to stop the current command by right-clicking in the work space and selecting Done from the pop-up menu.

Padstack editor

The Padstack Editor is used to define and modify padstacks (see Fig. 3.13). In OrCAD PCB Editor 17.2 it was updated to provide some new functionality. The

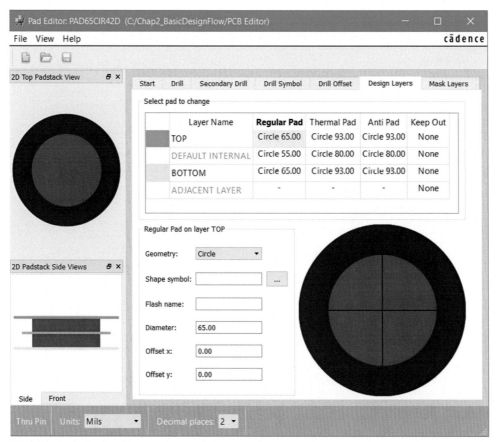

Figure 3.13 Padstack Editor.

padstacks can be from the PCB Editor library or within a PCB design only. You use `Padstack Editor` both during PCB layout work and separately during footprint development. `Padstack Editor` can be launched from within PCB Editor (from the `Tools` menu) or in stand-alone mode from the Windows `Start` menu. Many examples of using the `Padstack Editor` are presented in Chapter 8, Making and editing footprints (footprint design examples), and Chapter 9, the PCB design examples.

Manufacturing artwork and drill files

Once the board is laid out and routed, the artwork must be generated. The information in the board design is separated into specific data files (Gerber files), which are used by the board manufacturer to make the different parts of your board. The `Artwork Control Form` (Fig. 3.14) is used to specify all the different types of layers for which Gerber files will be created and the format of the files. You can add as many layers as necessary to fully define your board design. Chapter 10, Artwork development and board fabrication, describes how to set up the `Artwork Control Form` based on a board layout example.

Figure 3.14 Artwork Control Form dialog box (Film Control tab).

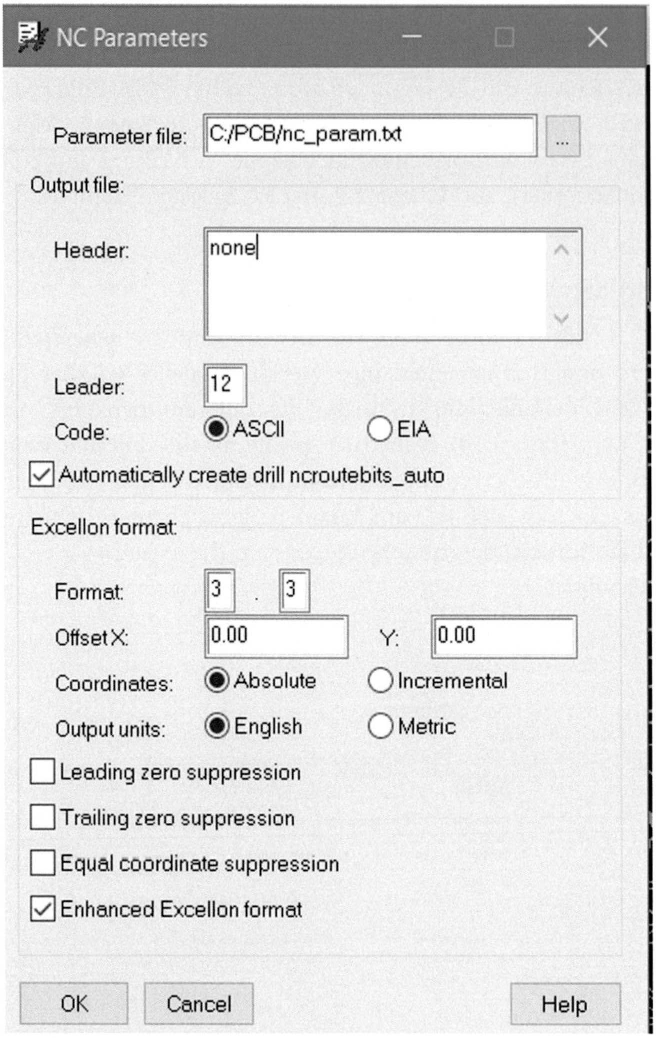

Figure 3.15 NC (drill) parameters dialog box.

The drill files are generated by selecting `Export → NC Drill…` from the menu. The drill files are generated from drill settings that you specify in the `NC Parameters` dialog box (shown in Fig. 3.15). The use of this dialog box is discussed further in Chapter 9, Printed circuit board design examples.

The number and types of files generated by the `Artwork Control Form` depends on the complexity of your board and the requirements of your board manufacturer. Most of the artwork files correspond to the layers with which you work in the layout environment, but some of the files do not. For example, files are generated for etch and silk-screen layers and drill files, all of which depend on the output format you chose.

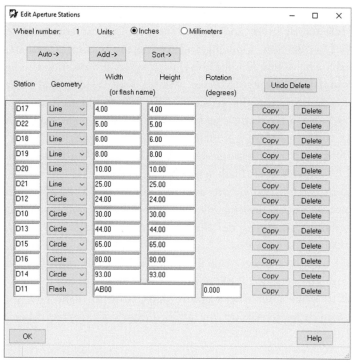

Figure 3.16 Edit Aperture dialog box.

The most common type of output formats is RS-274X. But your board manufacturer should tell you which specific files it needs and which format it prefers. Artwork instructions are set up from the Export → Gerber Parameters menu, and we will look at specific settings in Chapter 9, Printed circuit board design examples.

Understanding the documentation files

Some of the terminology used in the PCB fabrication business is left over from the early PCB manufacturing days. *Drill tapes* and *apertures* are such terms. Nowadays, a drill tape is just an electronic file that describes drill holes and sizes just like any of the other files describes its data. However, originally the drill tape was actually a role of paper or Mylar with holes punched into it that described hole sizes and locations for early computer numerical control machines, but it is still sometimes called a *drill tape*. Here is another example. In Chapter 1, Introduction to printed circuit board design and computer-aided design, the current technology of photoplotting and laser direct imaging was discussed. In contrast the older technology used a xenon flash lamp and a shutter to expose photosensitive film or glass plates. The shutter controlled the exposure and an aperture controlled the size of the exposed area. Any shape or drawing could be made by having apertures of different sizes and shapes. Light would be shone through the selected aperture and moved in the $\pm x$ direction, and the film was moved in the $\pm y$ direction. Opening and closing the shutter in one area (to make a pad) was called a *flash*, while holding the shutter open and moving the light source or film in the x and y directions (to make a trace) was called a *draw*. The technology today is more advanced, but the concept is the same, and the same terminology is used. Fig. 3.16 shows a list of the apertures the `Artwork Control Form` uses, where you will notice D codes and flash geometries (circles, rectangles, and flash symbols). A Gerber D code is a just a chronological number that specifies the size and shape of the aperture used on a given layer. Flash symbols (e.g., D code D11 in the figure) are used to produce thermal reliefs, which are used to connect plated through holes to copper planes. The procedure for drawing thermal flash symbols is described in Chapter 8, Making and editing footprints, and the use of them is described in Chapter 9, the PCB design examples.

You should now be familiar with the PCB design flow and PCB Editor tool set. In the following chapters, we look at how to assign footprints to parts from within Capture, how to make your own Capture parts, how to make new footprints with PCB Editor, and how to set up the PCB Editor's Constraint Manager to route complex boards.

CHAPTER 4

Introduction to industry standards

Contents

In the previous chapters, we looked at how to use the OrCAD® tools to design circuit schematics in Capture, route PCBs with PCB Editor, and postprocess the design for fabricating the board. We have not, as of yet, taken a look at how to design the PCB itself. Chapters 4–6 provide an introduction to PCB design. This chapter introduces industry standards related to PCB design.

Complete PCB Design Using OrCAD® Capture and PCB Editor
DOI: https://doi.org/10.1016/B978-0-12-817684-9.00004-7

Not every PCB that is manufactured makes it into service. The ratio of the number of PCBs that enter service to the number of PCBs manufactured is called *yield* (in percent). The higher the yield the better it is because failed boards cost time and money and produce waste. There are several failure points that can be addressed to increase yield. To have high yield, we need three things. First, the board has to be manufacturable, second, it has to perform properly (signal integrity and quality), and third, it has to be reliable (it has to work for the full length of its expected life span).

Being manufacturable means two things. The bare board has to be able to be *fabricated* given standard fabrication allowances (SFAs) and also has to be able to be *assembled*, that is, parts need to be able to be attached to the board with proper solder joints without damaging the parts or the board.

Performance refers to both mechanical and electrical considerations. Mechanically, the board must physically fit into its enclosure, and it must be able to handle its environment with respect to ambient temperature, vibration, and humidity. Electrical performance refers to whether it meets the operational design constraints (power, I/O, etc.), is immune to outside interference, and does not cause interference to neighboring equipment.

Being reliable means it meets the abovementioned considerations over the expected life of the device. If it is designed correctly it should not fail before the expected end of life unless the user exceeds specified operational design criteria. If the user stays within the operational guidelines and the unit still fails, then there was a bare board defect (a manufacturing defect that was not caused by your design), or there was a circuit or PCB design defect that resulted in a poor assembly process, which resulted in damage to components or board, or the circuit was operating too close to limits, which stressed the system, and failed early due to accelerated stress aging. If reliability problems exist, then a failure analysis is conducted, and the board is redesigned.

Introduction to the standards organizations

When you begin a new PCB design, you may be asking how big and what shape should the board outline be, where should the parts be placed and in what order, what kind of layer stack-up should be used, how wide and far apart should the traces be routed, and what grounding and shielding techniques should be used? Is there a "right" way to do it, and who says so?

There are several standards related to the PCB design. The organizations mentioned later set standards that may be guides, rules for certification, or even laws. To cover all aspects of these standards would fill an entire book by itself. Some of the standards organizations that you may have heard of or will hear of are listed next with a brief description of who they are and what they do. The discussion in this chapter is limited to the basic standards for PCB design. A listing of applicable design standards is presented in Appendix A.

IPC—Association Connecting Electronics Industries

The IPC is a global trade association consisting of more than 2300 member companies. It is an organization made up of contributors from industry and includes designers, board manufacturers, assembly companies, suppliers, and original equipment manufacturers. Contributing members bring lessons learned and known good practices to the table, and they document and disseminate the knowledge base through industry-accepted standards. Over the past several years the IPC standards have replaced many of the military standards (MIL-STD) and are sources that you should be familiar with. A list of IPC standards is provided in Appendix A. You can also visit the IPC website (www.ipc.org).

Electronic Industries Alliance

The Electronic Industries Alliance (EIA) was a national trade organization comprising over 1000 US manufacturers and high-tech associations and companies. Its primary focus was promoting the market development and competitiveness of the US high-tech industry in the global economy through domestic and international policy efforts. It had influence on design standards set by contributing groups, which has included the following:

• CEA—the Consumer Electronics Association
• ECA—the Electronic Components, Assemblies, and Materials Association
• GEIA—the Government Electronics and Information Technology Association
• JEDEC—the JEDEC Solid State Technology Association
• TIA—the Telecommunications Industry Association
• EIF—the Electronic Industries Foundation
• ISA—the Internet Security Alliance

JEDEC Solid State Technology Association

The JEDEC Solid State Technology Association is an independent semiconductor engineering trade organization and standardization body. It is an association of several hundred organizations, which represents all areas of the electronics industry. Its primary focus with regard to PCB design is in standardizing discrete and integrated circuit semiconductor devices and packages. You need to know the package specifications in order to design footprints for your PCB. You can access many of the standards online at www.jedec.org. A list of package specifications is provided in Appendix B.

International Electrotechnical Commission

The International Electrotechnical Commission (IEC) is an international standards organization that prepares and publishes International Standards for all electrical, electronic and related technologies – collectively known as "electrotechnology". You can find out more about the IEC (including a free online educational program) at www.iec.ch.

Military standards

MIL-STD are maintained by Defense Supply Center Columbus (DSCC), which is a field activity of the Defense Logistics Agency, whose purpose is to provide logistics and contract management support to the US armed forces. The Department of Defense develops and procures an incredible amount of material and engineering services through private contractors. MIL-STD set and communicate standards on how things are to be designed, built, and tested in a controlled, known, and acceptable manner so that all who bid on contracts know exactly what is expected of them, so that they can be successful and competitive. In recent years, specialized MIL-STD have been replaced by commercial standards such as IPC standards for PCB design and manufacturing and many other fields of engineering and manufacturing (ASME, etc.). Some of the MIL-STD are becoming obsolete, but new commercial standards are based on and update the old MIL-STD. For example, the newer IPC-2221B PCB standard originated from MIL-STD-275. You can obtain many of the old MIL-STD for free at the https://quicksearch.dla.mil website.

American National Standards Institute

The American National Standards Institute (ANSI) is a private, nonprofit organization that administers and coordinates voluntary consensus standardization and conformity assessment in the United States. Organizations may become accredited by ANSI, which signifies that their procedures meet ANSI's requirements for due process. There are a couple of PCB and electronic design standards produced through a joint effort between ANSI and other standards organizations (see the "Institute of Electrical and Electronics Engineers" section). You can find out more about ANSI at www.ansi.org.

Institute of Electrical and Electronics Engineers

The Institute of Electrical and Electronics Engineers (IEEE) is a developer of technology standards that are designed to build consensus in an open-based process with input from interested parties. IEEE is a central source of standardization in fields, such as telecommunications and power generation.

Examples of IEEE standards related to schematic design and PCB layout include IEEE/ANSI 315-1975 and IEEE-1445-1998. Visit IEEE at www.ieee.org.

Classes and types of printed circuit boards

The design approach for a PCB depends on many factors including its intended end use, design and fabrication complexity, acceptable fabrication allowances, and type of component and attachment technology. Standard classifications have been established to aid designers, fabricators, and consumers in communicating with each other on these issues. The classifications include performance class, producibility level, and type of construction.

Performance classes

PCBs can fall into any of three end-use performance classes. Throughout many of the IPC standards, (IPC-7351B, Section 1.3; IPC-CM-770E, Section 1.2.1; IPC-D-330, Section 1.1.42.6) material performance and tolerance levels are determined by the class rating. Performance classes are based on the following things: allowed variation in copper-plating thickness, feature location tolerance, and hole diameter tolerance (plated and unplated), to name a few. The three classes are as follows:

- Class 1, General Electronic Products, includes general consumer products such as televisions, electronic games, and personal computers that are not expected to have extended service lives and are not likely to be subjected to extensive test or repairability requirements.
- Class 2, Dedicated-Service Electronic Products, includes commercial and military products that have specific functions, such as communications, instrumentation, and sensor systems, from which high performance is expected over a longer period of time. Since these items usually have a higher cost, they are usually repairable and must meet stricter testing requirements.
- Class 3, High-Reliability Electronic Products, includes commercial and military equipment that has to be highly reliable under a wide range of environmental conditions. Examples include critical medical equipment and weapons systems. They typically have more stringent test specifications and possess greater environmental robustness and reworkability.
- Class 3/A, Military and/or space avionics circuits.

Producibility levels

Producibility levels are described in detail here. The levels are not a set of explicit requirements but a way of describing how complex a design is and the precision required to produce the particular features of a PCB or PCB assembly. Smaller features (trace widths, etc.) require stricter tolerances, which increase the design complexity. The IPC standards (IPC-7351B, Section 1.3.1; IPC-CM-770E, Section 1.2.2; IPC-D-330, Section 1.1.42.6) provide several tables that assist the designer in determining the complexity as it relates to SFAs. For example, issues, such as tolerances for interconnecting lands and conductor width tolerances, are described in the standards. The three producibility levels are as follows:

- Level A, general design—preferred complexity
- Level B, moderate design—standard complexity
- Level C, high design—reduced producibility complexity

Fabrication types and assembly subclasses

PCB fabrication types (IPC-CM-770E, Section 1.2.3) are indicated by a number; the higher the number, the greater the sophistication required to make the board. Issues that are related to the fabrication type are the number of copper layers (e.g., single layer, two layer, or multilayer) and the types of vias used to connect the layers, etc. The six fabrication types defined by IPC are as follows:

- Type 1, single-sided printed board
- Type 2, double-sided printed board
- Type 3, multilayer printed board without blind or buried vias
- Type 4, multilayer printed board with blind and/or buried vias
- Type 5, multilayer metal-core printed board without blind or buried vias
- Type 6, multilayer metal-core printed board with blind and/or buried vias

Each PCB type can be further defined by an assembly subclass, which describes how components will be attached to the board (IPC-CM-770E, Section 1.2.2). The subclasses are as follows:

- Subclass A, through-hole devices (THD) only
- Subclass B, surface-mounted devices (SMD) only
- Subclass C, mixed THD and SMD (simple)
- Subclass X, complex THD/SMD, fine pitch, ball grid array packages
- Subclass Y, complex THD/SMD, ultrafine pitch, chip-scale packaging
- Subclass Z, complex THD/SMD, fine pitch, flip-chip packaging

Typically more sophisticated types and subclasses require stricter tolerances (producibility levels). Higher performance class boards are made more reliable by using stricter producibility levels and lower (easier) fabrication types and assembly classes.

IPC land pattern density levels

The density levels are used to gauge PCB footprint (land pattern) designs with regard to how densely a board can be populated and with regard to the difficulty of trace routing and fabrication. More will discussed in Chapter 5, Introduction to design for manufacturing, as to how this relates to footprint design in PCB Editor. The three IPC land pattern density levels (IPC-7351B, Section 1.4) are as follows:

- Density Level A, most land protrusion (largest courtyard and least density)
- Density Level B, nominal land protrusion (median courtyard and median density)
- Density Level C, least land protrusion (smallest courtyard and highest density)

Introduction to standard fabrication allowances

No manufacturing process is perfect, and it is therefore subject to tolerance limitations. PCB manufacturing is no exception. Design tolerances include drill-hole location and diameter, copper plating and etching, and soldermask resolution, to name a few. Manufacturing tolerance becomes increasingly important as the number of layers increases and line widths and spacing decrease. The tolerance errors can add up at each manufacturing step and result in a scrapped board.

It is important to be aware of manufacturing limitations so that you stay within the boundaries of the manufacturer's capabilities. Industry standards exist to set minimum performance and process guidelines. Individual manufacturers also have their own capabilities, and you need to be aware of them as well. Just because your design meets certain minimum industry standards does not mean that every manufacturer has the ability to manufacture or assemble the PCB as designed. The following discussion covers the major design issues to look out for and references to the appropriate standards.

Registration tolerances

As described in Chapter 1, Introduction to printed circuit board design and computer-aided design (see also Coombs, 2001, p. 17.1.4), many steps and design files are required to fabricate a multilayer PCB. The design parameters in each step have to line up with the next, or misregistration can occur, which can result in manufacturing defects and a nonoperational board. One of the PCB's most vulnerable spots is the plated through hole (PTH) because it requires accurate alignment of many layers performed over several manufacturing steps.

Breakout and annular ring control

Fig. 4.1 shows how fabrication allowances result in a final hole tolerance. Fig. 4.1A shows the ideal hole, which has a specified diameter and location. Fig. 4.1B shows the uncertainty of the final hole diameter due to drilling tolerances. Fig. 4.1C shows the uncertainty of the hole location. Fig. 4.1D shows the desired hole compared to the possible hole after considering the combined uncertainties. Feature dimensions on each layer of a PCB have uncertainty, which, when combined, can result in a bad board if allowances are not designed into the board.

In addition to hole tolerances, there are also tolerances on trace/land location and size due to plating thickness and variation in etch rate and consistency. The combination of these uncertainties can result in problems such as loss of annular ring control and subsequent land breakout as shown in Fig. 4.2 (Coombs, 2001, p. 42.2). PTHs can often function with breakout, but reliability is greatly reduced. By knowing limitations of fabrication processes and following design guides you can greatly reduce the occurrence of defects and increase yield.

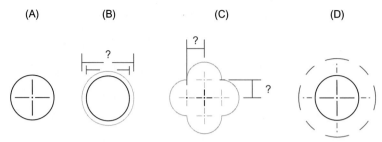

Figure 4.1 Example of fabrication tolerance and dimensional uncertainty. (A) ideal hole, (B) drilling tolerance, (C) hole location uncertainty, (D) desired hole tolerance.

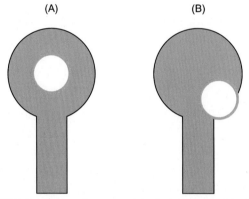

Figure 4.2 Breakout of PTH due to misalignment: (A) ideal PTH and (B) PTH breakout. *PTH*, Plated through hole.

Printed circuit board dimensions and tolerances

The following sections briefly introduce design limitations that you should be aware of.

Standard panel sizes

There are 16 industry standard board panel sizes (per ANSI/IPC-D-322; IPC-2221B, Fig. 1, p. 5—41). The boards are identified by a letter and a number, which represent the x and y dimensions. Sizes range from A1 to D4 as shown in Table 4.1. An example of a size C2 panel is listed in the table, where C2 is 7.1 × 6.7 in.

Being aware of standard panel sizes may help reduce costs since smaller PCB designs can be panelized onto one large panel. If you have flexibility in specifying the size of your board, you can do so in a way that will allow you to maximize the number of boards on one panel. This helps reduce cost by minimizing the number of parts being handled and the amount of waste generated. This is not always an option as

Table 4.1 Standard copper clad panel sizes.

Letter	Number			
	1	**2**	**3**	**4**
A	2.4 × 3.2	2.4 × 6.7	2.4 × 10.2	2.4 × 13.8
B	4.7 × 3.2	4.7 × 6.7	4.7 × 10.2	4.7 × 13.8
C	7.1 × 3.2	7.1 × 6.7	7.1 × 10.2	7.1 × 13.8
D	9.5 × 3.2	9.5 × 6.7	9.5 × 10.2	9.5 × 13.8

Sizes are given in inches.

PCB size is often driven by design constraints that are out of your control. Small boards may have to be panelized, though, if they are to be assembled and soldered by automated processes since automated machinery has minimum size limitations (and maximums as well) on the size of PCBs that they can process.

Tooling area allowances and effective panel usage

When manufacturers make PCBs, they need to have an area around the PCB outline to place tooling holes and certain manufacturing data marks. This area is called a *tooling area*. The required tooling area ranges from 0.375 to 1.5 in. (typically 1.0 in.) as measured from the edge of the board outline to the panel boundary (ANSI/IPC-D-322). On panelized designs the distance between board outlines is typically 0.1–0.5 in. The goal is to utilize as much of a panel as possible without having to go to the next larger size (IPC-D-330, Section 2, Table 2-6, p. 9).

Most of the time, these considerations are handled by the board manufacturer and are transparent to the board designer, but being aware of the issues may allow you to optimize the board layout and reduce production costs.

Standard finished printed circuit board thickness

As described in Chapter 1, Introduction to printed circuit board design and computer-aided design, a PCB is an assembly of one or more cores joined together by sheets of partially cured epoxy (called *prepreg*). By stacking the combinations of various core thicknesses and sheets of prepreg, a wide variety of finished board thicknesses can be achieved. Table 4.2 lists the typical board thicknesses in the industry. The following sections discuss standard core, prepreg and copper thickness, and tolerances. Unless you are designing controlled-impedance PCBs, you may not be immediately concerned about the discussion of each of the thicknesses described later. The information is presented here for completeness and as a reference for the curious and those who are interested in designing controlled impedance PCBs and can be used when working

Table 4.2 Typical finished board thicknesses.

Inches	Mils	Millimeters
0.020	20	0.51
0.030	30	0.76
0.040	40	1.02
0.062	62	1.6
0.093	93	2.4
0.125	125	3.2
0.250	250	6.4
0.500	500	12.7

with PCB Editor's Cross-Section Editor, which is discussed in Chapter 9, Printed circuit board design examples.

Core thickness

Cores are made up of substrates (i.e., fully cured laminate epoxy), which are then plated with copper on one or both sides. Table 4.3 shows typical laminate epoxy thicknesses without copper cladding or foil (see also Coombs, 2001, Table 5-5, p. 5−12; IPC-4101E, Table 3-7). Copper (foil and plated cladding) thicknesses are described next.

Prepreg thickness

The prepreg that is used to join the plated cores during PCB stack-up comes in sheets of various thicknesses. Table 4.4 shows the various prepreg types and their sheet thicknesses before curing (Coombs, 2001, Table 6-3, p. 6.8, and Table 10-2, p. 10.4). Board manufacturers stack-up combinations of sheets to obtain the desired board thickness. The actual thickness of a sheet once it is in a board and cured depends on whether it is between plane layers or signal layers, because signal layers tend to sink into the prepreg, which results in a thinner end thickness. The dielectric constant of the prepreg also varies by manufacturer and is discussed in Chapter 6, Printed circuit board design for signal integrity.

Copper thickness for plated through holes and vias

Electroless copper plating is used to plate holes to make PTHs and vias. While drilled holes are being plated some of the external surfaces are also plated. The thickness of the plating is typically 20−100 μin. both in the holes and in the surfaces depending on the board manufacturer's processes (Coombs, 2001, p. 28.8; MIL-STD-275, Table 4-3).

Table 4.3 Typical laminate core thicknesses (without copper).

Mils		Millimeters	
Range	Avg	Range	Avg
0.98−4.69	2.8	0.025−0.119	0.072
4.72−6.46	5.6	0.120−0.164	0.142
6.50−11.8	9.1	0.165−0.299	0.232
11.8−19.6	15.7	0.300−0.499	0.400
19.7−30.9	25.3	0.500−0.785	0.643
30.9−40.9	35.9	0.786−1.039	0.913
40.9−65.9	53.4	1.040−1.674	1.357
65.9−101	83.4	1.675−2.564	2.120
101−141	121	2.565−3.579	3.072
141−250	195	3.580−6.350	4.965

Table 4.4 Standard prepreg thicknesses.

Prepreg type	Thickness range	
	(mil)	(mm)
106	1.5−2.3	0.038−0.058
1080	2.3−3.0	0.058−0.076
2313	3.5−4.0	0.089−0.102
2116	4.5−5.3	0.114−0.135
2165	5.0−6.8	0.127−0.173
2157	5.8−6.5	0.147−0.165
7628	7.0−7.8	0.178−0.198

Later in the fabrication process other plating processes are used to "finish" the board. After finishing processes are completed, the external surfaces of a PCB can be much thicker, but the wall thickness of a PTH is usually 1 mil (25 μm) or less. Most of the time, it does not enter into the drill size calculations unless the hole is very small (most finished hole sizes are 8 mil or larger), but the information is presented here for completeness. Table 4.5 shows the minimum PTH wall thickness as presented in the literature (MIL-STD-275, Table 4-3).

Copper cladding/foil thickness

When you order your PCB, the manufacturer will need to know how thick you want the copper. The thickness of the copper depends on how much current the trace will be required to carry and the required impedance of the traces (for controlled-impedance PCBs). The thickness also plays a role in how narrow the

Table 4.5 Minimum electroless plating thicknesses (surfaces and holes).

Minimum thickness	Classes 1 and 2		Class 3	
	(mil)	(mm)	(mil)	(mm)
Average	0.79	0.020	0.98	0.025
Thin	0.71	0.018	0.79	0.020

Source: IPC-2221B, Table 4-2 (partial data).

Table 4.6 Nominal and finished copper thickness by weight and gauge ($\pm$ 10%).

Area wt. (oz/ft^2)	Nominal thickness		Internal minimum finished thickness		External minimum finished thickness	
	(mil)	(mm)	(mil)	(mm)	(mil)	(mm)
0.148 (1/8)	0.20	0.005	0.12	0.0031	0.91	0.0231
0.25 (1/4)	0.34	0.009	0.24	0.0062	1.03	0.0262
0.35 (3/8)	0.47	0.012	0.37	0.0093	1.15	0.0293
0.50 (1/2)	0.68	0.017	0.45	0.0114	1.32	0.0334
0.75 (3/4)	1.01	0.026	0.76	0.0193	1.62	0.0410
1	1.35	0.034	0.98	0.0249	1.89	0.0479
2	2.70	0.069	2.19	0.0557	3.10	0.0787
3	4.05	0.103	3.41	0.0866	4.32	0.110
4	5.40	0.137	4.63	0.118	5.49	0.139
5	6.75	0.171	5.92	0.150	6.32	0.160
6	8.10	0.206	7.13	0.181	7.28	0.185
7	9.45	0.240	8.35	0.212	8.22	0.209
10	13.5	0.343	12.0	0.305	10.9	0.277
14	18.9	0.480	16.9	0.428	14.3	0.364

traces can be because thicker copper takes longer to etch and can result in variations in trace width and etchback effects as described later. As mentioned above, various finishing processes (etching and plating) alter the final cladding or foil thickness. The processes are described in detail in the IPC standards and Coombs and are not described here. Table 4.6 shows the copper thicknesses by weight and gauge as described in the literature (Coombs, 2001, Tables 5-4 and 5-11; IPC-4101E, Table 1-2, p. 3; IPC-D-330, Section 2, Table 2-16). The values listed here are for reference only.

Copper trace and etching tolerances

When copper is etched (instead of milled), the edge of the copper trace is neither a completely smooth nor a vertical wall. The roughness (called the *edge definition*) occurs because of mask resolution limitations, nonuniformity of the acid circulation, gas bubbles during etching, etc. The wall will have a slight angle to it because, as the acid begins to work its way into the exposed copper, a sidewall begins to form, which also is attacked by the acid. As Fig. 4.3 shows, the copper near the etch resist begins to be removed under the mask. This effect is called *etchback* or *undercutting*. If the etching process is stopped as soon as the last bit of copper cladding or foil is removed from the surface of the board, the trace width at the bottom will be the initial size of the mask width (which is defined by the Gerber files). For this reason the wider value (W, not w', in Fig. 4.3) is used for design calculations when trace width dimensions are required.

Width tolerances range from ± 4.0 to ± 0.6 mil for 1.5 oz copper depending on producibility level and plating (MIL-STD-275). The designer should be aware of the width variations when calculating trace widths for controlled impedances and for current handling ability. For general design considerations, traces should be made as wide as practical. Per IPC-2221B the minimum trace width and spacing is 3.9 mil. Individual board manufacturers may have their own etching and spacing tolerances (MIL-STD-275). Typical minimum trace widths are 4—8 mil. It is a good idea to call and ask what their capabilities are or check their websites.

Standard hole dimensions

Holes are drilled into PCBs by various techniques including twist bits, router bits, lasers, and plasmas. The capabilities with regard to placement and size accuracy and the speed can vary considerably. But the board designer needs to know what size of drill hole to specify in a padstack. Standard drill bit sizes are specified in ANSI standard B19.11M. However, not all board manufacturers carry every size bit. When you specify a particular hole size manufacturers have different ways of "adjusting" it to fit their capabilities. Some may round (up or down) to the nearest drill size available, or they may always round up to the next largest drill bit to make sure that the hole is never too small. This affects annular ring width, which can lead to breakout as described earlier. IPC-D-330 (Section 2, Table 2-4) lists minimum PTH sizes by board thickness and class.

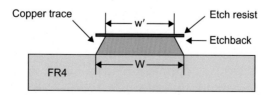

Figure 4.3 Result on finished trace width due to etchback.

When laminate materials are drilled, they become soft due to frictional heating. The softened laminate then smears during the drilling process, which coats the surface of the copper and can prohibit good plating (Coombs, 2001, p. 10.7). To solve this problem the laminate inside the hole is etched back after being drilled to desmear the hole. Etchback enhances the plating of PTHs (Coombs, 2001, p. 48.30), but excessive etchback can cause partial delamination and internal shorts when combined with misregistration and may make the hole larger than requested. This is one reason why adequate clearance is required between the PTH and the ground plane (Coombs, 2001, p. 17.9).

After the hole is drilled and desmeared, it is plated. Depending on the manufacturer's plating methods and processes, the plating can add as much as 1 mil of thickness on all surfaces, which means that the diameter of the hole could be as much as 2 mil smaller than the drill tool diameter. The padstack drill size in PCB Editor is the finished hole size. In most cases the variance in available drill bit size and plating is not a problem. But if your holes are very small and/or the lead diameters are very close to the size of the drill hole, the hole variations can cause a lead to not fit into its through hole. Padstack calculations that account for plating widths and tolerances are described in the next chapter. Another problem that can occur is that if the hole size is very small compared to the thickness of the board, the plating process may not be satisfactory. The "proper" hole diameter to PCB thickness is called the *aspect ratio* (AR).

AR is the ratio of the plated hole diameter to the PCB thickness. It is recommended that the AR be between 3:1 and 5:1 for Level A boards (IPC-2221B, Table 5-1, p. 40; also IPC-CM-770E; and IPC-D-330, Table 6-30), with 3:1 being a good target (Coombs, 2001, p. 42.3).

The AR for Level B boards is 6:1−8:1, and Level C boards may be as high as 9:1 or higher. Check with your board manufacturer to know what its capabilities are, because if the AR is too high, plating problems can occur inside the hole, which can lead to open circuits from incomplete plating and barrel cracking.

Padstack design, which includes all of the issues described earlier, is covered in detail in Chapter 5, Introduction to design for manufacturing, and Chapter 8, Making and editing footprints.

Soldermask tolerance

Due to photolithography misregistration and swell of the soldermasks, lands can be partially covered. Soldermasks that are patterned on solder-plated lands may be damaged when the solder reflows during soldering operations and can adversely affect solderability. This may be especially troublesome on very small parts (SOT23). To reduce these risks the soldermask openings are usually larger than the lands (oversized). There are two basic categories of soldermask materials: liquid screen printed and photoimageable masks. The recommended oversizing for liquid-screen-printed coatings is 16–20 mil, and the recommended oversize of photoimageable masks is 0–5 mil (IPC-2221B). Many of the PCB Editor footprints have clearances of between 0 and 10 mil. Many fabricators will adjust the soldermask as necessary for their processes. However, some do not and expect (or assume) that you will do it. Check with your board house to find out if it requires oversizing and, if so, how much and who it expects to do it.

References

ANSI B94.11M-1993. *Twist drills.*–ANSI.org.

ANSI/IPC-D-322. *Guidelines for selecting printed wiring board sizes using standard panel sizes.* Lincolnwood, IL: IPC/Association Connecting Electronic Industries.

Coombs, C. F., Jr. (Ed.), (2001). *Coombs' printed circuit handbook* (5th ed.). New York: McGraw-Hill.

IEEE-1445-1998. *IEEE standard for digital testing interchange format.*–standards.ieee.org.

IEEE/ANSI 315-1975. *Graphic symbols for electrical and electronics diagrams.*–standards.ieee.org.

IPC-2221B. *Generic standard on printed board design.* Northbrook, IL: IPC/Association Connecting Electronic Industries. Easy reading, ~160 pages, lots of pictures.

IPC-4101E. *Specification for base materials for rigid and multilayer printed boards.* Northbrook, IL: IPC-Association Connecting Electronics Industries.

IPC-7351B. *Generic requirements for surface mount design and land pattern standard.* Northbrook, IL: IPC/Association Connecting Electronic Industries. Easy reading, ~100 pages, lots of pictures.

IPC-CM-770E. *Guidelines for printed board component mounting.* Northbrook, IL: IPC/Association Connecting Electronic Industries.

IPC-D-330. *Design guide manual.* IPC/Association Connecting Electronic Industries.

MIL-STD-275. Printed Wiring for Electronic Equipment (similar to and superseded by IPC-2221B) *(free download).*–https://quicksearch.dla.mil.

Further reading

Highly suggested reading for new PCB designers is given below. A more thorough list is given in Appendix B.

47CFR15. *FCC rules on radio frequency devices (including unintentional radiators).* Free download.–Federal Communications Commission (FCC).

ASME B18. *Series standards for mounting hardware (includes tables for standard screw sizes and recommended drill hole sizes, etc.).*–ANSI.org.

MIL-HDBK-1861A. *Selection and use of electrical and electronic assemblies, boards, cards, and associated hardware.* Free download.–https://quicksearch.dla.mil.

MIL-HDBK-198. *Selection and use of capacitors.* Free download.– https://quicksearch.dla.mil.

MIL-HDBK-199. *Selection and use of resistors.* Free download.– https://quicksearch.dla.mil.

MIL-HDBK-5961A. *List of standard semiconductor devices.* Free download.–https://quicksearch.dla.mil.

CHAPTER 5

Introduction to design for manufacturing

Contents

Introduction to printed circuit board assembly and soldering processes

For a PCB to be manufacturable the bare board has to be able to be fabricated within standard fabrication allowances, and the board has to be able to be assembled given the different component technologies. As discussed in the previous chapters, there are many steps to designing and fabricating a PCB. When a design is complete and submitted to a board house, the house must be able to perform the manufacturing steps that the design calls for. Whether a PCB is manufacturable or not really begins with and includes parts creation and schematic entry in Capture, padstack and footprint design, parts placement and trace routing, and artwork production in PCB Editor.

But it does not end there. Once the board has been fabricated, it is of little use without the functional parts. Those parts need to be able to be attached to the board without them or the board being damaged. Parts attachment encompasses positioning the parts (which depends on proper parts placement and orientation) and making reliable solder joints between the component's leads and the board's mounting pads (which depends on good padstack and footprint design). This chapter provides the information necessary for padstack and footprint design and component placement for the design of manufacturable PCBs.

Complete PCB Design Using OrCAD® Capture and PCB Editor
DOI: https://doi.org/10.1016/B978-0-12-817684-9.00005-9
83

Assembly processes

PCBs may be manually assembled or assembled by automated machinery. Assembly processes depend on the class of component technology (Classes A through Z as described earlier in Chapter 4: Introduction to industry standards) and the number of boards to be assembled at one time. Some companies may both fabricate and assemble boards under one roof, while some companies may specialize in PCB fabrication only, and others in PCB assembly only. The method of assembly plays a role in how you lay out your PCB because of clearance and orientation issues and soldering processes. A brief summary of assembly processes is given here along with component placement, orientation, and spacing considerations.

Manual assembly processes

Manual assembly is typically used for prototype and low-volume work and in postautomated assembly for odd-form components. Both surface-mount technology (SMT) and through-hole technology (THT) components may be assembled manually. In low-volume work an assembly line of several assemblers may be used, in which each person is responsible for attaching specific components. The assembly processes may be interrupted several times to functionally test sections of the PCB as it is assembled. Manual assembly may involve both manual placement and soldering or a mixture of manual placement and automated soldering (described below).

Manual assembly can be tedious work. Consistent component placement and orientation in the PCB project can aid manual assemblers. For example, orienting polarized components (capacitors and diodes) in the same direction and orienting integrated circuits (ICs) so that pin 1 on all ICs is located in the same direction can significantly reduce assembly defects and increase yield.

Automated assembly processes (pick and place)

Automated insertion processes (pick and place) exist for both surface-mounted devices (SMDs) and through-hole devices (THDs), which include both radial- and axial-leaded devices. Automated machines are programmed to extract parts from reels or bins and place the components on the PCB in the correct location and orientation. THDs can typically be populated at rates from 20,000 components per hour (CPH) to 40,000 CPH (Coombs, 2001, Section 41.2.5, p. 41.10). Data for programming automated placement machinery can be supplied by output from PCB Editor and other computer-aided manufacturing programs.

THDs are usually packaged as roles or strips of components, which are taped together by their leads. Through-hole components are usually placed only on the top side of the board so that the leads can be wave soldered, and the components themselves are not exposed to molten solder. Soldering processes are described below.

The typical automated process step for THDs is the insertion of dual inline packages first, then axial-leaded devices, then radial-leaded devices, and finally odd-form devices. After the components have been inserted, the board is most often wave soldered (wave soldering is described below) but can be reflow soldered (intrusive reflow; see Coombs, 2001, p. 43.10), also described below.

SMDs are commonly packaged in tubes, matrix trays, tape and reel, and bulk. SMDs may be mounted on one or both sides of a PCB. When attached to the top side only, solder paste is screen printed onto the PCB's solder pads. The parts are then placed onto the board by the pick-and-place equipment with the component lead terminations set into the paste, temporarily holding the parts in place. The board is then run through a reflow oven, which melts and then cools the solder, thereby attaching the part to the board. SMDs can typically be populated at rates from 10,000 to 100,000 CPH (Coombs, 2001, Section 41.3.2, p. 41.15).

When SMDs are placed on both sides of a PCB or when a board contains both surface and THDs, a sequential, reflow/wave soldering process is employed. First the top-side SMDs are attached to the board using the solder paste and reflow process described below. Next, THDs are inserted from the top and held in place by clinching (bending) the leads on the bottom, by gluing the part to the top, or in some cases by friction between the lead and the hole. The board is flipped over and adhesive dots are applied to the bottom of the board by an automated dispenser. The bottom-mounted SMDs are then positioned manually or by pick-and-place machines onto the glue dots. The adhesive holds the bottom-side SMDs in place until the solder joint has been completed. The assembly is run through an oven to cure the adhesive. The board, with through-hole components on the top and SMDs on the top and bottom, is then run through a wave-soldering station, which solders the through-leads and the bottom-mounted SMDs. The previously soldered top-mounted SMDs remain soldered on the top.

When a PCB has only SMDs but has them on both sides, a two-step reflow soldering process is sometimes used. The top-mounted SMDs are attached to the board first using a high-temperature solder paste, and then the board is run through a high-temperature reflow oven. With the top-side components securely in place the board is flipped over and the bottom-mounted SMDs are attached with the lower temperature paste and reflow process as described below. Very often the same type of solder paste is used for the both sides of PCB. Then it's important for the designer to place all of the heavy components on one side of PCB that will be assembled in the second step, to prevent them falling down into the oven during the second reflow process.

Soldering processes

Soldering is used both to attach components physically to the PCB and to provide electrical conductivity between the component's leads and the PCB traces. For the soldering process to be successful an intermetallic compound, or alloy, must be formed between the solder and the base material (the leads and traces). To protect the solder joint areas from oxidation, contact areas on new PCBs receive a surface finish by being dipped in a solder bath and hot-air solder leveled or are plated by some other plating process such as electroless nickel—gold or palladium (Coombs, 2001, p. 32.1). Just prior to or during soldering, the surfaces to be joined are cleaned (deoxidized) with flux so that the solder can flow over and wet the surfaces.

There are two general soldering methods: mass soldering (which includes wave, oven reflow, vapor phase reflow, and conduction reflow) and directed energy (which includes hot gas, hot bar, laser, iron, and pinpoint torch) (see Coombs, 2001, p. 43.10; IPC-CM-770E, p. 34). Only manual, wave soldering, and oven-reflow soldering are discussed here and only briefly.

Manual soldering

Manual soldering is used for a wide variety of applications from complete PCB assembly to simple repair work and touch-up. There are several types of manual soldering tools available, including but not limited to hot-air pencils, soldering irons, and induction coils. Other than slower soldering speed, the biggest drawbacks to manual soldering are the increased risk of electrostatic discharge during handling and thermal gradients caused by localized heating of the board and parts. Parts placed on the board that will be manually placed and soldered require no special layout consideration as far as spacing and orientation other than the basics described below. However, it is helpful to the assembler if parts placement affords room to work and similar parts are aligned and oriented in a consistent manner as described above.

Wave soldering

During wave soldering the board is held by its edges on a conveyor, fluxed, and preheated as shown in Figs. 5.1 and 5.2. The conveyor moves the board past a standing wave of molten solder so that only the bottom side of the board is exposed to the solder. Wave soldering can be used for both THDs and SMDs, but reflow is preferred for SMDs. The through-hole components are placed on the top with the leads protruding out the bottom of board, which is prefluxed. As the conveyor moves the PCB into the solder wave, solder wicks up the barrel and creates fillets on the top and the bottom. SMDs are glued to the board, fluxed, and run through the wave. Very small components or large tantalum caps can be problematic with wave soldering. The small parts cause problems during the gluing process because in some cases the glue dot is larger than the component and the glue can ooze over onto the solder pads. Large SMDs cause problems because of thermal stresses that can lead to component cracking.

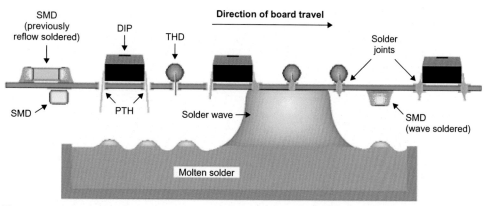

Figure 5.1 Wave soldering (side view): *DIP*, dual inline package; *PTH*, plated through hole.

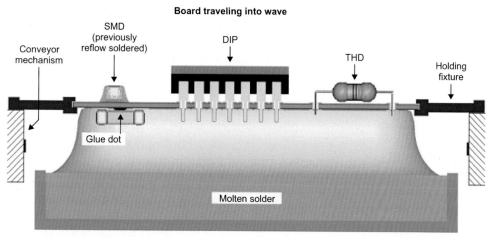

Figure 5.2 Wave soldering (rear view).

When SMDs are wave soldered (i.e., mounted to the bottom side of the board), the designer needs to know (or specify) the direction the board will travel through the wave. The components should be placed on the board as shown in Fig. 5.3 so that smaller parts are not shadowed by larger parts, which can cause poor solder joints on the smaller parts, and so that solder bridging does not occur across the leads.

Solder bridging can occur because of the fine lead pitch of some SMDs. As each pin leaves the solder wave, it tends to draw some of the solder from the pin beside it as the molten solder attempts to reduce its surface tension (for the orientation shown in Fig. 5.3). This causes a problem at the last two rows of pins as there are no pins that follow to draw the excess solder away from them, which can result in the last two pins on a side being bridged by the excess solder. One method to minimize the

Direction of board travel

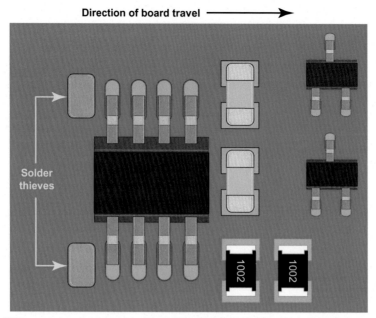

Figure 5.3 SMD component orientation for wave soldering. *SMD*, Surface-mounted devices.

bridge is to place solder thieves on the trailing edges of SMD ICs as shown in Fig. 5.3. The solder thieves are extra pads placed after the last pads that pull the excess solder away from the pads. Solder thieves can also be made by simply making the last pads a little larger and extending farther back than the typical pads on that device.

If SMD parts are on top of the board and are reflow soldered prior to wave soldering, fan-out vias need to be located away from lands to prevent solder migration away from the SMD lands and down into the via. This can occur as heat conducts from the solder wave up via barrels and re-reflows the solder, which then draws the solder down the via by capillary action. The suggested spacing is ≥ 20 mil (IPC-7351B) between the edge of the via and the edge of the pad. A default spacing can be set in PCB Editor using the Fanout tab in the Automatic Router Parameters dialog box.

Some boards may be too large or too small to be wave soldered. Large boards sag as they are heated unless special holding fixtures are used. Very small boards may need to be panelized with breakaways (tab routes or V-scores) so that they can be handled by automated equipment without having to make specialized board holders.

Reflow soldering

There are various types of reflow soldering, but the discussion here is limited to oven-type reflow soldering. Reflow soldering is most often used for surface-mount devices, but through-hole components can also be soldered this way

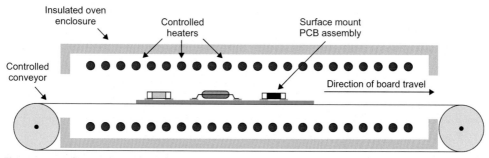

Figure 5.4 Reflow soldering oven (side view).

(called *pin-in-paste* or *intrusive reflow soldering*). A schematic diagram of a reflow oven is shown in Fig. 5.4.

Solder paste (which contains solder and flux in a viscous paste) is applied to a bare board with a stencil. The components are placed into position (usually by pick-and-place machines) so that the leads sit in the solder paste. The assembly is placed on a conveyor and run through a temperature-controlled oven that heats the boards and parts evenly. As the assembly heats up the flux is activated and the solder reflows (melts). Surface tension of the molten solder tends to self-align the parts. However, if parts are not in proper alignment and thermal gradients do not melt the solder on all pads at the same time, the component may stand up on one end (called *tombstoning*). Once the assembly has reached the proper temperature and the solder has reflowed, the assembly exits the oven and is cooled to solidify the solder.

Solderable components need to be selected for both wave and reflow soldering. It is also important not to place all of the thermally massive parts in the same area of the board, as this can cause thermal gradients across the board and poor solder joint formation in the cooler areas (Coombs, 2001, p. 43.12; for board design, see Coombs, 2001, p. 43.25).

Component placement and orientation guide

A PCB assembly consists of the bare board, the attached components, and connectors. The board design can have a significant impact on how easily components can be placed on and attached to the board and how reliable the end product is.

The board topology (class and level), the component technology (SMD or through hole), and the soldering method used to attach the parts (reflow vs wave soldering) play significant roles in how parts should be placed and spaced on the board.

Component placement and orientation depend on the type of component (THT or SMT), the assembly method (manual, wave soldered, or reflow), and the electrical performance requirements (electrical discussed below). These by themselves do not make the board function better, but they do make it easier to assemble, inspect, test, and troubleshoot. General guidelines are

1. Components should be placed so that they are neat and organized with uniform spacing and alignment.

2. Components should be oriented such that component edges are parallel to the board edges (with the exception noted below for wave soldering).

3. If a board is machine soldered, through-hole components should be mounted on a single side that is opposite the solder whenever possible.

4. When placing components on both sides of the board and when mixing through-hole and SMD technologies, keep in mind that multiple assembly phases may be required to place all of the components, which increases cost and potential failure points and makes rework more difficult.

5. Do not put plastic-leaded chip carriers or large tantalum capacitors on the bottom side of board (i.e., do not wave solder them), as they can easily crack due to thermal stresses.

6. A 100-mil (2.5-mm) grid should be used if possible, but a 20-mil (0.5-mm) grid or even 2-mil (0.05-mm) grid can be used if required for component leads that are not on standard grids (IPC-2221B, Section 5.4.2 and Section 8.1.2). When a metric grid is used, a grid of 2.54 mm (0.100 in.), 1.27 mm (0.05 in.), 0.64 mm (0.025 in.), or 0.50 mm (0.020 in.) should be used (IPC-7351B).

7. A 0.100-in. (2.54-mm) grid should be used for boards that will undergo bed-of-nails testing.

8. Polarized capacitors and diodes should all be oriented consistently throughout the board for ease of inspection and testing.

9. When machine–vision-assisted assembly processes are used, add fiducials (global and local) to aid in component placement.

10. If the design allows, connectors should be placed on the short side of the board when using automated soldering processes.

11. Allow adequate space along board edges during component placement for handling the board and to accommodate mounting hardware.

12. Components that weigh more than 5.0 g per lead should be mechanically supported if the board will experience vibration (IPC-2221B, Section 5.2.7, p. 43).

13. Thermal management during soldering processes and circuit operation should be considered throughout the design process.

14. Electrical considerations usually have priority over mechanical considerations when there is a conflict between the two, unless it will result in a mechanical failure of the board.

15. For mixed-signal (analog/digital) PCBs, components should be segregated to minimize the effect of switching noise on analog circuits. High-power circuits should also be segregated from low-power and low-noise circuits.

Component spacing for through-hole devices

The following tables provide minimum recommended spacing guidelines for discrete and IC THDs.

Discrete through-hole devices

For discrete THDs, see Tables 5.1 and 5.2.

Table 5.1 Minimum recommended spacing for discrete, axial through-hole devices.

Parameter		Mils	Millimeters	PCB Editor default
Side to PCB edge *End to PCB edge*	(a) (b)	$a = 75$ $b = 90$	$a = 1.9$ $b = 2.29$	Depends on pad to track space setting. No DRC error occurs as long as place outline does not cross the package keep-in outline
End to end		100	2.54	DRC error occurs if place outlines overlap
Side to side When body diameters are <100 mil (2.54 mm)		100	2.54	DRC error occurs if place outlines overlap
Side to side When $D_2 > D_1 > 100$ mil	(a) D_1 (b) D_2	$a = 70 + (1/2)D_1$ $b = 10 + (1/2)$ $D_1 + (1/2)D_2$ (a and $b \geq 100$ mils minimum)	$a = 1.78 + (1/2)D_1$ $b = 0.25 + (1/2)$ $D_1 + (1/2)D_2$ (a and $b \geq 2.54$ mm minimum)	DRC error occurs if place outlines overlap
Side to end When one or more body diameters are >100 mil	D_1	$95 + (1/2)D_1$	$2.41 + (1/2)D_1$	DRC error occurs if place outlines overlap

DRC, Design rule checker; *PCB*, printed circuit board. Reference: IPC-2221B, Fig. 7–1. p. 68.

Table 5.2 Minimum recommended spacing for discrete, radial through-hole devices.

Parameter		Mils (millimeters)	PCB Editor default
PCB edge		$r = 1/2$ the diameter of the device or $1/2$ the height, whichever is greater, *and* $r \geq 60$ mil (1.52 mm)	No DRC error occurs as long as place outline does not cross the package keep-in outline
Others parts		$r = 1/2$ the diameter of the device or $1/2$ the height, whichever is greater	DRC error occurs if place outlines overlap

DRC, Design rule checker; *PCB*, printed circuit board.

Integrated circuit through-hole devices

For IC THDs, see Table 5.3.

Table 5.3 Minimum recommended spacing for through-hole mounted integrated circuits.

Parameter		Mils	Millimeters	PCB Editor default
Side to PCB edge		$a = 100$	$a = 2.54$	No DRC error occurs as long as place outline does not violate the package keep-in outline
End to edge		$b = 75$	$b = 1.91$	
End to end		200	5.08	No DRC error occurs as long as place outlines do not overlap. Could violate IPC standard and not cause DRC error
Side to side		100	2.54	

DRC, Design rule checker; *PCB*, printed circuit board.

Mixed discrete and integrated circuit through-hole devices

For mixed discrete and IC THDs, see Table 5.4.

Table 5.4 Minimum recommended spacing between through-hole discretes and integrated circuits.

Parameter		Mils	Millimeters	PCB Editor default
	$a\ (D_1 > 100)$ $b\ (D < 100)$ $c\ (D < 100)$ $d\ (D_1 > 100)$	$115 + (1/2)D_1$ 200 100 $40 + (1/2)D_1$	$2.91 + (1/2)D_1$ 5.08 2.54 $1.02 + (1/2)D_1$	No set rule. DRC error occurs if place outlines overlap

DRC, Design rule checker; *PCB*, printed circuit board.

Holes and jumper wires

For holes and jumper wires, see Table 5.5.

Table 5.5 Minimum recommended spacing for holes and jumper wires.

Parameter		Mils (millimeters)	PCB Editor default
Hole to hole (plated or nonplated)		1. So as not to violate pad spacing rules 2. So that residual laminate is >20 mils (0.5 mm) between holes	DRC error occurs if pad-to-pad spacing rules are violated
Jumper wires (any direction)		100 mils (2.54 mm)	DRC error occurs if pad-to-pad spacing rules are violated

DRC, Design rule checker; *PCB*, printed circuit board.

Component spacing for surface-mounted devices

The following tables provide minimum recommended spacing guidelines for discrete and IC SMDs.

Discrete surface-mount devices

See Table 5.6 for discrete SMDs.

Table 5.6 Minimum recommended spacing for discrete surface-mounted devices.

Parameter		Mils	Millimeters	PCB Editor DRC
Side to PCB edge and/or end to PCB edge		60	1.5	DRC error occurs if space from pad to edge of board outline is less than pad-to-track spacing rule or if place outline crosses package keep-in outline
End to end and/or side to side		Size 0603 or larger		Spacing determined by relationship of place outline to pads and body. DRC error occurs if place outlines overlap
		20	0.50	
		Smaller than 0603		
		12	0.30	
Pad to via		20	0.50	DRC error occurs if distance between edge of pads is less than pad-to-pad spacing rule

References: IPC-2221B, p. 73; IPC-7351B. Tables 3.5–3.8, Fig. 3.15.
DRC, Design rule checker.

Integrated circuit surface-mount devices
For IC SMDs, see Table 5.7.

Table 5.7 Minimum recommended spacing for integrated circuit surface-mounted devices.

Parameter		Mils	Millimeters	PCB Editor default
Component side to PCB edge and/or end to PCB edge		60	1.5	DRC error occurs if space from pad to edge of board outline is less than pad-to-track spacing rule or if place outline crosses package keep-in outline
End to end (body)		20	0.50	Spacing determined by relationship of place outline to pads and body. DRC error occurs if place outlines overlap

(Continued)

Table 5.7 (Continued)

Parameter		Mils	Millimeters	PCB Editor default
Side to side (pad to pad)		20	0.50	

DRC, Design rule checker; PCB, printed circuit board. References: IPC-2221B, p. 73; IPC-7351B, Tables 3.2—3.22.

Mixed discrete and integrated circuit surface-mount devices

Use the greater of any of the preceding spacing rules for the components involved.

Mixed through-hole devices and surface-mounted devices spacing requirements

Use the greater of any of the preceding spacing rules for the components involved (usually the THT spacing).

Footprint and padstack design for printed circuit board manufacturability

Many footprints are included with the PCB Editor software, but you will need to make your own at some point. Chapter 8, Making and editing footprints, describes the PCB Editor tools that are used to design the footprints, and the following describes the design considerations and industry standards related to designing footprints. Footprint design for SMDs and THDs are significantly different, but both require consideration of PCB manufacturability and assembly. A footprint (land pattern) in PCB Editor consists of padstacks, silk-screen elements, and a place-boundary outline (see Chapter 8: Making and editing footprints, for specifics).

There are some useful component spacing constraints in PCB Editor based on distances between the component boundaries and other objects. The footprint design (the size of the place-boundary outline) determines the spacing and therefore the maximum board density possible [without incurring design rule checker (DRC) errors]. Per the IPC standards, there are several spacing recommendations, which are dependent on the board classification and package types. PCB Editor's footprints do not necessarily fall within specific IPC design standards. So if you need to produce a Level C board, you will need to modify or make new footprints that fall within the specific guidelines. Or if you want footprints that specifically meet Level A requirements to increase yield and reliability, then you may also need to design special footprints or modify existing ones. The following discusses some of the design issues and provides references to the industry standards.

Land patterns for surface-mounted devices

When you are doing a board layout and need a part that is not in the PCB Editor footprint library, there are several things you can do to construct one. If PCB Editor has a footprint that is similar to the one you need, but it has the wrong number of pins, you can save a copy of the existing part with a new name and use it as a pattern to add pins and resize the place outline and silk screen as necessary. If PCB Editor has no footprint that you can use as a pattern, the next step is to check with the manufacturer's data sheet to see if it has a suggested land pattern. If it does not provide the necessary information, then you may be able to find the design parameters from the IPC Land Pattern Viewer available now as the free software tool named PCB Library Expert. If all else fails you will need to design one from scratch using the package information from the component's data sheet or one of the JEDEC package standards.

Component data sheets and the JEDEC package standards provide information about the package dimensions, but not the PCB land pattern, typically. The IPC standards provide guidance on the land pattern dimensions relative to the lead dimensions but does so in a way that requires additional calculations to construct the padstacks and footprints in PCB Editor. When designing footprints in PCB Editor, you need to know the size of the padstack and the distances between the centers of the pads. The IPC standards typically provide spacing information relative to the edges of the pads.

So for convenience, a way of translating the JEDEC package dimensions directly into PCB Editor footprint dimensions while maintaining compliance with IPC standards is needed.

When you look up a part's data sheet, the dimensions are usually with regard to the package. An example of an eight-pin small outline integrated circuit (SOIC) is shown in Fig. 5.5. The part manufacturer may also provide the JEDEC specification. Fig. 5.6 illustrates an example of JEDEC standard MS-012 for an SOIC package. Fig. 5.6 is reprinted with permission of JEDEC.

From the manufacturer's information and/or the standards, we need to be able to determine the padstack width and height and the spacing between the pads in both the x and the y directions as shown in Fig. 5.7, in order to create a new footprint in PCB Editor.

We begin by looking at the design of the padstack followed by footprint design using the established padstacks.

Surface-mounted device padstack design

A good padstack promotes the best possible solder joint between a component termination (lead) and the PCB. The padstack must allow for component dimensional variations, PCB fabrication tolerances, placement tolerances, and solder fillet specifications. THDs are relatively large and therefore more forgiving of these tolerances, but SMDs are

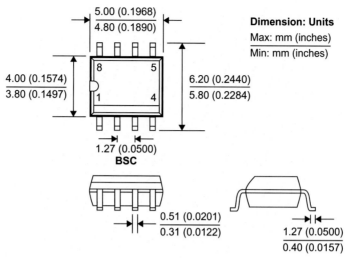

Figure 5.5 Data sheet package dimensions (typical convention).

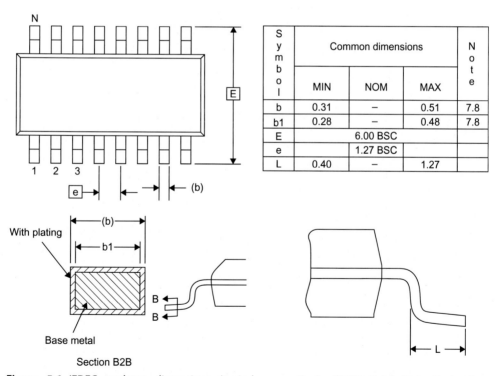

Figure 5.6 JEDEC package dimensions (typical convention). *JEDEC*, Solid State Technology Association. *From JEP95, MS-012. Copyright JEDEC. Reproduced with permission by JEDEC.*

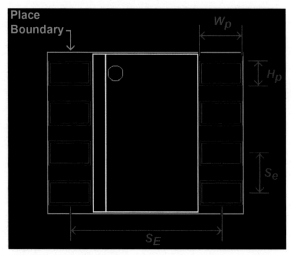

Figure 5.7 Footprint dimensions (typical convention).

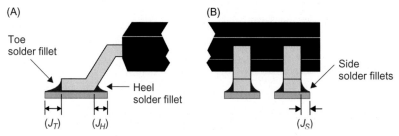

Figure 5.8 SMD padstack requirements: (A) side view, (B) toe view. *SMD*, Surface-mounted device.

typically much smaller and are therefore more sensitive to manufacturing and placement variations. IPC-7351B (which superseded IPC-SM-780/2) is the standard for surface-mount land pattern design for both padstack and footprint design.

As shown in Fig. 5.8, the solder pad (padstack) needs to be larger than the component lead to allow a proper solder joint. The required pad oversize is defined in IPC-7351B (pp. 10−21), where the term J_T defines the distance from the end of the pad to the toe of the lead, J_H defines the distance from the end of the pad to the heel of the lead, and J_S defines the distance from the sides of the pad to the sides of the lead.

The values for J_T, J_H, and J_S depend on the type of component and the desired density level (A through C); nominal values are provided in Tables 5.8−5.10.

To design a padstack in PCB Editor, we need to know the width of the padstack, W_P, and the height of the padstack, H_P, to be able to fill in the required values in the PCB Editor's Padstack Editor tool (Fig. 5.9).

Table 5.8 Nominal toe solder fillet values (J_T) by package type.

Package type	Nominal density	
	Mils	Millimeters
Gull wing (SOG)	14	0.35
J lead (SOJ)	14	0.35
Chip components (0603 and larger)	14	0.35
Chip components (smaller than 0603)	4	0.10
Small outline (SO)	12	0.30
Tantalum capacitors	6	0.15
Leadless chip carrier	22	0.55
MELF	16	0.40
Butt joints	31	0.80

MELF, Metal electrode face.

Table 5.9 Nominal heel solder fillet values (J_H) by package type.

Package type	Nominal density	
	Mils	Millimeters
Gull wing (SOG)	14	0.35
J lead (SOJ)	−8	−0.20
Small outline (SO)	0	0.00
Chip components (all)	−2	−0.05
Tantalum capacitors	20	0.50
MELF	4	0.10
Leadless chip carrier	6	0.15
Butt joints	31	0.80

MELF, Metal electrode face.

Table 5.10 Nominal side solder fillet values (J_S) by package type.

Package type	Nominal density	
	Mils	Millimeters
Gull wing (SOG) (pitch greater than 0.625 mm)	1	0.03
J lead (SOJ)	1	0.03
Gull wing (SOG) (pitch less than 0.625 mm)	−1	−0.02
Chip components (0603 and larger)	0	0.00
Chip components (smaller than 0603)	0	0.00
MELF	2	0.05
Small outline (SO)	−2	−0.04
Tantalum capacitors	−2	−0.05
Leadless chip carrier	−2	−0.05
Butt joints	8	0.20

MELF, Metal electrode face.

Figure 5.9 Pad design using Padstack Editor. (A) copper layers, (B) mask layers.

Using the parameters from the component data sheet or the JEDEC standard, Eqs. (5.1) and (5.2) can be used to determine maximum sizes for W_P and H_P.

$$W_{P(MAX)} = E_{MIN} - (E_{MAX} - 2L_{MIN}) + 2J_T + 2J_H$$
$$+ \sqrt[2]{(E_{TOL(\Delta)})^2 + F^2 + P^2} \tag{5.1}$$

where $W_{P(MAX)}$ is the maximum pad width (see Figs. 5.7 and 5.9); $E_{(MIN\ and\ MAX)}$ is the distance between the ends of the leads per the JEDEC dimensions (Fig. 5.6); $E_{TOL(\Delta)}$ is the tolerance of E per the JEDEC dimensions or calculated from $(E_{MAX} - E_{MIN})$ (Fig. 5.6); L is the length of the lead that will be soldered to the pad per the JEDEC dimensions (Fig. 5.6); J_T and J_H are solder fillet allowances as described in Tables 5.8 and 5.9, which are derived from IPC-7351B; F is the PCB fabrication tolerance (IPC-2221B, 0.1 mm or 4 mil typically); and P is the placement tolerance of pick-and-place machines (depends on the machine, 0.15 mm or 6 mil is typical); and

$$H_{P(MAX)} = b_{MIN} + 2J_S + \sqrt{(b_{TOL(\Delta)})^2 + F^2 + P^2} \tag{5.2}$$

where $H_{P(MAX)}$ is the maximum pad height (see Figs. 5.7 and 5.9), b_{MIN} is the minimum lead width, J_S is the solder fillet allowances as described in Table 5.10, $b_{TOL(\Delta)}$ is the tolerance of b per the JEDEC dimensions or calculated from $(b_{MAX} - b1_{MIN})$ (Fig. 5.6), and F and P are as described for Eq. (5.1).

These equations were derived from various tables in IPC-7351B, but the actual standard also includes rounding factors, which are not included in Eqs. (5.1) and (5.2). Tables 5.8−5.10 were also derived from various tables in IPC-7351B, but the source standard includes additional data for greater and lesser density levels, while only the nominal density levels are included here. Please see the source for full dimensional considerations or download the IPC Land Pattern Viewer (a demo version is available for free from www.pcblibraries.com/).

Once the pad is designed (i.e., you have calculated the width and height), use the procedures given in Chapter 8, Making and editing footprints, to create the pad with `Padstack Editor`.

Surface-mounted devices footprint design

Once the padstacks are designed, they need to be correctly located to complete the footprint design. An example of how IPC-7351B defines land pattern parameters is shown in Fig. 5.10 for an eight-pin SOIC. The component outline defines the outermost boundary of the IC including both the edge of the package and the end of the leads. Next, a courtyard is defined around the part that includes the body and the basic land pattern. The courtyard is adjusted outward from the part outline to protrude into and consume PCB real estate. The amount of protrusion (called *courtyard excess*) determines the minimum separation between components during placement. The greater the courtyard excess, the less densely the PCB can be populated. Density levels are categorized by Levels A through C, as listed in Table 5.11 for various SMD packages (from IPC-7351B, Tables 3-2—3-22).

PCB Editor does not use the courtyard concept used by the IPC-7351B standard for its SMD footprints. Fig. 5.11 shows the eight-pin SOIC place-boundary outline in PCB Editor overlaid onto the IPC protruded courtyard outline. Notice that the PCB Editor place-boundary outline does not include the excess space at the ends of the body or around the leads.

For most of the PCB Editor footprints included with the software library, the courtyard excess is nonexistent. Also in OrCAD® PCB Editor's Constraint Manager there are some package spacing constraints that you can set, they are located in its

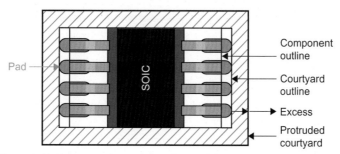

Figure 5.10 IPC-7351B land pattern description of an eight-pin SOIC. *SOIC*, Small outline integrated circuit.

Table 5.11 Courtyard excess (protrusion) by density level for various surface-mounted devices.

Package type	Density Level A		Density Level B		Density Level C	
	Mils	Millimeters	Mils	Millimeters	Mils	Millimeters
Gull wing (SOG)	20	0.50	10	0.25	4	0.10
J lead (SOJ)	20	0.50	10	0.25	4	0.10
Small outline (SO)	20	0.50	10	0.25	4	0.10
Chip components (0603 and larger)	20	0.50	10	0.25	4	0.10
Tantalum capacitors	20	0.50	10	0.25	4	0.10
MELF	20	0.50	10	0.25	4	0.10
Leadless chip	20	0.50	10	0.25	4	0.10
Chip components (smaller than 0603)	8	0.20	6	0.15	4	0.10
Butt joints	59	1.50	31	0.80	8	0.20

MELF, Metal electrode face.

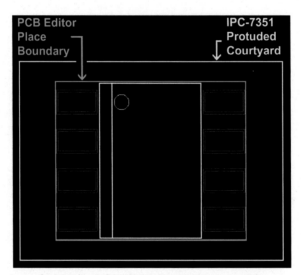

Figure 5.11 Comparison of PCB Editor place boundary to IPC courtyard.

`/Manufacturing/Design for Assembly/Spacing/` subfolder. The much more powerful constraint table exists in higher level tier—Allegro Venture PCB Editor, which is not discussed in this book. If you didn't set up the "Manufacturing" constraints in OrCAD PCB Editor, the DRC looks for violations only where place outlines cross. So satisfying density level requirements requires that you either modify the footprint definitions or set the place grid conservatively and manually check each of the components individually using Tables 5.6 and 5.7 as guides.

Standardized footprint design for THDs is perhaps not as widely known as the SMD standards. This may be partly because of the greater variety of package styles of the THDs and partly because their typically larger size makes them less sensitive to many of the design and manufacturing issues related to SMDs. As a result, land pattern standards or guides for THDs are harder to come by for many devices and may not exist for others. In this section, we look at how to design footprints and padstacks for THDs.

Land patterns for through-hole devices
Footprint design for through-hole devices

THDs fall generally into one of two categories: axial leaded or radial leaded. An example of a radial–leaded capacitor is shown in Fig. 5.12. The footprint design for this type of device is determined strictly by the construction of the device. Clearly the padstacks have to be located where the leads extend from the body. Radial-leaded devices include pin grid arrays and many discrete transistor devices such as TO-220 and TO-92 packages. The only variable is the padstack design with regard to the lead diameters. Padstack design is described below.

Footprint designs for components with axial leads are highly variable compared to the radial–leaded devices, but the IPC standards (IPC-CM-770E, Section 11.1.8) and

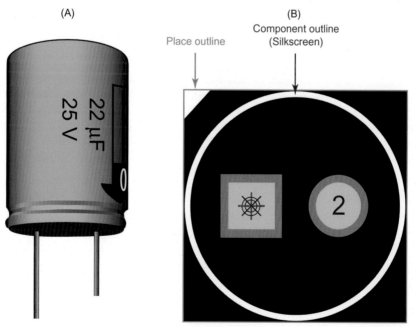

Figure 5.12 Radial-leaded through-hole device: (A) radial-leaded capacitor, (B) PCB Editor radial footprint.

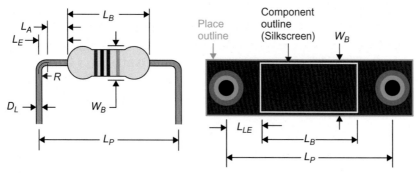

Figure 5.13 Generic footprint design parameters for axial-leaded components.

Table 5.12 Bend radius and lead extension allowances.

D_L (mil)	R (mil)	L_A (mil)
$D_L < 31$	$1.0 \times D_L$	31
$31 \leq D_L \leq 47$	$1.5 \times D_L$	D_L
$D_L > 47$	$2.0 \times D_L$	D_L

other sources in the literature provide general guidance on footprint design. Fig. 5.13 illustrates the design parameters for an axial-leaded component (in this case carbon resistor). The location of the padstacks depends on the length of the body and the location of the lead bends (lead form). The minimum required distance between the padstacks is calculated using the following equation:

$$L_P = L_B + 2(R + L_A), \qquad (5.3)$$

where L_P is the pad spacing (center to center), L_B is the length of the body, R is the bend radius allowance (as read from Table 5.12), and L_A (as read from Table 5.12) is the length of the lead extension from the end of the body to the beginning of the bend. Note that R and L_A are dependent on the diameter of the lead, D_L, and that L_E is the sum of L_A and R. Padstack calculations are described in the next section. Once this minimum padstack spacing is calculated, the padstacks should be placed on the closest standard grid location. Standard grid spacing is described under Component Placement and Orientation Guide. In most cases a 100–mil grid is used. The total length of both leads should not exceed 1 in. (25 mm) unless the component is mechanically supported.

Padstack design for through-hole devices

The biggest vulnerability of PCB fabrication is the plated through hole (PTH) because of registration errors, aspect ratio, plating, etc. The likely failure points are breakout (annular ring control) due to misalignment and open due to thermal stress and cycling during soldering processes (assembly and rework). But if we follow the standards and fabricator's design guides, then we will usually have little trouble.

In designing the PTHs for through-hole components, we need to consider the lead-diameter/drill-hole relationship, the PCB-thickness/drill-hole relationship (aspect ratio), the width of the copper ring around the hole (annular ring), the clearance of plane layers from the PTH, and the capabilities of the board manufacturer (see Figs. 4.1 and 4.2). Both Coombs and the IPC standards are consulted here to define a process for designing PTHs in `Padstack Editor`.

Hole-to-lead ratio

The size of the PTH should be large enough so that the lead can easily slip into the hole, but it should not be so large that it prevents capillary action during soldering operations. There are two methods for determining the required hole size relative to the diameter of the lead that will be soldered into it. The first method is derived from a combination of IPC-2221B and Coombs (Section 42.2.1, p. 42.3). This method takes into account the plating thickness (either assumed or known) and allows the designer to define a tolerance factor. This method produces a clearance between the lead and the PTH that varies with the diameter of the lead. The drill tool diameter is calculated using the following equation:

$$D_H = (D_L + 2T_P) \times k, \tag{5.4}$$

where D_H is the diameter of the drill tool in `Padstack Editor`, D_L is the diameter of the lead (per the component's data sheet or by measurement), T_P is the thickness of the plating inside the hole (if not known, use $T_P = 1$ mil), and k is a user-defined tolerance factor, for which $1.05 < k \leq 3.0$ (1.5 is recommended). The finished hole diameter in `Padstack Editor` is $2 \times T_P$ smaller than D_H.

As an example, if the diameter of a component lead, D_L, is 32 mil, then the drill bit should be $D_H = (32 + 2 \times 1) \times 1.5 = 51$ mil, and the finished hole is 49 mil. With the exception of the variable k, this method is explicit and easy to use. However, as lead diameters become larger, the clearance between the lead and the hole becomes wider, and at some point the clearance may become too large for proper capillary action to take place. The point at which this occurs is not well documented in the literature.

The second method is to use a lookup table in which the finished drill-hole size is dependent on the lead diameter and the desired producibility level (A—C) as shown in Table 5.13 (derived from IPC-2222A, Table 9-5, p. 25). So as an example using this method, if we use the same 32-mil lead diameter and the data sheet specifies a 10% tolerance on the lead diameter, then the maximum lead diameter is 35.2 mil and the

Table 5.13 Hole-to-lead size relationship by producibility level.

	Level A		Level B		Level C	
Hole size	Mils	Millimeters	Mils	Millimeters	Mils	Millimeters
Minimum = max lead diameter +	10	0.25	8	0.20	6	0.15
Maximum = min lead diameter +	28	0.70	28	0.70	24	0.60

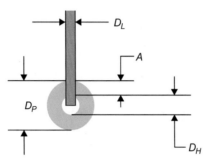

Figure 5.14 PTH design parameters. *PTH*, Plated through hole.

minimum is 28.8 mil. And if we want a producibility level of A, then the hole diameter would be between 45.2 mil (35.2 + 10) and 56.8 mil (28.8 + 28).

If you search the Internet, you can find other algorithms for calculating the "proper" hole-to-lead ratio. Which method you choose is up to you. The two methods here were derived from the IPC standards and other sources, but they are not hard and fast requirements.

Plated through hole land dimension (annular ring width)

Once the size of the drill hole has been determined, the next step is to determine the diameter of the pad (also called the *land*). The difference between the pad diameter and the drill-hole diameter is called the *annular ring*. This is the copper area where the solder joint is made with the component lead. The larger the pad diameter is, the larger the solder joint will be—to a point. Too large a pad does not necessarily make a better solder joint and only increases the amount of heat required to make the solder joint. Too small a pad can cause a weak solder joint, and the pad can be easily damaged by heat or mechanical stress and lift off from the board. The pad diameter recommended by IPC–2221B (p. 96) is calculated using the following equation:

$$D_P = a + 2b + c, \tag{5.5}$$

Table 5.14 Annular ring requirements for internal and external rings.

	Internal	External
Mils	1	2
Millimeters	0.025	0.05

Table 5.15 Standard fabrication allowances for plated through hole design.

	Level A	Level B	Level C
Mils	16	10	8
Millimeters	0.40	0.25	0.20

where D_P is the pad diameter (see Fig. 5.14), a is the finished hole size ($a = D_H - 2T_P$ from Fig. 5.14), b is the minimum annular ring requirements from Table 5.14 (IPC-2221B, Table 9-2), and c is the standard fabrications allowances from Table 5.15 (IPC-2221B, Table 9-1).

As an example, if we use the 32-mil lead from the example above (using a hole size of 50 mil) and we are calculating the size of an external pad (TOP or BOTTOM in Padstack Editor) and we want a Level A producibility level, then the pad size would be $(50 - 2 \times 1) + (2 \times 2) + 16 = 68$ mil.

Clearance between plane layers and plated through holes

PTHs and vias typically have lands on all layers even when there is no connection to it on that layer; these are called *nonfunctional lands*. IPC-2222A (Section 9.1.4, p. 23) states that nonfunctional lands should be used on internal layers when possible, but nonfunctional lands are not required on every layer if the board is over 10 layers thick, and they are not required on plane layers. When a via or PTH passes through a plane layer, a minimum amount of space is required between the plane and the plated hole or its nonfunctional land to allow for fabrication allowances, to minimize wicking of plating solution into the laminate from the through hole (resulting in internal shorts to internal plane layer), and to meet voltage withstanding (insulation) requirements.

IPC-2221B (Section 6.3.1, p. 57) states that the trace-to-trace spacing requirements (see also Table 6.8 in Chapter 6: Printed circuit board design for signal integrity, of this text) apply to the clearance between the plane edge and the PTH or its land. Furthermore, per IPC-2222B (p. 22, Fig. 9-1), the minimum clearance between the plane edge and the edge of internal nonfunctional lands or plated holes is to be a minimum of 10 mil (0.25 mm), that is, the clearance diameter is to be 20 mil (0.51 mm) larger than the drill diameter.

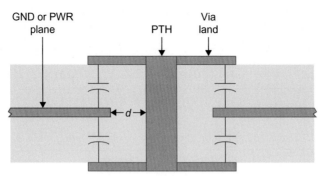

Figure 5.15 PTH or via with oversized lands overlapping plane. *PTH*, plated through hole.

Normally, the padstacks in PCB Editor do not contain nonfunctional lands on plane layers, so the clearance diameter should be 20 mil (0.51 mm) larger than the drill diameter (i.e., the edge of the plane should be 10 mil from the edge of the plated through hole barrel). Padstacks that are not designed this way may cause a problem during fabrication or operation of the board.

If the land (pad) is designed per Eq. (5.5) and the minimum clearance between the hole and the plane is used, the clearance diameter will typically be several mils larger than the lands. If the lands are larger than the clearance diameter, however, there will be an overlap of the lands and the plane layers as shown in Fig. 5.15, even if the clearance (dimension *d*) meets IPC standards. This should be avoided as there will be capacitive coupling between the pad and the plane, which can alter the characteristic impedance of the trace and could cause cross-talk problems at high frequencies. Characteristic impedance is described in greater detail in Chapter 6, Printed circuit board design for signal integrity.

Another thing to pay attention to is that the clearance diameter should be no larger than necessary, as large slots in the ground (return) plane can result if several pins are close together (as in a board or socket connector), which can cause problems with high-speed digital and high-frequency analog circuits when signal lines pass between the pins. These forces return signals on the ground planes to go around the open slot in the ground plane and increase the loop inductance of the circuit. We will look at loop inductance and ground planes in Chapter 6, Printed circuit board design for signal integrity.

Soldermask and solder paste dimensions

The soldermask opening is typically *larger* than the pad (land) diameter in `Padstack Editor`. Board manufacturers sometimes require a particular oversized opening (5 mil is not uncommon). If they do not automatically adjust the size for you, then you will need to modify the soldermask opening before you create the PCB artwork.

The solder paste dimensions are usually the same size as the external lands (TOP and BOTTOM layers). Most of the PCB Editor footprints that come with the default library do not have a solder paste defined. If you will need the solder paste definitions make sure that they are properly specified for your specific assembly needs.

References

Coombs, C. F., Jr (Ed.), (2001). *Coombs' printed circuit handbook* (5th ed.). New York: McGraw-Hill.

IPC-2221B. November 2003. *Generic standard on printed board design.* Northbrook, IL: IPC/Association Connecting Electronic Industries.

IPC-2222A. December 2010. *Sectional design standard for rigid organic printed boards.* Northbrook, IL: IPC/Association Connecting Electronic Industries.

IPC-7351B. June 2010. *Generic requirements for surface mount design and land pattern standard.* Northbrook, IL: IPC/Association Connecting Electronic Industries.

IPC-CM-770E. January 2004. *Guidelines for printed board component mounting.* Northbrook, IL: IPC/Association Connecting Electronic Industries.

IPC-SM-780/2. July 1998. *Component packaging and interconnecting with emphasis on surface mounting.* Northbrook, IL: IPC/Association Connecting Electronic Industries.

CHAPTER 6

PCB design for signal integrity

Contents

Complete PCB Design Using OrCAD® Capture and PCB Editor
DOI: https://doi.org/10.1016/B978-0-12-817684-9.00006-0
111

Desirable electrical characteristics of a circuit and its PCB include low noise, low distortion, low cross talk, and low radiated emissions, to name a few. The purpose of this chapter is to introduce the issues that cause PCB performance problems and how to route the PCB to minimize them and maximize signal integrity.

Circuit design issues not related to PCB layout

Circuit design constraints are primarily the responsibility of the circuit design engineer and will not be covered in detail here, but some of the issues will be mentioned briefly since the symptoms of poor circuit design can be confused with PCB design problems.

Noise

Noise generally refers to any signal that interferes with or degrades a signal of interest. It is often used with an adjective for problems, such as phase noise, switching noise, cross talk noise, and reflection noise. In this text we will limit the term *noise* to mean random or pseudo-random, natural signals, which are generally not a result of the PCB design. Functional problems such as cross talk or ringing (which are PCB-related problems) will be named as such. From this perspective there are two basic categories of noise: background noise and intrinsic component noise. These noise problems are generally addressed by the circuit designer, not the PCB designer, but are discussed here briefly for completeness.

Background noise

Background noise is an uncontrolled signal that originates from the system or environment your board is working in. For example, if your circuit is an audio amplifier that is supposed to amplify a speaker's voice as he or she speaks into a microphone, but a crowd of people is talking around the speaker or a jet plane flies overhead, both the speaker's voice and the background sounds will be amplified and the signal would be considered noisy or said to suffer from a low signal-to-noise ratio. There is nothing you can do about it from the PCB design perspective. Sensors may also be noisy because of their sensitivity, but that is also a circuit design issue and needs to be handled long before the PCB is laid out.

Intrinsic noise

There are four basic types of intrinsic noise: thermal noise, shot noise, contact noise, and popcorn noise. Thermal noise (also known as *Johnson noise*) is due to the motion of electrons in a conducting material. It is present in any material that exhibits a resistance to current flow and is a function of temperature. It is white noise (is constant over frequency) and is prominent in resistors and semiconductor devices. Shot noise is also white noise and is due to potential barriers and is also prevalent in semiconductor devices.

Contact noise (also called *excess noise in resistors* and *1/f noise*) is due to imperfect connections at contact junctions or interfaces. It is not constant over frequency and can be fairly large at low frequencies. Your best defense against this type of noise is good-quality connectors and good solder joints.

Popcorn noise (also called *burst noise*) is typically proportional to $1/f^2$ and is worse in high-impedance circuits. It is caused by manufacturing defects in semiconductors and ICs.

Distortion

Distortion is an issue more related to analog circuitry because of the nature of continuous signals. In analog circuitry, all voltages between the power supply rails may be of significance. Digital signals are not continuous: they are either HI or LO and usually nothing in between matters. As long as voltage levels meet threshold specifications, there is no ambiguity and therefore no quality issues. Ringing on the rising and falling edges of a square wave might be considered distortion, but that is handled differently, as described below. Distortion of a sinusoidal signal (which normally has a single spike on a frequency spectrum) begins to occur in amplifiers as the sine waves either are clipped or experience a phase reversal. Op amps have amplitude limits imposed by the power supplies, their drive capabilities, and their frequency response. If the amplitude of a sinusoidal output signal (as determined by the input signal times the gain) exceeds the output capability of the op amp, then the output signal will be clipped off and begin to resemble a square wave. Square waves are composed of many sine waves, which are primarily odd harmonics of the fundamental frequency of the square wave. The dominant harmonic is typically the third one, so as a sine wave begins to clip the onset of third harmonic distortion is observed.

If the input signal exceeds the op amp's input limits (as imposed by the power supply rails), the output signal will also be distorted. Some amplifiers simply clip the signal (causing third harmonic distortion), while other op amps experience phase inversion, which also causes harmonic distortion.

These problems are caused by the circuit design and component selection and are not the fault of the PCB design. These effects are mentioned because, if you are not used to them or do not know about them, they can be confused with PCB layout problems. Along with harmonic distortion ringing will produce unwanted frequency components, which can be seen with a spectrum analyzer and may be confused with other forms of distortion or noise. Ringing is caused by reflections, which in turn are caused by impedance mismatches on PCB traces, which *is* a function of the PCB design.

Frequency response

Both analog and digital circuits have frequency limits. In digital circuitry, if frequency limits are exceeded the signal level may rise and fall before a gate has a chance to switch states. This may give the appearance that the signal is attenuated or that the receiving gate is "not seeing" the signal. This too is a circuit design problem and not a PCB problem. In circuit design, we need to make sure that the components selected are within design constraints.

When signals exceed the frequency limits of analog circuitry, the output signal will also be attenuated, and distortion will result if the sine wave begins to look like a triangle wave at the output of the frequency-limited component. This is a function of the amplifier's slew rate, -3 dB BW, and gain bandwidth product. Again these issues need to be handled at the circuit design level, well before the PCB design stage.

Issues related to PCB layout

Electromagnetic interference and cross talk

There are three goals in designing PCBs for electrical performance and signal integrity: (1) The PCB should be immune from interference from other systems, (2) it should not produce emissions that cause problems for other systems, and (3) it should demonstrate the desired signal quality. A common factor relating these three issues is electromagnetic waves. As Fig. 6.1 shows, "noise" can be introduced into your PCB from outside sources, and it can produce noise that is radiated to other systems and to itself.

When electromagnetic waves get into your system, this is referred to as *electromagnetic interference* (EMI). On the flip side, your PCB can be the source of EMI and cause

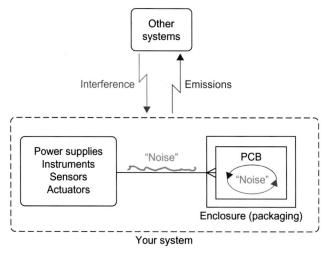

Figure 6.1 The enemy—electromagnetic interference.

problems for other systems. The ability for systems to "play nice together" is referred to as *electromagnetic compatibility* (EMC). The FCC has established rules for many types of systems regarding EMI and EMC, which, depending on your application, your PCB may have to abide by. Properly laying out your PCB can greatly reduce EMI and improve EMC. In this section we take a look at how to minimize EMI and its effects. There are many good books available that address these issues in greater detail. The material in this chapter is not intended to duplicate those works but to provide an overview of the issues and provide insight on how to design PCBs with PCB Editor with regard to signal integrity issues.

The method by which systems and circuits can "reach out and touch" another circuit is inductive and capacitive coupling of electromagnetic fields.

In the 1820s Faraday and Henry showed that an electric current could be produced in a conductor by changing the current in another, nearby conductor (Serway, 1992, p. 806). And years later, Maxwell showed that changing electric fields also produce magnetic fields. These fields are the source of many woes in PCB design. We begin by looking at magnetic fields and inductive coupling.

Magnetic fields and inductive coupling

As shown in Fig. 6.2 a magnetic field vector, **B**, is developed around a conductor when current flows through the conductor, into the "X" end and out of the "Dot" end of the conductor. The right-hand rule (from Ampere's law) is used to determine the direction of the field: If the thumb of the right-hand points in the direction of conventional current flow (movement of positive charges), then the magnetic field

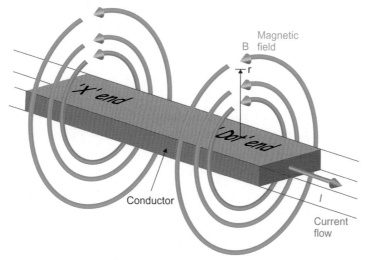

Figure 6.2 Magnetic field caused by a current-carrying conductor.

curls in the direction of the fingers. This is defined mathematically by a cross product called the *Biot—Savart law*. By applying some calculus, which will not be shown here, an equation can be derived for the scalar (nonvector) magnitude, B, of the magnetic field vector, **B**, near the conductor.

The magnitude of the magnetic field a distance, r, from a long straight conductor (Serway, 1992, p. 838) is given by Eq. (6.1),

$$B = \frac{\mu_0 I}{2\pi r^2},\tag{6.1}$$

where B is the magnitude of the magnetic field in Wb/m², μ_0 is the permeability of free space ($\mu_0 = 4\pi \times 10^{-7}$ Wb/A m), I is the current in amps (A), and r is the distance from the conductor.

The total magnetic field within or through an area is called magnetic flux, Φ, which has units of Wb and is described by Eq. (6.2) (Serway, 1992, p. 849; Wb/m² $\times$ m² = Wb):

$$\Phi = BA\cos(\theta),\tag{6.2}$$

where B is the magnetic field magnitude per unit area (Wb/m²), A is the area intersected by the magnetic field (m²), and θ is the angle between B and A.

Magnetic flux expands or contracts in proportion to changes in current flow. As the flux expands or contracts around the conductor, we see from Faraday's Law of Induction, given in Eq. (6.3) (Serway, 1992, p. 877), that a voltage is induced into the conductor. This is known as *self-induced electromotive force* (emf):

$$E = -\frac{d\Phi}{dt}.\tag{6.3}$$

The minus sign in Eq. (6.3) is a result of Lenz's law, which states that the emf induced into the conductor produces a current in the conductor that creates a magnetic flux that will oppose the changing magnetic flux. This effect is called *self-inductance*. The self-inductance tends to limit how fast the current can change in a conductor. This is what makes an inductor have inductance and oppose AC currents in analog circuits and fast rise times in digital circuits. The magnitude of the inductance, L, is shown in Eq. (6.4):

$$L = \frac{N\Phi_m}{I}.\tag{6.4}$$

The inductance is directly proportional to N, the number of turns on a coil ($N = 1$ for a PCB trace and its return path), and the magnetic flux and is inversely proportional to the current, I.

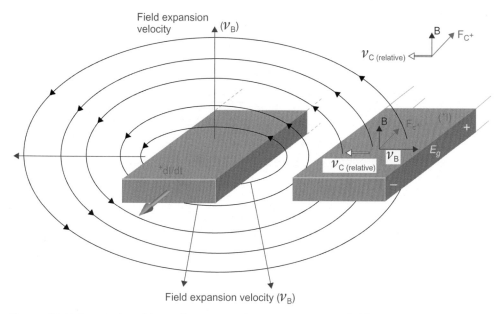

Figure 6.3 Voltage induced into adjacent trace by varying magnetic fields.

If a secondary conductor is near a primary conductor, which is carrying current as shown in Fig. 6.3, some of the flux will be felt by the secondary conductor. If the current through the primary conductor changes with respect to time (as in the case of a rising edge of a digital signal or an AC signal) the magnetic field (and therefore the magnetic flux) also changes with respect to time as it increases in strength and expands outward (or decreases in strength and contracts inward) from (or toward) the primary conductor.

When the flux that is impinging on the secondary conductor changes with respect to time, as it expands or contracts, we see again from Faraday's Law of Induction in Eq. (6.3) that a voltage, E_g, is induced into the secondary conductor. Using the right-hand rule for electric generators, the direction of the induced current can be determined. To use the right-hand rule point the thumb in the direction of the motion of the conductor (relative to the expanding/shrinking flux—ν_C), point the first finger in the direction of the B field, and the middle finger will point in the direction of the force applied to the positive charges (F_c^+) and therefore the positive current flow. The resultant voltage and current are shown in Fig. 6.3. Notice that the current in the secondary conductor flows in the direction opposite that of the current in the primary conductor as a result of the induced voltage. This produces a secondary magnetic field that opposes and partially cancels the primary magnetic field (by Lenz's law).

If the induced current flow in the secondary conductor is changing with respect to time (which it is because the primary conductor is causing it to) and it is in close proximity to the first conductor (which it is), then the secondary conductor will also induce an emf back into the primary conductor. The magnetic flux that goes back and forth between the two conductors is called *mutual inductance*.

The emf (voltage) generated into the primary conductor (by the secondary conductor) will be in a direction that aids the original current flow in the primary conductor as the secondary flux tries to oppose the primary flux. If the original flux is partially canceled, then the self-inductance is also partially canceled and the changing current in the primary conductor is not as limited (i.e., it sees less inductance).

It would seem that unlimited current could flow in the primary inductor since the secondary conductor aids it, but the secondary inductor also experiences its own self-inductance and is therefore limited. Also the amount of mutual inductance between the two conductors is limited by how much of the flux couples the two conductors. This is similar to the coupling coefficient in transformers and in both circumstances is never 100%.

When a trace on a PCB induces a voltage into an adjacent signal trace we call that *cross talk*, which is bad because it generates noise in the adjacent signal trace. But if the second conductor is the PCB's ground plane, that is good because it reduces the trace's inductance and therefore the overall *loop inductance* of the circuit. For the time being we will stick with using the term *return plane*, rather than *ground plane*, for reasons described later.

Loop inductance

In Fig. 6.3 only segments of the traces were shown. Of course for current to flow through the circuit, a closed path must exist, as shown in Fig. 6.4. Any conductor in the circuit that carries current will produce a magnetic field as indicated by the circular arrows.

An equation for inductance is given in Eq. (6.5) (Serway, 1992, p. 905),

$$L = \mu_0 n^2 A \ell, \tag{6.5}$$

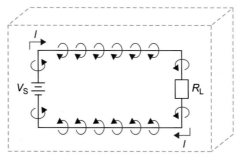

Figure 6.4 Loop inductance of a closed circuit.

where n is the number of turns (1), $A\ell$ is the volume that the circuit occupies, and μ_0 is the relative permeability of the material in which the circuit exists. For most materials used in PCBs, $\mu_0 = 1$. So the inductance is a function of the volume of the inductor and the number of turns of the conductor around the space. Therefore, inductance is dependent on the circuit geometry, by which a smaller volume results in a smaller circuit or loop inductance.

If we look at the closed–loop circuit shown in Fig. 6.4 with respect to its volume, we can see that, if the circuit is physically large and makes a large loop, it will have what is referred to as high loop inductance.

If on the other hand we arrange the circuit as shown in Fig. 6.5, we can see from Eq. (6.5) that the loop inductance will be less since there is less volume. If you notice the direction of the arrows (the magnetic fields) you can see that the source current and the return current magnetic fields oppose each other, thereby reducing the flux and inductance.

Fig. 6.6A shows the resultant magnetic field of two conductors in close proximity where the currents are in the same direction. As indicated the magnetic fields circulate in the same direction and aid each other. This is the case for an inductor in which the turns are wound in the same direction and build up an overall strong magnetic field.

Fig. 6.6B shows the resultant magnetic field when the currents of the two conductors are in opposite directions. The magnetic fields oppose each other and result in

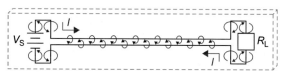

Figure 6.5 Closed-loop circuit with low loop inductance.

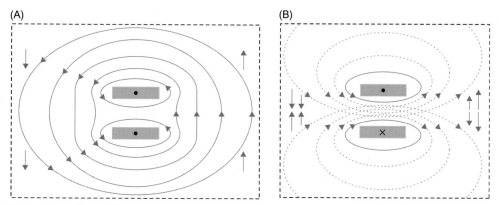

Figure 6.6 Aiding and opposing magnetic fields: (A) aiding fields and (B) opposing fields.

partial flux cancellation. The amount of cancellation depends on the amount of mutual inductance, which in turn depends in part on the distance between the conductors.

The situations in Figs. 6.4–6.6 are steady-state DC circuits, but if we apply the concept in Fig. 6.3 and Eq. (6.3) for time-varying currents, then we can see that, if the return trace on a circuit board is in close proximity to the signal trace, then the inductance of the traces is reduced; that is, the loop inductance of the closed-loop circuit is reduced. If the loop inductance is reduced in an AC circuit, then by Eq. (6.6) it can be seen that there will be less inductive reactance (X_L), less voltage drop, and less cross talk (fewer EMI problems):

$$X_L = 2\pi f\, L \qquad\qquad (6.6)$$

To maintain a small X_L (and low loop inductance), we need to have the return path as wide as possible (low self-inductance) and as close as possible to the signal path wherever possible (maximum coupling and small cross-sectional areas). The easiest way to do this is by using a plane layer as the return path. The return plane has historically (and most often inappropriately) been called the *ground plane*, but it is being referred to more often as an *image plane* or a *return plane*. In PCB design a return plane has low inductance (and therefore low self-inductance) and it is everywhere the signal trace is and therefore allows for maximum coupling between the signal trace and the plane for any and all widths of the signal trace.

From this discussion then we can say that one of the most important functions of the return (image) plane is to reduce loop inductance. Reducing loop inductance (and the magnetic fields related to it) provides a low impedance return path for power and signal lines and reduces unwanted cross talk to nearby conductors. It should also be stated that cross talk between unrelated conductors is also reduced by keeping them farther apart (i.e., *r* is large).

Electric fields and capacitive coupling

We saw in the previous paragraphs that keeping signal and power lines close to their return paths provides flux cancellation and reduces loop inductance, which is beneficial in all respects. But what happens with the electric fields under these circumstances and what is the effect on the circuit? Fig. 6.7A shows electric field lines for a single charged conductor, which can represent a signal trace that is a long way away from its return path. Fig. 6.7B shows electric field lines between two oppositely charged conductors, which can represent a signal or power line close to its return path. As shown, the solitary conductor radiates electric field lines in all directions, while the coupled conductors contain (or at least concentrate) the electric field between them.

The Ampere–Maxwell law (Serway, 1992, p. 851) states that "magnetic fields are produced both by conduction currents and by changing electric fields." So, to

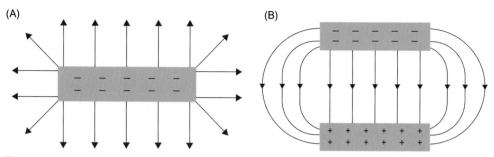

Figure 6.7 Electric fields on conductors: (A) a single charged conductor and (B) a field between oppositely charged conductors.

minimize cross talk, it would seem to be in our best interest not to allow traces to radiate electric fields in an uncontrolled manner but to keep the signal (and power) traces close to their return paths. This is true for both magnetic and electric fields.

But what happens to the capacitance of the traces relative to the return plane when we do this? The equation of a parallel plate capacitor (in farads) in air is given in Eq. (6.7) (Serway, 1992, p. 712),

$$C = \frac{\varepsilon_0 A}{d}, \tag{6.7}$$

where C is capacitance (in farads), ε_0 is permittivity of free space, A is the area common to the parallel plates, and d is the distance between the plates. As indicated, as the plates become closer together or as the area becomes larger, the capacitance increases. This also holds for the capacitance between a trace and the return plane, although as we will see later the equation is slightly different and the units are in F/in. to make it easier to calculate capacitance for various trace lengths.

From Eq. (6.8) we can see that as the capacitance, C, increases the capacitive reactance decreases:

$$X_C = \frac{1}{2\pi f C}. \tag{6.8}$$

By combining Eqs. (6.7) and (6.8) as shown in Eq. (6.9), we can see that, by keeping traces and power planes close to their return planes (small d), the capacitive reactance between the trace and the return plane is reduced (coupling is increased):

$$X_C = \frac{d}{2\pi f \varepsilon_0 A}. \tag{6.9}$$

And by keeping unrelated signal traces farther apart (large d), the reactance between the traces is higher and the coupling (cross talk) is reduced. With both magnetic and electric fields, the wider the return path (the areas of the conductor), the better the coupling, and the closer the signal or power conductor is to the return plane, the better the coupling.

Ground planes and ground bounce

What Ground is and what it is not

Ground Symbol

In the previous discussion, the term *return plane* was used instead of the term *ground plane*. Unless the return plane is physically connected to the earth by some means, it really has nothing to do with "ground." The ground symbol, ⏚, has long been used on schematics (in academia and in practice) to indicate a connection to the point to which all closed–circuit currents must return. An example is shown in Fig. 6.8. This gives the impression that ground, somehow, is an omnipresent, unfaltering current sink and equipotential reference. Equipotential means that the voltage is the same everywhere, regardless of how much current is flowing through it. This is a myth.

Although the depiction of the ground connection shown in Fig. 6.8 is convenient to use on the schematic, in reality there has to be a physical, real-world connection. And, just so you know, the official ground symbols per IEEE Std 315-1975 (ANSI Y32.2-1975) are given in Table 6.1. The IEEE disclaims any responsibility or liability resulting from the placement and use in the described manner. However, it is quite common that the ground and return symbols are used otherwise. Since the ground concept has been around a long time, it is unlikely that its use will change overnight. The important thing is that it is clear what the symbol means. The standard states that the symbols can be given supplementary information (such as names) on the schematic to specifically annotate the symbol's purpose or function.

There are two basic ground and source connection schemes, parallel and series connections, as shown in Fig. 6.9. A parallel ground system is shown in Fig. 6.9A. A parallel ground system is also called a *separate ground system*, since the current flow in each branch is supplied by and returns to the source through completely separate paths. A series-connected ground system is shown in Fig. 6.9B. The series-connected

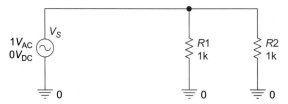

Figure 6.8 Typical depiction of "ground."

Table 6.1 IEEE/ANSI standard ground symbols.

Symbol	Name	Purpose/usage
	Earth GND	A direct connection to the earth.
		A direct connection to a vehicle's or an airplane's frame that serves the same function as earth ground
	Noiseless GND	Used to indicate a low or noiseless earth ground
	Safety GND	Used to indicate a ground connection that serves a safety function against electric shock
	Chassis GND	A connection to a chassis, or frame, or similar connection of a *printed circuit board* and may be completely different from earth ground
	Return	Used to indicate common return connections

Source: Adapted and reprinted with permission from IEEE, ©2003.

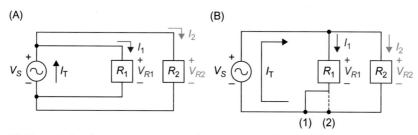

Figure 6.9 Typical signal and return connection schemes: (A) parallel connected and (B) series connected.

ground system is also referred to as a *common ground system* or *daisy chain*, since the current flows in the two branches share a common path.

At the schematic level the circuits in Fig. 6.9 are identical; mathematically the circuits are also identical, that is, $I_T = I_1 + I_2$ and $V_{R1} = V_{R2} = V_s$. Furthermore, as indicated in Fig. 6.9B, you could connect R_1 to the common return path at either point (1) or point (2) without changing the meaning of the circuit description in any way. However, on the PCB, the two circuits shown in Fig. 6.9 can be significantly different. There may even be significant differences between points (1) and (2) in Fig. 6.9B.

It was said earlier that any conductor that carries current will produce a magnetic field. Even if there is very tight coupling between the signal conductor and its return conductor, some inductance will always exist because the coupling is not 100.0% complete; that

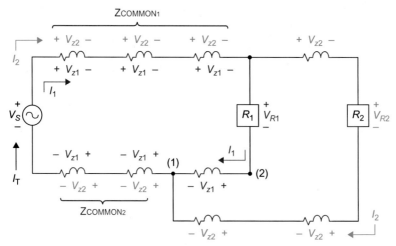

Figure 6.10 The actual circuit—the hidden schematic.

is, there is less than perfect mutual inductance from the primary trace to the secondary trace (the return plane) and from the secondary trace back to the primary trace.

With this understanding we can see from Fig. 6.10 that there is an "unseen" schematic within the PCB. As a result, although the relationship $I_T = I_1 + I_2$ is still true, it should be realized that $V_{R1} = V_{R2} = V_s$ is no longer true because there are voltage drops along the shared and individual impedances between the source and each of the loads. Furthermore points (1) and (2) and any other points along the return path are not equipotential. The ideal, equipotential ground plane does not exist in practice.

Common impedance is particularly troublesome when high-power or noisy signals share common return paths ($Z_{COMMON1}$ and $Z_{COMMON2}$) with low-power signals. High-speed digital signals operating near low-level analog signals are an example.

Clearly then, it would seem that the best return system is the parallel system shown in Fig. 6.9A. However, a potential problem arises when we try to implement this approach on the PCB. Fig. 6.11 shows the routing scenarios in which Fig. 6.11A is the parallel connection, and Fig. 6.11B is the series connection. As described above and shown in Fig. 6.11A, the current paths do not interfere with each other in the parallel connection since their paths are completely separate, and since each signal path is directly above its return path, there is tight coupling between the signal and the return, and therefore the inductance is minimized. The problem is that it would become incredibly cumbersome to route a PCB using this approach with even a few additional components and worse on a high-density, multilayer PCB. If components are moved, movement of the signal and return paths would have to be carefully coordinated, resulting in many opportunities for routing "errors" that could be difficult to detect.

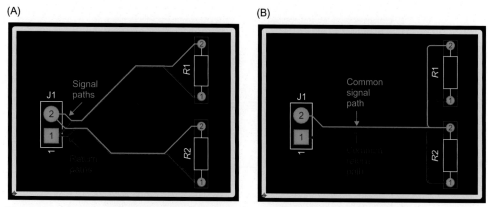

Figure 6.11 Signal and return connection schemes in PCB Editor: (A) parallel (separate) connected and (B) series (common) connected.

The series connection in Fig. 6.11B is obviously much easier to route, but it loses the benefit of the separate signal and return paths. Even with the signal and return paths tightly coupled, the common paths (impedances) could be problematic for circuits operating at high frequency or with fast rise times.

Ground (return) planes

Above, we said that we want the return path to be as wide as possible and as "everywhere" as possible, which, when taken to the extreme, causes the return path to become a plane. But since (at first glance) it appears that a plane is potentially a common return path (a common impedance), the question arises as to whether this is really the best solution to overcome the inconvenience of the routing problem. If we reroute the circuit of Fig. 6.8 as shown in Fig. 6.12, in which the return path is the GND plane in PCB Editor (where the thermal reliefs indicate a connection to the GND plane), we can analyze the situation. If the signal path is relatively close to the return path, the return signal will automatically flow through the GND plane directly below the signal trace. The reason this happens is that by doing so the loop inductance of the circuit is minimized. As is commonly known in DC circuits, current follows the path of least resistance. Perhaps not as commonly known, AC currents will follow the path of least impedance and, particularly on PCBs, the path of least inductance. The only way for that to happen is if the return current travels directly under the signal trace on its way back to the source. This is true no matter what kind of crazy path the signal trace makes, as long as no discontinuities exist in the return plane.

So (in Fig. 6.12) even if R_2 was directly between R_1 and the PCB connector, the return paths would not be common as long as the signal traces to R_1 and R_2 did not overlap (which can happen if the signal traces were routed on different layers) or

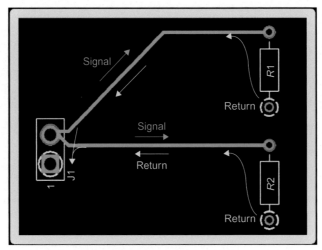

Figure 6.12 Pseudo-common return path using a "ground" plane.

become "too" close. We will look at the appropriate trace separation (what *too close* means) in the routing section discussed later.

Ground bounce and rail collapse

In a typical PCB the power distribution system contains one or more power and ground planes. The power and ground planes are like very wide traces (have little inductance) and are usually adjacent to each other (high capacitance). This is exactly what we want for the power distribution system. However, despite these advantages, a problem occurs in high-speed digital systems when gates switch from one state to another. The problem in general is known as *switching noise*.

Fig. 6.13 shows a representative CMOS logic gate. The capacitor, C_L, represents all of the capacitances related to construction of the CMOS transistors Q1 and Q2. When the gate switches from one state to another, C_L has to charge (or discharge) before the gate can reach its steady-state value. For example, at a logic state of 0, Q1 is off and Q2 is on, the output (and VC_L) is at V_{SS} (0 V). When the gate tries to switch to a high state, logic level 1, Q1 turns on, Q2 turns off, and C_L begins to charge to V_{DD}. During this transition the gate consumes significant power because for a brief moment Q1 and Q2 are both partially on. A short circuit exists from V_{DD} to V_{SS} through Q1 and Q2 and through C_L (which has low impedance while it charges). Because a high current results (even if only briefly) the voltage at the V_{DD} pin tends to drop until the switching is complete and C_L is fully charged. A similar thing happens when the gate tries to change from a logic 1 to a logic 0 state, except that as C_L tries to discharge through Q2 (which is turning on as Q1 turns off) the voltage at the V_{SS} pin tends to rise until the switching is complete and C_L is fully discharged.

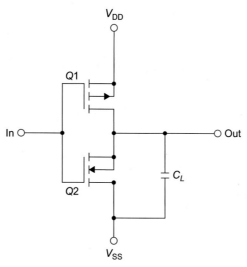

Figure 6.13 A CMOS logic gate.

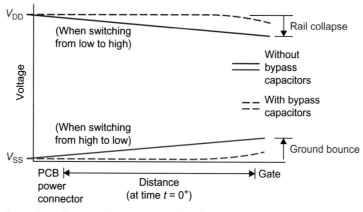

Figure 6.14 Illustration of ground bounce and rail collapse.

Because the power and ground planes are not superconductors, there is a drop in voltage between the supply pins of the gate and where power is connected to the PCB (same for the return plane). Remember that there is always some amount of resistance and inductance—even on the so-called ground plane. This is shown in Fig. 6.14, in which we see the supply voltages across the PCB while the gate is switching. The drop in the positive rail is called *rail collapse* and the rise in ground potential is called *ground bounce*. Note that, since there is really nothing magical about the so-called ground plane, the term *rail collapse* can refer to the ground rail rising as well as the supply rail dropping.

The solid line in Fig. 6.14 shows that a relatively linear drop in positive rail voltage (or rise in ground potential) occurs along the distance from the connector to the switching gate when none of the ICs on the PCB have bypass capacitors. The worst voltage drop occurs at the gate itself, but any gates located between the switching gate and the connector will "feel" the voltage drop as well.

The dashed lines in Fig. 6.14 indicate the voltage drop (or rise in ground potential) when all of the ICs have bypass capacitors. The capacitors act as current reservoirs and help hold up the positive rail voltage and keep down the return (ground) rail voltage except at close proximity to the gate that is switching (although it is minimized there as well).

The primary purpose of bypass capacitors in digital circuits is to promote a stable PCB power distribution system and prevent rail collapse and ground bounce—that is, to keep switching noise off from the rails. Conversely, the purpose of analog bypass capacitors is to keep any power supply noise that does exist from getting to the analog circuitry; that is, the bypass capacitors act as low-pass filters and short out power supply transients (noise) before they get to the amplifiers.

Since analog circuits (particularly amplifiers) are usually designed to operate strictly between (and a safe distance from) the rails, they rarely cause rail collapse but are usually the victims of it. The purpose of amplifier circuits is to amplify small signals, which are often in the millivolt or microvolt range; thus a very quiet circuit environment is highly desirable. Since digital switching noise can be as much as 100 mV or more, making analog and digital systems work together on the same PCB can be a significant challenge.

Even when a PCB's power distribution system is well-designed (low inductance planes and lots of bypass caps) switching noise can be a significant problem for analog circuits (and other digital circuits for that matter) as the switching currents surge across the PCB planes. This is particularly true if the analog circuitry is between the PCB connector and the noise digital circuitry. Any voltage drops along the path will be seen as noise by the analog system.

It was said earlier (see Fig. 6.12) that, as long as traces did not overlap or if they were not too close, the return currents would tend to stay directly under the signal traces and cross talk would be minimized. However, in very high-frequency analog or high-speed digital systems, return currents (whether signal or power return) may deviate from the ideal path because of imperfections in the PCB's copper plating and variations in the laminate materials. As a result, high-frequency analog and high-speed digital return currents may actually spread across a return plane "looking" for the path of least inductance. This occurs particularly when a signal leaves one layer and goes to another through a via and the return currents do not have an easy path from the one return plane to another that is closer to the new signal layer.

Split power and ground planes

The solution to the problem of digital noise being injected into analog circuitry through the supply planes is to segregate the analog components from the digital ones and eliminate common return paths. Segregating the components is straightforward; the components are physically placed in different places on the board. Eliminating common return paths can be accomplished by splitting the ground and power planes into separate areas. The various planes are shown in Fig. 6.15. A typical plane is one continuous sheet (an entire layer dedicated to a single power or ground connection) as shown in Fig. 6.15A. But it is possible and advantageous to break up the plane into sections or to create completely separate planes for the digital and analog areas. Fig. 6.15B shows a split plane that provides isolated areas on a single layer while providing an electrically common reference point. This configuration is common where power electronics are placed over the plane area that is close to the board connectors, and analog and digital electronics are placed over their respective return planes. This allows all circuits on the board to be referenced to a common ground but forces the specific return currents to stay within their own areas. This is demonstrated in the PCB Design Examples.

Return planes can also be completely separate areas by using moats as in Fig. 6.15C or distinct continuous planes as shown in Fig. 6.15D. Moated planes are sometimes used as local ground or reference planes for high-speed clocks or small

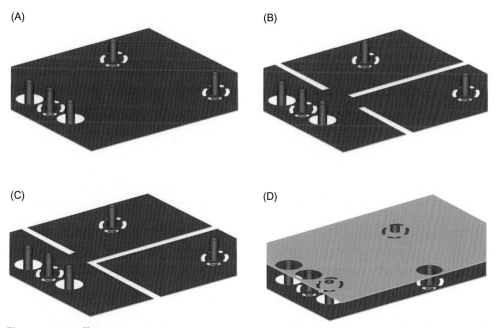

Figure 6.15 Different types of power/ground planes: (A) continuous plane, (B) split plane, (C) moated plane, and (D) isolated, continuous planes.

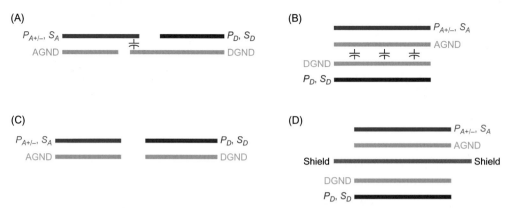

Figure 6.16 Power and ground plane stack-up scenarios: (A) coupling between overlapped planes, (B) coupling between parallel planes, (C) nonoverlapping split planes, and (D) shielded isolated planes. *P*, power; *S*, signal; *A*, analog; *D*, digital.

sections of a circuit that require their own regulated supply or ground potential, and isolated planes are used where parts of the system do not share a common ground reference or power supply system.

Care must be exercised when using split or multiple isolated ground and power planes. Even if the planes are separated physically, noise can be capacitively coupled from one plane to the next as shown in Fig. 6.16A and B. To minimize noise coupling between analog and digital planes, split planes on different layers should be prevented from overlapping each other, as shown in Fig. 6.16C, or should be separated with a shield plane, as shown in Fig. 6.16D. This is demonstrated in Example 3 in Chapter 9, PCB design examples.

On rare occasions the analog and digital return or reference planes of a PCB may be on different layers (and not overlapping or separated by a shield plane) but must be referenced to a common point (for example, when working with analog-to-digital and digital-to-analog converters). So the question becomes how to keep them physically separated but electrically connected. The easiest way is by using the isolated planes (Fig. 6.15C) and then connecting the planes at a point using a plated through hole as shown in Fig. 6.17A or a shorting bar. Moated planes (Fig. 6.15C) can also be connected using the shorting bar. Both of these methods can be used in PCB Editor and are demonstrated in the PCB Design Examples.

PCB electrical characteristics

Characteristic impedance

It was stated earlier that to minimize cross talk we want to minimize trace (loop) inductance and maximize the capacitance to the return plane. What does this do to the characteristic impedance, Z_0, of a trace?

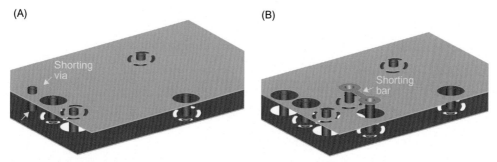

Figure 6.17 Methods of shorting together separate plane layers: (A) a via as a short and (B) a copper trace (strip) as a short.

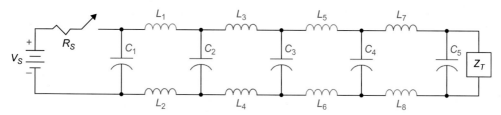

Figure 6.18 A lumped-element transmission line model.

Perhaps the first thing to do is to look at what characteristic impedance really is. For example RG58 is a coaxial cable that is often used as a shielded transmission line in 50-Ω systems. Actually, RG58 is about 52 Ω, not 50 Ω. But even so, what does that mean? If you use an ohmmeter to measure the resistance from the center conductor to the shield, you will see that it is neither 52 nor 50 Ω. So how is its characteristic impedance 52 Ω?

Fig. 6.18 shows a model of a transmission line, which consists of series inductors and parallel capacitors. This is called the *lumped-element model*, which assumes that the series resistance is negligibly small and that the transmission line is infinitely long (or at least long enough to watch what happens). Each *LC* "lump" represents a finite section of the transmission line, and the sum total of the elements is representative of the total inductance and capacitance of the transmission line.

We begin the analysis with all of the capacitors discharged and all currents at zero. At time $t = 0$ seconds the switch is shut, which applies the source voltage, V_S, to the transmission line through the source resistance R_s as shown in Fig. 6.19. Initially C_1 acts as a short circuit so $I = V_S/R_S$. Current, I, begins to charge capacitor C_1, and a return current will also flow out of the bottom of C_1 back to the source (note that this is a *displacement* current as postulated by Maxwell rather than a *conduction* current as defined by Ampere). The instantaneous impedance is $Z_{C1} = V_{\text{line}}/I$.

As C_1 charges (no longer acting like a short), current begins to flow into L_1. Each inductor pair (L_1 and L_2, L_3 and L_4, etc.) is mutually coupled, so the magnetic field of L_1 induces the return current in L_2.

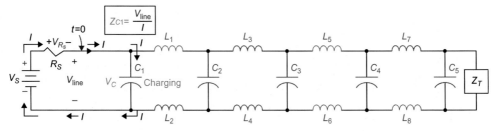

Figure 6.19 A signal applied to the transmission line.

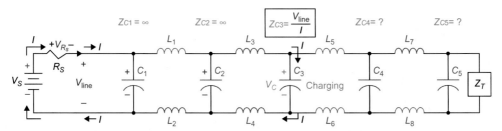

Figure 6.20 The instantaneous impedance propagates along the transmission line.

As current flows past L_1, C_2 begins charging positively on the topside; and as L_2 forces return current to flow back to the source (due to mutual inductance), C_2 begins charging negatively on the bottom side (relative to its top lead).

At some point C_1 becomes fully charged to a value of $V_{C1} = V_{line} = V_S - I \times R_S$ and then the displacement current no longer flows through C_1, so the instantaneous impedance at C_1 is $Z_{C1} = \infty$ and $Z_{C2} = V_{line}/I$. Current continues down the line, charging up each capacitor in turn to a value of $V_{Cn} = V_{line}$.

As each capacitor along the way is charging, the instantaneous impedance across the line is $Z_{Cn} = V_{line}/I_{Cn}$ as shown in Fig. 6.20. As each capacitor becomes fully charged, its impedance goes to infinity because the displacement current through it goes to zero. As seen from the source (V_S) the impedance of the line is $Z_{line} = V_{line}/I = V_{line}/I_{Cn}$ and is dynamic since it travels along the line. Furthermore, the impedance farther down the line is unknown.

The speed at which the instantaneous impedance travels along the line is dependent on the inductance and the capacitance of each section. It was said above it is desirable to have as little loop inductance as possible (which will never be zero) and as much capacitance as possible (which will never be infinite). Thus there will always be finite inductive reactance (X_1) and capacitive reactance (X_C) during any transient. However, the capacitors that are charged have nothing to do with the impedance (since they look like open circuits) and the inductors that have steady-state current flowing through them have nothing to do with the impedance (since they look like shorts). The capacitors and inductors farther down the line have nothing to do with the impedance either, since they do not see any action until the capacitors and inductors before them have approached a steady-state condition. Until the voltage (V_{line}) reaches the load, Z_T, the source actually has no idea the load, Z_T, even exists; neither does it know how many sections of L and C there are until all of the previous sections have reached steady state. If the impedance of each section is the same all along the line, then we call the instantaneous impedance the characteristic impedance of the transmission line and give it the special symbol Z_0.

Before we consider what happens to the current flow and line voltage in Fig. 6.20 once all of the capacitors are charged and the line voltage and current reach Z_T, we need to take a closer look at the behavior of the transmission line. From the above discussion we see that it takes a finite amount of time for the applied voltage (minus the voltage drop, V_{RS}) to propagate down the line, and, as the applied voltage propagates, it essentially behaves as a wave. In fact the effects described here are due to wave properties and not directly due to electrons flowing (at least not like we normally think of them). The key to understanding Z_0 (and reflections and ringing, as we will see shortly) is in understanding how and at what speed the waves travel.

If you ask an average person how fast electricity travels, you will usually get the answer that it travels at the speed of light. Except in one particular case, that answer is not correct. If we think of electricity as flowing electrons, then electricity actually travels at only about 1 cm/s (Bogatin, 2004, p. 211), pretty slow really. This seems counterintuitive since when we turn on a light switch the lights come on seemingly immediately, as if the "electricity" traveled at the speed of light from the switch to the light bulb. But what does travel at (almost) the speed of light is the electromagnetic wave that is launched into the wiring by the switch closing.

Fig. 6.21 can be used to explain the difference between the speed of electrons and the electromagnetic wave velocity. The figure shows a copper tube, which contains marbles that are separated by small springs. If an additional marble (No. 5) is shoved into the tube, marble 4 is shoved further into the tube, compressing the spring between it and marble 3. Note that in this early stage marbles 2 and 1 have no idea what is going on yet. As No. 5 is shoved into No. 4's place the rest of the marbles must "do the wave" to make room for it. Eventually all of the marbles have slid over by one marble space and marble 1 pops out the other end.

Notice now that all of the marbles have moved a distance of only one marble space, but the effect of this movement (a wave) is felt at the end of the tube in about the same amount of time. The speed of the wave is determined for the most part by the value of the spring constants and partly by the momentum of the marbles.

So in a transmission line the electrons travel slowly, but the electromagnetic (EM) waves travel fast. The speed of the EM wave is determined by how quickly the magnetic fields in the inductors and the electric fields in the capacitors can be built up or dissipated, which is influenced by the material properties and geometry of the PCB through which the wave travels.

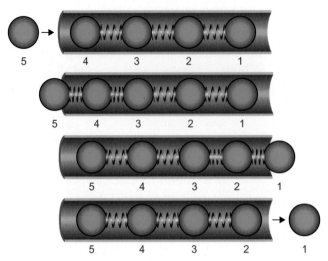

Figure 6.21 Wave velocity versus particle velocity.

The velocity of an EM wave through a medium is described by Eq. (6.10),

$$v_{EM} = \frac{1}{\sqrt{\varepsilon_0 \varepsilon_r \mu_0 \mu_r}}, \tag{6.10}$$

where v_{EM} is the velocity of the EM wave in a given material, ε_0 is the permittivity of free space (8.89×10^{-12} F/m), ε_r is the relative permittivity (dielectric constant) of the material (a unitless constant relative to ε_0), μ_0 is the permeability of free space ($4\pi \times 10^{-7}$ H/m), and μ_r is the relative permeability of the material (a unitless constant relative to μ_0).

You may recall that the speed of light, c (a special EM wave), in free space is

$$c = \frac{1}{\sqrt{\varepsilon_0 \mu_0}}. \tag{6.11}$$

So we can rewrite Eq. (6.10) as

$$v_{EM} = c \times \frac{1}{\sqrt{\varepsilon_r \mu_r}}. \tag{6.12}$$

As stated, the terms ε_r (relative permittivity) and μ_r (relative permeability) are unitless. Furthermore μ_r is equal to 1 in. free space and in most polymers (including FR4 laminate), so we can further simplify Eq. (6.12) as shown in Eq. (6.13):

$$v_{EM} = c \times \frac{1}{\sqrt{\varepsilon_r}}. \tag{6.13}$$

From Eq. (6.13) we see that the velocity of an EM wave (which comprises both electric and magnetic fields) in a PCB varies inversely with the relative permittivity, ε_r.

Relating this observation with Eq. (6.7), we can state (without rigorous proof) that the capacitance of a transmission line is determined by the geometry of the transmission line and the relative dielectric constant (ε_r) within the transmission line. And the inductance of a transmission line (specifically the loop inductance) is determined by the geometry, but μ_r falls out since it is equal to 1 [see Eq. (6.5) and Figs. 6.4 and 6.5)].

In practice, calculating the characteristic impedance, capacitance, and inductance can be fairly complex, depending on the geometry of the circuit, but fortunately that has been done for us for the most common transmission line configurations. The equations are shown in Tables 6.2−6.5. The PCB designer has full control over the trace width (w) and partial control over the trace thickness (t) by selecting the ounces per square foot but may have little or no control over the thickness of the laminate (h). These equations are solved for w and presented later in this chapter (in Tables 6.6 and 6.7).

Reflections

So the next question is, What happens when the voltage "wave front," V_{line}, reaches the termination impedance, Z_T? The answer is that it depends on what Z_T is.

Let's assume for a minute that Z_T is an open circuit. When the last capacitor, C_5, in Fig. 6.18 is charged and V_{line} reaches Z_T (which equals infinity), then all capacitors are charged along the line (and their impedance equals infinity), so all current stops— or at least it would like to. But it cannot, because all of the inductors have current, I_{line}, through them and they will not allow I_{line} to stop instantly. As the magnetic fields of L_7 and L_8 begin to collapse to try to maintain their current (remember that they are mutually coupled and influence each other), they continue to shove current into C_5, raising its voltage a bit more (we will see later what *a bit more* means).

The magnetic fields of each inductor pair (L_3 and L_4, L_1 and L_2, etc.) will collapse, one after the next, back toward the source and raise the voltage of its nearest capacitor, all the way down the line. This new voltage front (V_{line} + a bit more) propagates back from Z_T toward the source with the collapsing magnetic fields until all of the magnetic fields have collapsed and all of the capacitors have this new charge on them. An analogy of a reflection from a high-impedance termination is shown in Fig. 6.22, which shows a person launching a wave into a rope. If the rope experiences little or no friction, the wave will propagate down the rope unattenuated. If the end of the rope is loose (a high impedance), the wave will be reflected back toward the person, who will feel an identical wave returned.

In the example above the reflected wave has the same polarity and amplitude of the transmitted (incident) wave. In reality the rope would not be in a frictionless

Table 6.2 Surface microstrip transmission lines.

Microstrip transmission lines	$Z_0(\Omega)$	C_0 (pF/in.)
Surface	$Z_0 = \dfrac{k}{\sqrt{\varepsilon_r + 1.41}} \ln\left(\dfrac{5.98h}{0.8w + t}\right)$	$C_0 = \dfrac{0.67(\varepsilon_r + 1.41)}{\ln\left(5.98h/(0.8w + t)\right)}$
Surface differential	$Z_{\text{diff}} = 2Z_0\left[1 - 0.48 \times e^{(-0.964(d/h))}\right]$ Z_0 same as surface microstrip	

$L_0 = (Z_0^2 \times C_0)/12$ in nH/in, where $k = 87$ for $15 < w < 25$ mil and $k = 79$ for $5 < w < 15$ mil. Restrictions: $0.1 < w/h < 3.0$ and $1 < \varepsilon_r < 15$ (typically 4.0–4.5 for FR4).

Table 6.3 Embedded microstrip transmission lines.

Microstrip transmission lines	$Z_0(\Omega)$	C_0 (pF/in.)
Embedded	$Z_0 = \dfrac{k}{\sqrt{\varepsilon_r + 1.41}} \ln\left(\dfrac{5.98h}{0.8w + t}\right)\left(1 - \dfrac{h_1}{0.1}\right)$ OR $Z_0 = \dfrac{k}{\sqrt{\varepsilon_{r,\text{eff}} + 1.41}} \ln\left(\dfrac{5.98h_2}{0.8w + t}\right)$ $\varepsilon_{r,\text{eff}} = \varepsilon_r\left(1 - e^{-1.55H/h_2}\right)$ $Z_{\text{diff}} = 2Z_0\left(1 - 0.48 \times e^{-0.964/(2h-t)}\right)$ $h = h_1 = h_2$ Z_0 same as embedded microstrip	$C_0 = \dfrac{1.41\varepsilon_{r,\text{eff}}}{\ln\left(5.98h/(0.8w + t)\right)}$
Embedded edge coupled differential		

$k = 87$. Restrictions: $0.1 \, w/h < 3.0$, $1 < \varepsilon_r < 15$, and $Z_0 < Z_{\text{diff}} < 2Z_0$.

Table 6.4 Balanced stripline.

Stripline transmission lines	$z_0(\Omega)$	C_0 (pF/in.)
Balanced (symmetric)	$Z_0 = \dfrac{60}{\sqrt{\varepsilon_r}} \ln\left[\dfrac{1.9(H + h + t)}{0.8w + t}\right]$ OR $Z_0 = \dfrac{60}{\sqrt{\varepsilon_r}} \ln\left[\dfrac{1.9(2h + t)}{0.8w + t}\right]$	$C_0 = \dfrac{1.41\varepsilon_r}{\ln\left(3.81h_f(0.8w + t)\right)}$
Differential (edge coupled)	$Z_{\text{diff}} = 2Z_0\left(1 + 0.347 \times e^{-298/(2h+t)}\right)$ Z_0 same as symmetric stripline	

Restrictions: $w/(h - t) < 0.35$ and $w/h < 2.0$, $t/h < 0.25$, and $0.005 < w < 0.015$ in.

Table 6.5 Unbalanced stripline.

Stripline transmission lines		$Z_0(\Omega)$	C_0 (pF/in.)
Unbalanced (asymmetric)		$Z_0 = \dfrac{80}{\sqrt{\varepsilon_r}} \ln\left[\dfrac{1.9(H+h+t)}{0.8w+t}\right]\left[1 - \dfrac{h}{4(H+h+t)}\right]$	$C_0 = \dfrac{2.82\varepsilon_r}{\ln\left[(2(h-t))/(0.268w + 0.335t)\right]}$
Differential (broadside coupled)		$Z_{\text{diff}} = \dfrac{82.2}{\sqrt{\varepsilon_r}} \ln\left(\dfrac{5.98H}{0.8w+t}\right)\left(1 - e^{-0.6h/t}\right)$ Z_0 same as unbalanced stripline	

Note: Unless otherwise noted, $L_0 = (Z_0^2 \times C_0)/1000$ in nH/in.

Source: Brooks, D. (2003). Signal integrity issues and printed circuit board design (p. 203). Upper Saddle River, NJ: Pearson Educational; IPC–2141A. (2004). Controlled impedance circuit boards and high speed logic design. Northbrook, IL: IPC–Association Connecting Electronics Industries; IPC–2221B (2012). Generic standard on printed board design. Northbrook, IL: IPC–Association Connecting Electronics Industries; Montrose, M.I. (1999). EMC and the printed circuit board: Design, theory, and layout made simple (2nd ed.; pp. 171–177) New York: IEEE Press.

Table 6.6 Microstrip transmission line configurations.

Microstrip transmission lines	Topology	Characteristics	Intrinsic propagation delay
Surface	Topology (diagram: w, t, ε_r, h)	*Characteristic impedance* $Z_0 = \dfrac{k}{\sqrt{\varepsilon_r + 1.41}} \ln\left(\dfrac{5.98h}{0.8w + t}\right)$ ohms $k = 87$ for $15 < w < 25$ mil (IPC–2141A, 2004; IPC–2251, 2003, p. 32; Montrose, 1999, p. 172; Montrose, 2000, p. 101) $k = 79$ for $5 < w < 15$ mil (Montrose, 1999, p. 172; Montrose, 2000, p. 101) *Restrictions* $0.1 < w/h < 3.0$ (IPC–2251, 2003, p. 32) $1 < \varepsilon_r < 15$ (IPC–2251, 2003, p. 32) *Design equations* Trace routing width to use in PCB Editor $w = 7.475h \times e^{(-Z_0\sqrt{\varepsilon_r + 1.41})/k} - 1.25t$ (use $k = 87$, then check against width rules, use $k = 79$ if necessary)	$t_{\mathrm{PD}} = 84.75\sqrt{0.475\varepsilon_r + 0.67}$ (ps/in.)
Surface differential	Topology (diagram: w, d, w, t, ε_r, h)	*Characteristic impedance* $Z_0 = \dfrac{87}{\sqrt{\varepsilon_r + 1.41}} \ln\left(\dfrac{5.98h}{0.8w + t}\right)$ ohms (Same as surface microstrip) (IPC–2251, 2003, p. 36; Montrose, 1999, p. 177; Montrose, 2000, p. 107) *Differential impedance* $Z_{\mathrm{diff}} = 2Z_0\left[1 - 0.48e^{(-0.96(d/h))}\right]$ ohm *Restrictions* None specifically noted except as applies to the surface microstrip *Design equations* Trace routing width to use in PCB Editor $w = 7.475h \times e^{(-Z_0\sqrt{\varepsilon_r + 1.41})/k} - 1.25t$ Trace separation: $d = \ln\left(2.08\dfrac{-1.04Z_{\mathrm{diff}}}{Z_0}\right)\left(\dfrac{-h}{0.96}\right)$	$t_{\mathrm{PD}} = 84.75\sqrt{0.475\varepsilon_r + 0.67}$ (ps/in.)

Embedded	Topology		
		Characteristic impedance $Z_0 = \dfrac{87}{\sqrt{\varepsilon'_r + 1.41}}\ \ln\left(\dfrac{5.98h}{0.8w + t}\right)$ (ohms) (IPC-2141A, 2004; Montrose, 1999, p. 177; Montrose, 2000, p. 103) where $\varepsilon'_r = \varepsilon_r\left[1 + e^{\left(\frac{-1.55H}{h_2}\right)}\right]$ OR $Z_0 = \dfrac{87}{\sqrt{\varepsilon_r + 1.41}}\ \ln\left(\dfrac{5.98h}{0.8w + t}\right)\left(1 - \dfrac{h_1}{0.1}\right)$ (ohms) (IPC-2251, 2003, p. 32) *Restrictions* (IPC-2251, 2003, p. 33; Montrose, 2000, p. 103) $0.1 < w/h_2 < 3.0$ $1 < \varepsilon_r < 15$ Line widths: 0.127(5 mil)–0.381 mm (15 mil) Dielectric thickness: 0.127 (5 mil)–0.381 mm (15 mil) $40 < Z_0 < 90\ \Omega$ *Design equations* Trace routing width to use in PCB Editor $w = 7.475 h_2 \times e^{-x} - 1.25t$ where $x = \dfrac{-Z_0 \sqrt{\varepsilon'_r + 1.41}}{87}$ for Z_0 from (IPC-2141A, 2004; Montrose, 1999, p. 177; Montrose, 2000, p. 103) OR $x = \dfrac{-Z_0 \sqrt{\varepsilon_r + 1.41}}{87\left(1 - (h_1/0.1)\right)}$ for Z_0 from (IPC-2251, 2003, p. 32)	$t_{\mathrm{PD}} = 84.75\sqrt{\varepsilon'_r}$ (ps/in.) OR $t_{\mathrm{PD}} = 84.75\sqrt{0.475\varepsilon_r + 0.67}$ (ps/in.) (IPC-2251, 2003, p. 32)

(Continued)

Table 6.6 (Continued)

Microstrip transmission lines	Topology	Characteristics	Intrinsic propagation delay
Embedded differential		*Characteristic impedance* $Z_0 = \frac{87}{\sqrt{\varepsilon_r + 1.41}} \ln\left(\frac{5.98h}{0.8w + t}\right)\left(1 - \frac{h_1}{0.1}\right)$ (ohms) (IPC–2251, 2003, p. 36) *Differential impedance* $Z_{\text{diff}} = 2Z_0\left[1 - 0.48e^{\left(-0.96\left(d/\left(h_1 + h_2 + t\right)\right)\right)}\right]$ (ohms) *Restrictions* (IPC–2251, 2003, p. 36) Same as embedded microstrip *Design equations* Trace routing width to use in PCB Editor $w = 7.475h_2 \times e^{-x} - 1.25t$ where $x = \frac{-Z_0\sqrt{\varepsilon_r' + 1.41}}{87\left(1 - \left(h_1/0.1\right)\right)}$ (IPC–2251, 2003, p. 32) Trace separation of pair: $d = \ln\left(2.08 - \frac{1.04Z_{\text{diff}}}{Z_0}\right)\left[\frac{-\left(h_1 + h_2 + t\right)}{0.96}\right]$	$t_{\text{PD}} = 84.75\sqrt{0.475\varepsilon_r + 0.67}$ (ps/in.) (IPC–2251, 2003, p. 36)

Table 6.7 Stripline transmission line configurations.

Stripline transmission lines	Characteristics	Intrinsic propagation delay
Balanced (symmetric)	Topology *Characteristic impedance* $Z_0 = \frac{60}{\sqrt{\varepsilon_r}} \ln\left(\frac{1.9H}{0.8w + t}\right)$ (ohms) [10, 11] *Restrictions* Line widths: 0.127 (5 mil)–0.381 mm (15 mil) Dielectric thickness: 0.127 (5 mil)–0.381 mm (15 mil) $40 < Z_0 < 90 \ \Omega$ *Design equation* Trace routing width in PCB Editor: $w = 1.25\left[1.9H \times e^{\left(-Z_0\sqrt{\varepsilon_r}/60\right)} - t\right]$	$t_{\mathrm{PD}} = 84.75\sqrt{\varepsilon_r}$ (ps/in.)
Unbalanced (asymmetric)	Topology *Characteristic impedance* $Z_0 = \frac{80}{\sqrt{\varepsilon_r}} \ln\left[\frac{1.9(2h_2 + t)}{0.8w + t}\right]\left(1 - \frac{h_2}{4(H)}\right)$ (ohms) (IPC–2251, 2003, p. 33; Montrose, 2000, p. 105) *Restrictions* $\frac{w}{h_2} - t < 0.35$ $\frac{t}{h_2} < 0.25$ *Design equation* Trace routing width in PCB Editor $w = 2.375(2h_2 + t)e^{-x} - 1.25t$ where $x = \frac{Z_0\sqrt{\varepsilon_r}}{80(1 - h_2/4H)}$	$t_{\mathrm{PD}} = 84.75\sqrt{\varepsilon_r}$ (ps/in.)
Broadside coupled differential stripline (symmetric) (IPC–2251, 2003, p. 35)	Topology *Characteristic (between conductors) impedance* $Z_0 = \frac{82.2}{\sqrt{\varepsilon_r}} \ln\left(\frac{5.98d}{0.8w + t}\right)\left(1 - e^{(-0.6h)}\right)$ ohms *Restrictions* None given in the references *Design equations* Trace routing width to use in PCB Editor $w = 7.475d \times e^{-x} - 1.25t$ Where $x = \frac{Z_0\sqrt{\varepsilon_r}}{82.4(1 - e^{-0.6h})}$	$t_{\mathrm{PD}} = 84.75\sqrt{\varepsilon_r}$ (ps/in.)

(Continued)

Table 6.7 (Continued)

Stripline transmission lines	Topology	Characteristics	Intrinsic propagation delay
Edge coupled differential stripline (symmetric) (IPC–2251, 2003, p. 35) For symmetric ($h_1 = h_2$) or asymmetric ($h_1 \neq h_2$)		*Characteristic impedance* For symmetric ($h_1 = h_2$) $Z_0 = \frac{60}{\sqrt{\varepsilon_r}} \ln\left(\frac{1.9H}{0.8w+t}\right)$ (ohms) For asymmetric ($h_1 \neq h_2$) $Z_0 = \frac{80}{\sqrt{\varepsilon_r}} \ln\left[\frac{1.9(2h_2+t)}{0.8w+t}\right]\left(1 - \frac{h_2}{4(H)}\right)$ (ohms) (IPC–2251, 2003, p. 33; Montrose, 2000, p. 105) *Differential impedance (both)* $Z_{\text{diff}} = 2Z_0\left[1 - 0.374 e^{(-2.9(d/H))}\right]$ *Design equations* Trace routing width to use in PCB Editor For symmetric ($h_1 = h_2$) $w = 1.25\left[1.9H \times e^{(-Z_0\sqrt{\varepsilon_r}/60)} - t\right]$ For asymmetric ($h_1 \neq h_2$) $w = 2.375(2h_2 + t) \times e^{-x} - 1.25t$ where $x = \frac{Z_0\sqrt{\varepsilon_r}}{80\left(1 - h_2/4H\right)}$ Trace separation in layer stack-up (for symmetric or asymmetric) $d = -0.347 \ln\left[2.67\left(1 - \frac{Z_{\text{diff}}}{2Z_0}\right)\right]$	$t_{\text{PD}} = 84.75\sqrt{\varepsilon_r}$ (ps/in.)

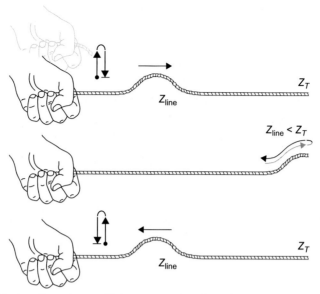

Figure 6.22 Positively reflected wave (Z_T is an open circuit).

environment (and Z_T would not be infinitely high). In that case the reflected wave would still have the same polarity as the incident wave but the amplitude would be less.

The magnitude and polarity of the reflected wave are described by the reflection coefficient, ρ (Greek letter *r*), as shown in Eq. (6.14):

$$\rho = \frac{Z_T - Z_{\text{line}}}{Z_T + Z_{\text{line}}}. \tag{6.14}$$

The reflection coefficient can have values between -1 and $+1$. If $Z_T > Z_{\text{line}}$ (i.e., as Z_T approaches ∞, as in the example above), then

$$\rho \cong \frac{Z_T}{Z_T} = 1,$$

which means that the reflected wave will be exactly the same amplitude and have the same polarity as the incident wave.

Next we consider what happens if Z_T (in Fig. 6.18) is a short circuit instead of an open circuit. At first the exact same thing occurs as described above when the switch is shut. That is (assuming the same initial conditions as above, all caps are discharged, etc.), the capacitors and inductors take their turn charging up and building up magnetic fields, V_{line} is applied to the line, current I_{line} flows, and $Z_{Cn} = V_{\text{line}}/I_{Cn}$. So the instantaneous line impedance is equivalent to Z_{Cn}. A different result occurs at the end of the transmission line. Since $Z_T = 0\ \Omega$ and inductors L_7 and L_8 again want to maintain their current flow, I_{line} flows straight through the short, Z_T.

Since the current through L_7 and L_8 is maintained (even for just an instant) and since the voltage drop across an inductor with a constant current flow is zero, capacitor C_4 sees the short and begins to discharge through L_7 and L_8 (helping to maintain their current flow) and on through the short, Z_T. A short moment later C_4 is at the same potential as C_5 and Z_T (0 V), while L_7 and L_8 have managed to maintain their current. The capacitors continue to discharge one after the other (C_3 then C_2, etc.) and each inductor pair maintains its current until finally all capacitors are shorted (and all the inductors look like a short if they have the same constant current). In the final analysis, $V_{\text{line}} = V_{ZT} = 0$ V and therefore $Z_{\text{line}} = 0/I_{\text{line}} = 0\ \Omega$ and $I_{\text{line}} = V_s/R_s$.

A mechanical analogy of a wave reflected negatively from a "dead short" is shown in Fig. 6.23. If a positive wave is launched into a rope that is rigidly fixed at the end, the wave will be negatively reflected. In a perfectly lossless environment, the reflected wave will be of the same magnitude but opposite polarity as the incident wave.

In the electrical example above, the negatively reflected wave has the same magnitude as but opposite polarity to the voltage stored on the capacitors. As the negative wave hits each capacitor, it is forced to give up its charge (as current), which helps maintain the current flow through the nearest inductors and all the way down the line through the short at the end of the line. This negatively reflected wave is again represented mathematically by the reflection coefficient [Eq. (6.14)], but in this case since

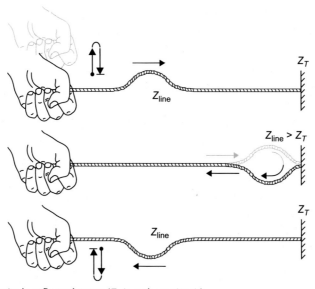

Figure 6.23 Negatively reflected wave (Z_T is a short circuit).

$Z_T < Z_{\text{line}}$ (i.e., as Z_T approaches 0), $\rho = -1$, as shown in Eq. (6.15) the following equation:

$$\rho = \frac{-Z_{\text{line}}}{Z_{\text{line}}} = -1. \tag{6.15}$$

Now let's say for argument that the characteristics of our transmission line are such that when we calculate $Z_{Cn} = V_{\text{line}}/I_{Cn}$ at each capacitor/inductor section, $Z_{Cn} = 50\ \Omega$. Let us also set R_s to $50\ \Omega$. Now what happens when $Z_T = 50\ \Omega$? As you can suppose by this time, at the moment the switch is shut, the capacitors take their turn getting charged (and the inductors are building their fields). Since each $Z_{Cn} = 50\ \Omega$, then Z_{line} is also $50\ \Omega$. Since R_s and Z_{line} are equal (and act as a voltage divider), $V_{\text{line}} = 1/2V_s$. Once the wave front has propagated down the line and reaches Z_T, which is also $50\ \Omega$, I_{line} continues to flow into Z_T as if nothing different has occurred and $V_{ZT} = V_{\text{line}} = 1/2V_s$. As long as Z_T is purely resistive, then everything is at steady state and $Z_{\text{line}} = Z_T = 50\ \Omega$. Also no voltage is reflected back toward the source because no change in voltage occurred on the capacitors and no current change occurred in the inductors.

In this case, since $Z_T = Z_{\text{line}}$ the reflection coefficient is 0 ($\rho = 0$) as shown in Eq. (6.16):

$$\rho = \frac{0}{Z_T + Z_{\text{line}}} = 0. \tag{6.16}$$

The mechanical analogy is shown in Fig. 6.24, in which none of the wave energy is reflected but is perfectly absorbed into the load at the end of the line.

From these examples we can conclude that Z_{line} is in effect only during voltage transitions and is the result of the voltage and the current transients that flow to charge the line capacitance to the new voltage and to build the magnetic fields in the inductors. We can also see that, if the impedance Z_T is not the same as Z_0, then a reflection will occur, but if the impedance Z_T is the same as Z_0, then no reflection will occur. Furthermore it takes a finite amount of time for a wave front to propagate from one end of a transmission line to the other, and if a reflection does occur it takes another finite amount of time for the reflection to propagate back to the source. What happens at the source is the next topic.

Ringing

When $\rho \neq 0$ between any adjacent impedances, reflections will occur. This is true both from driver to transmission line and from transmission line to load (and back). If there is little or no loss along the transmission line, the reflected waves will bounce back and forth between the driver and the load if they are not matched to the transmission

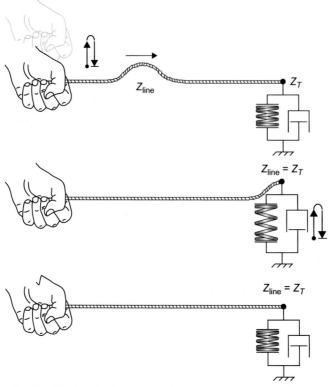

Figure 6.24 No reflection (Z_T absorbs wave energy).

line (or if the transmission line is not matched to them). When viewing a particular point along the path, for example at the output pin of a gate or amplifier, the repeated reflections will be evident as ringing. Ringing is a direct result of reflections, which in turn are due to impedance mismatches.

One of the problems with ringing is that the voltage at any point along the line is effectively out of control, since ringing causes voltage overshoots and undershoots (see Fig. 6.26). Overshoots can actually damage active devices that have input voltage limitations and will radiate greater EMI than normal signals. Overshoots and undershoots can cause digital circuits to be falsely triggered if the reflected voltage swings across switching thresholds. In analog circuits the interactions between a continuous wave signal and its reflections creates standing and/or traveling waves that can degrade the signal of interest.

The magnitude and frequency of the ringing depend on the speed of the wave through the transmission line, the length of the line, and the reflection coefficient at each impedance discontinuity. We take a detailed look at ringing using the circuit shown in Fig. 6.25.

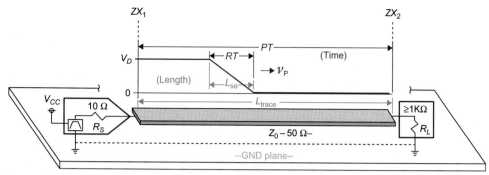

Figure 6.25 Representation of signal propagation on a PCB trace. *PCB*, Printed circuit board.

The circuit consists of a driver that is powered by V_{CC} and has a low output impedance, R_S (10 Ω); a transmission line with a characteristic impedance, Z_0 (50 Ω); and a receiver with a high input impedance, R_L (usually 1 kΩ or higher). The dashed lines indicate the interfaces of the mismatched impedances and are labeled as ZX_1, and ZX_2. The dimensions in green represent length, and the dimensions in blue represent time. The circuit in the figure can be used to represent an analog or digital circuit, but we will consider the digital application.

Consider the following:

- RT is the rise time, the time it takes for the output of a driver to transition from a minimum value to a maximum value. RT is specific to individual devices and is given in the data sheets.
- L_{trace} is the length of a trace (transmission line) on the PCB.
- v_P is the propagation velocity of a wave and is determined by Z_0 which is determined by ε_r and the transmission line dimensions (trace width and distance to the ground plane).
- PT is the propagation time, the time it takes for the transition to propagate from one end of the transmission line to the other.
- L_{SE} is the effective length of the rising edge [also called *transition distance* or the *spatial extent of the transition* (Bogatin, 2004, p. 215) or *edge length* (Johnson & Graham, 1993, p. 7)].
- Length and time are related by the propagation velocity of the wave, v_p (units of distance/time), where $PT = L_{\text{trace}}/v_p$ (units of time) and $L_{SE} = v_P \times RT$ (units in distance).
- If length of the trace, L_{trace}, is longer than the spatial extent of the rising edge, L_{SE}, then the rising edge will fit entirely within the length of the trace and the reflection voltage will be an amplitude-scaled copy of the entire rising edge, for which the scaling is determined by the reflection coefficient, ρ. Another way of looking at the same thing is if the RT (rise time) is faster than the PT, then the rising edge will have time to be fully reflected.

Electrically long traces

The goal is to design PCB traces such that they do not allow conditions to exist under which PTs are too slow (compared to signal RTs) or a trace's length is too long (compared to a signal's spatial extent). When these conditions cannot be met, the trace is considered to be "electrically long" and must be treated as a transmission line. Proper treatment of a transmission line means controlling the impedance of the line over the entire length of the line and matching the impedance of the line with the source and load impedances so that reflections do not occur.

The obvious question is, when is a trace too long (or when is the RT too fast)? The magnitude of reflections and the ringing frequency are governed by the down and back (round trip) time of the reflections. Much of the literature states that the PT, should be less than one-half of the rise time (i.e., $PT < 1/2RT$) or that the length of the trace should be less than one-half of the special extent of a rising edge (i.e., $L_{trace} < 1/2L_{SE}$). These relationships define the limits, not the goal. The shorter trace lengths are or the slower the RT is, the better off you will be. The examples below illustrate this in greater detail. After the examples general design recommendations are provided.

Fig. 6.26 shows what happens when PT is too long compared to RT. The data in the figure were generated using the transmission line model found in PSpice, and the PT was set four times longer than the RT (instead of being $<1/2RT$). Refer to Figs. 6.25 and 6.26 during the discussion.

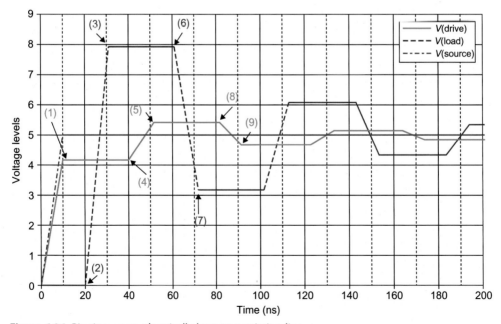

Figure 6.26 Ringing on an electrically long transmission line.

1. At time $t = 0$ ns the logic gate output (V_{source}) switches to $V_{CC} = 5\,V_{DC}$ and begins to increase in voltage at the output (V_{drive}) (start of rising edge). The first capacitor in the transmission line is uncharged and acts like a short to GND.

2. At $t = 10$ ns the gate has finished switching and the voltage at the output of the driver, V_D, is $V_D = V_{CC}(Z_0/(Z_0 + R_S)) = 5[50/(50 + 10)] = 4.17$ V because of the voltage divider established by R_s and Z_0. At this point the beginning of the rising edge is halfway to the load, and the tail end of the rising edge is just leaving the load side of R_s.

3. At $t = 20$ ns the beginning of the rising edge reaches the load resistor, R_L. Since there is an impedance mismatch between Z_0 (50 Ω) and R_L (1 kΩ), there is a positive reflection that begins immediately to head back to R_S. The reflected voltage is added to the rest of the rising edge as it continues to arrive at R_L. The reflection coefficient looking into the load from the transmission line is $\rho = (1000 - 50)/(1000 + 50) = 0.90$.

4. By $t = 30$ ns the trailing end of the rising edge reaches the load. The voltage at the load (V_{load}) is now the sum of its previous voltage (0 V) plus the value of the incoming voltage (4.17 V) plus the reflected voltage (4.17 × 0.90 = 3.75 V) for a total of $4.17 + 3.75 = 7.92$ V. The reflected voltage (3.75 V) is well on its way back toward the source.

5. At $t = 40$ ns the rising edge that was positively reflected from the load begins to arrive at the source, R_S. The voltage at the source begins to rise (pts 4−5). However, since the impedance of the transmission line is greater than that of the source resistor, a negative reflection immediately begins to head back to the load. The reflection coefficient from the transmission line looking into R_S is

$$\rho = \frac{10 - 50}{10 + 50} = -0.67.$$

6. At $t = 50$ ns the 3.75 V reflected off from the load has completely reached R_s. The voltage at the load side of R_s is the sum of its original value (4.17 V) and the incoming reflected voltage (3.75 V) plus the voltage being rereflected back to the load (-0.667×3.75 V $= -2.50$ V) for a total of $4.17 + 3.75 + -2.5 = 5.42$ V. And the -2.50 V is on its way to the load.

7. At $t = 60$ ns the -2.50 V reaches the load and begins to lower the load voltage from its previous value of 7.92 V. Because of the impedance mismatch between the transmission line and the load, a reflection is immediately launched again. The reflection coefficient is still $+0.90$, so the reflection will have the same polarity as the incident wave. Since the incident wave is the -2.50 V reflected off from R_S, the load will reflect back a negative voltage. As the incoming -2.50 V runs into a positively reflected negative voltage, the overall voltage at the load (pts 6−7) drops significantly since ρ is high (0.90).

8. At $t = 70$ ns the -2.50 V reflected off from the source has completely reached R_L where the voltage is the sum of its original value (7.92 V) and the incoming reflected voltage (-2.50 V) plus the voltage being rereflected back to the load (-2.50 V $\times$ 0.90 $= -2.25$ V) for a total of $7.92 + -2.50 + -2.25 = 3.16$ V. And of course the -2.25 V is on its way to the source.

9. At $t = 80$ ns the negative voltage that was reflected (positively, i.e., leaving the sign intact) from the load begins to arrive at R_s. Since the incoming voltage is negative, the voltage at R_s begins to fall. But since there is still a negative reflection coefficient (-0.667) from the transmission line looking into R_s, the wave that immediately begins to bounce off from R_S is now positive and heads back to the load.

10. At $t = 90$ ns the -2.25 V reflected off from the load has completely reached R_s. Again the voltage at R_s is the sum of its previous value (5.42 V) and the incoming reflected voltage (-2.25 V) plus the voltage being rereflected back to the load (-0.667×-2.25 V $= +1.50$ V) for a total of $5.42 + -2.25 + 1.50 = 4.67$ V. And of course the $+1.50$ V is on its way to the load.

The reflections continue back and forth but decrease in value each trip. The losses occur because the energy that is not reflected at each impedance interface is absorbed into the source and load resistors. Eventually, the reflections become too small to notice and we say that it has reached steady state. The time to reach steady state is called the *settling time* and the shorter the settling time, the better.

If the length of the trace, L_{trace}, is much shorter than the special extent of L_{SE} (as represented in Fig. 6.27), then the rising edge will not fit within the length of the trace and will reach the driver before the driver has even completely reached is steady-state value. If the trace is very short, the reflection voltage will be reflected many times and repeatedly fold back onto itself as the driver output climbs to its steady-state value. Since the voltage at an interface is the sum of its existing voltage, the incoming reflection, and the reflected reflection, the effects of the reflections become "smeared" into each other. By the time the driver has fully reached its final value most of the reflections have come and gone and only the last, smaller overshoots and undershoots are evident.

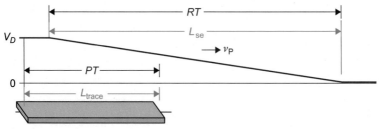

Figure 6.27 Representation of an electrically short trace.

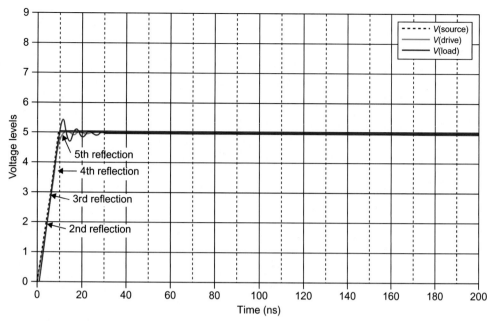

Figure 6.28 Negligible ringing on electrically short traces.

Fig. 6.28 shows what happens when PT is much shorter than RT on an electrically short trace. The reflections occur as previously described, but as shown in the graph many of the reflections have occurred by the time the driver reaches its full output level. Recall that the voltage at each impedance interface is the sum of its previous voltage plus the incoming voltage plus the reflected voltage, but since the reflections happen fast (relative to the rise time), each reflection never has a chance to reach its full voltage level (only a fraction of the rising edge at that time), so the reflections are much smaller and therefore the peaks (overshoots) and valleys (undershoots) are also much smaller (hardly noticeable), while the driver output is still rising. Also the ring frequency is higher with the shorter trace.

Critical length

As mentioned above an electrically short trace is one for which the PT is less than one-half of the rise time (i.e., PT < 1/2RT) or the length of the trace is less than one-half of the special extent of a rising edge (i.e., L_{trace} < 1/2L_{SE}). The length of a trace or transmission line for which these conditions are just barely met is called the *critical length*. Fig. 6.29 shows the voltage levels at critical length. Ringing still occurs, but the peaks never level off and the reflections settle sooner. But again the one-half rule is a limit and not a goal. Examples of determining critical length and designing

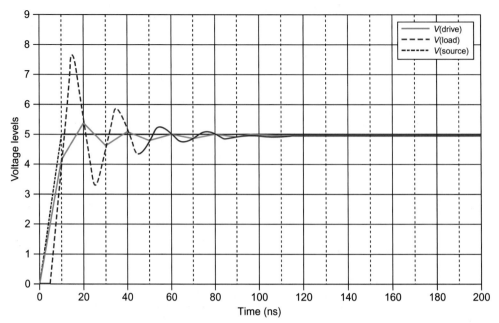

Figure 6.29 Reflections when PT = 1/2RT.

transmission lines are given in the design section below and in Example 4 in the PCB Design Examples.

Transmission line terminations

If we cannot make the rise time slower and/or the length of the trace shorter, then we will have noticeable reflections and ringing. The only other way to stop the reflections and ringing is to eliminate the impedance mismatches that are causing them by properly terminating the ends of the transmission line with the proper source and/or load resistors.

We can make $R_s = Z_0 = R_L$ by using a resistor in series with the source and a resistor in parallel with the load.

If the impedances are all matched then there will be no reflections, as shown in Fig. 6.30. However, only half the voltage will reach the load because a voltage divider results with R_S equal to R_L so $V_{load} = 1/2 V_{source}$. This lower voltage at the load may not reach required logic thresholds, preventing affected digital circuits from functioning.

An alternative is to put a resistor in series with the driver such that the impedance that the transmission line sees looking at the driver and series resistor is equal to Z_0. So if Z_0 is 50 Ω and the driver output resistance is 10 Ω, then the transmission line will be matched to the driver by putting a 40-Ω resistor in series with R_s. An example of

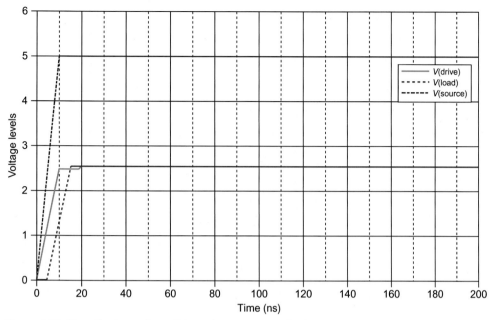

Figure 6.30 No reflections when all impedances are matched.

the result is shown in Fig. 6.31. Even when PT is $> 1/2RT$, only one reflection occurs (the flat section on V_{drive} between 10 and 20 ns), and it is absorbed into the 40-Ω resistor and R_s, so the reflection dies there. An advantage of this type of termination technique is that the voltage at the load is also much closer to the ideal voltage. The momentary hold on V_{drive} is usually not a problem, but it can be a problem in high-speed clock circuits for which the steady-state on (off) time is about the same duration as the rise time.

To match impedances between the source and the transmission line, place a resistor in series with the driver such that $R_{\text{series}} = Z_0 - R_s$.

To match impedances between the transmission line and the load, place a resistor in parallel with the load such that $R_{\text{parallel}} = (R_L Z_0)/(R_L - Z_0)$.

PCB routing topics

There are four areas for electrical considerations when routing your PCB: placing parts, PCB layer stack-up, bypass capacitors, and trace width and spacing width.

Parts placement for electrical considerations

Chapter 5, Introduction to design for manufacturing, addressed parts placement with manufacturability in mind. Here we consider parts placement with electrical

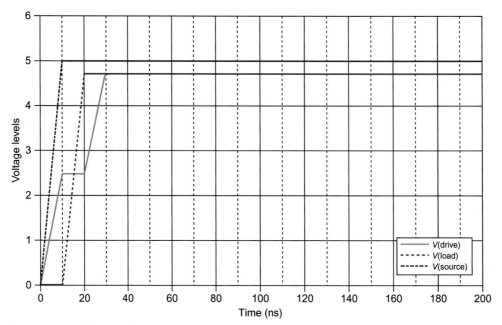

Figure 6.31 Reducing reflections by using a resistor in series with the source.

performance in mind. Usually the two goals complement each other, but occasionally they conflict. When conflicts do occur, an attempt should be made to resolve the conflict in a way that is mutually beneficial. If that is not possible, electrical considerations usually have priority over mechanical considerations unless doing so will result in a mechanical failure of the board. For example, it may be necessary to manipulate the assembly or soldering processes or place parts on both sides of the board in order to meet manufacturability requirements and meet electrical performance requirements. Whatever solution is chosen, the PCB needs to be manufacturable and operational.

Aside from manufacturability goals the first approach to placing parts for electrical considerations is usually determined by the function of the circuit. This is especially true for analog circuits where a signal enters the PCB, flows through the circuitry in more or less a single path (including feedback networks), and then leaves the board. Since analog circuits are susceptible to noise the goal is usually to place the parts to minimize the possibility of degrading the signals. This usually means keeping the parts as close together as possible so that the traces can be as short as possible and keeping the signal path as straight as possible (not zigzagging back and forth or from one side of the board to the other). This approach may increase the size of the board, however, and is not always possible.

With digital circuits it is also desirable to keep related parts close together and lines short, but because digital circuits often contain many parallel paths and branches and

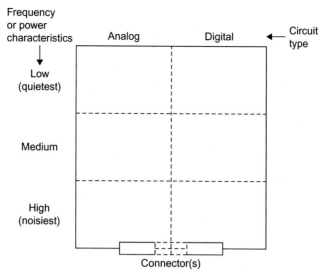

Figure 6.32 Board layout recommendations for noisy circuits.

may contain wide data busses it may be nearly impossible to do so. Sometimes the best that can be accomplished is to keep parts that are functionally related closer together or place parts together that have the highest speed clocks and rising edges in order to minimize the length of related signal lines.

Mixed signal boards are even more challenging, in that both analog and digital circuits and high-power circuits (such as switching regulators) exist on the same board. In these cases the PCB should be segregated into different areas as shown in Fig. 6.32. The topology may vary, but the idea is to keep the higher power and other noisy circuitry closer to the connector if possible. This limits the amount of return plane that is utilized by these circuits and therefore minimizes the amount of return plane that is common between them and the rest of the circuitry. Digital and analog circuits should also be kept apart from each other to minimize the effects of switching noise on analog circuitry. When dividing the PCB in this way, it is usually necessary (and beneficial) to set up split and isolated plane layers as shown in Fig. 6.15.

PCB layer stack-up

Although PCB Editor's Cross-section Editor dialog box allows you to define a stackup, the PCB layer stack-up and thicknesses are really assigned when the board is ordered from the manufacturer. But the PCB layer stack-up and thickness need to be defined early in the design stage before working in PCB Editor, since the stack-up will determine how many layers to enable and how many power and ground planes to establish. The strategy for stacking up a PCB depends on a number of things such as the capabilities of your board manufacturer, the circuit density (both routing and parts), the frequency (analog) and rise/fall times (digital) of the signals, and the acceptable cost of the board.

As circuits become more dense, additional routing and plane layers are required. Digital circuits commonly require more layers than analog circuits because digital circuits typically consist of parts with greater numbers of pins per chip and because they have a higher number of parallel interconnections. High-speed circuits (whether analog or digital) may require a greater number of layers even if they are not very dense. This is because multilayer boards can provide better impedance control and shielding since they can have additional ground planes. The more layers a board has, the more it costs, but once a PCB stack-up exceeds four layers, the cost increase per layer usually becomes less for additional layers (to a point). Additionally the benefits of shielding and impedance control can considerably outweigh the increased cost of extra layers.

There are many possible ways to design the stack-up. If the board will be operated at low speeds and in an isolated environment almost anything you can dream up will work. However, as signal speeds increase the layer stack-up and trace routing become increasingly important because a poorly designed stack-up can lead to reflections and excessive EMI radiation (affecting external circuits as well as causing self-inflicted cross talk). A well-designed layer stack-up (and proper routing) not only minimizes the energy it radiates, but also can make your circuit relatively immune from external sources of radiation. In designing a layer stack-up there are only a few guidelines to follow, but they are important. The following few paragraphs and the examples demonstrate the guidelines.

Since PCBs are constructed of double sided cores bonded together with prepreg, multilayer PCBs usually contain an even number of layers. Odd-layered boards can be made, but in most cases there is no cost benefit to adding only one layer instead of a layer pair (e.g., if you need five layers you may as well go with six). The extra layer can be an extra return/ground plane and the symmetry of an even number of layers helps minimize board warpage.

A signal layer should always be adjacent (and close) to a plane layer (preferably a return/GND plane) to minimize loop inductance, which minimizes electromagnetic radiation and cross talk. Power planes should also be adjacent (and close) to a return plane as this adds interplane capacitance, which helps minimize power supply noise and radiation. If you have to choose between a signal layer being adjacent to the return plane and a power plane being adjacent to the return plane, choose the signal layer. You can add more bypass and bulk capacitors between the power planes to make up for the loss in interplane capacitance.

The following stack-up examples are offered as a reference only. There are many more combinations possible, but only a few are shown here (please see the references for additional examples and details). The final board thickness in the examples is

0.093 in. and the copper is 1 oz (1.35 mil) thick. The availability of certain dimensions will depend on the manufacturer's capabilities and specific processes.

The thicknesses listed in the figures are given to provide perspective. You can use Tables 4.3−4.5 to get an idea of the different combinations of cores and prepreg types used to make up different layer thicknesses, but remember that the thicknesses from the tables are preassembly thicknesses, and finished thicknesses may vary since traces sink into the soft prepreg, while plane layers do not (see the Advanced Circuits Web site, www.4pcb.com, the ELLWEST Printed Circuit Boards Web site, www.ellwest-pcb. com, or other PCB suppliers websites, for examples of layer thicknesses).

A signal's return current will be on the ground or power plane that is closest to the signal line (if possible), and the relative dimensions determine the trace/plane routing pairs. Ideally all return currents will need to end up at the ground/return pin on the PCB's power connector. It may be beneficial to stack up and route your PCB to *try* to make it easier for the return currents of critical high-speed traces to be paired up with certain ground planes (so that they are not forced to go through capacitor leads or vias to get back to the ultimate return point), but there is no guarantee they will follow it. But remember that to an AC signal there is no real difference between the power plane and the ground plane because they are shorted together with bulk and bypass capacitors. So when signal traces transition from one layer to another, the return currents may not have an easy path from one ground plane to another and may choose to return to the source on a power plane until they find a convenient path back to the preferred ground plane through a bypass capacitor. While the return current is not flowing on the preferred plane, the impedance of a transmission line can be significantly different from what is expected. To assist the return current in staying with the signal line when it changes layers, return current bridges can be installed by using free vias connected to the ground planes or capacitors with fan-out vias near vias used to transition signals from one layer to another.

There are several symbols that need to be defined with regard to the stack-up figures. The H in the figures implies horizontal routing and the V implies vertical routing. In some of the figures, an R is used to imply nonspecific routing direction. Concentrating the routing strategies in horizontal or vertical directions makes the routing process more efficient in high-density designs and reduces the number of vias required to complete the connections. The symbol HS is used to imply high-speed signals that are usually buried between plane layers for extra shielding. The GND symbol is used to represent any ground or return plane (for signal or power). The symbols PWR or $\pm$PWR represent any power plane, such as $+5$ V V_{CC} for digital circuits or $+V$ and $-V$ for analog circuits.

There are three basic types of transmission lines: microstrip, stripline, and coplanar. You may not always need a transmission line, but no matter how you route your board one of these types will be represented (even if it is not a close resemblance). Other types of transmission line configurations can be realized by other types of

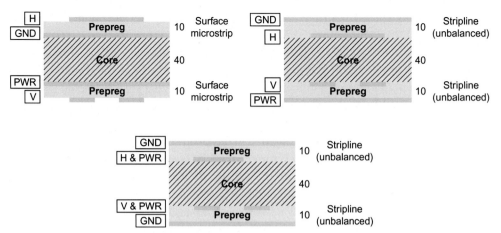

Figure 6.33 Typical four-layer stack-ups.

stack-ups, which are not shown here as they will depend on the capabilities and processes of your board manufacturer.

Fig. 6.33 shows three different four-layer stack-ups. Indicated thickness is in mils.

Fig. 6.33A is one of the most common four-layer stack-ups for simple digital or analog PCBs. The land patterns and traces are placed on the outside (top and bottom) of the board and the power and return planes are inside. Placing the traces on the outside enables postfabrication inspection and troubleshooting.

Fig. 6.33B and C places the power and return planes on the outside of the board and the traces on the inside. This arrangement helps shield the traces from outside EMI and helps contain self-generated radiation between the planes and consequently minimizes radiated EMI. The configuration in Fig. 6.33C can be used to route low-density analog circuits that require $\pm V$ for dual-supply op amps. Power is routed to the amplifiers with wide traces or copper pours and shares the layers used for routing signals.

Fig. 6.34 shows examples of six-layer PCB stack-ups. Fig. 6.34A is a typical stack-up for digital circuits or analog circuits that do not require dual-power supplies. This arrangement provides four routing layers and two plane layers. The inner routing layers can be used for the higher speed signals since they are shielded by the ground and power planes. For analog circuits that require dual-power planes, the stack-up in Fig. 6.34B provides a highly functional stack-up with two full routing layers, two partial routing layers (which share layers with the $+$ PWR and $-$ PWR layers), and ground planes that are adjacent to all signal and power layers and provide shielding for the inner signal traces. Alternatively the shared power and routing layers can be dedicated to power only, but this limits the board to two routing layers. The stack-up in Fig. 6.34C provides two well-shielded, balanced stripline routing layers for high-speed

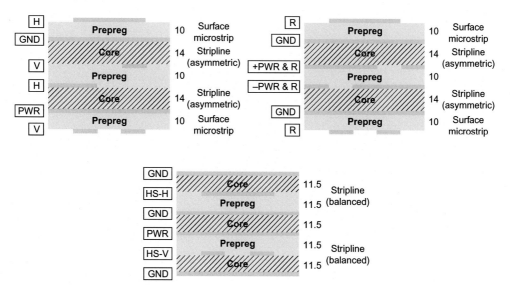

Figure 6.34 Six-layer stack-up examples.

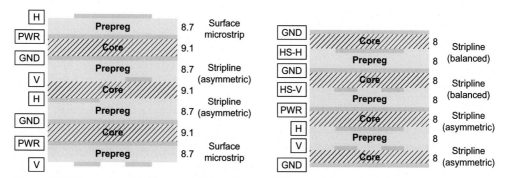

Figure 6.35 Examples of eight-layer PCB stack-ups. *PCB,* Printed circuit board.

digital circuits and adjacent ground planes for each of the signal and power layers. A limitation of this stack-up is that there are only two routing layers.

Figs. 6.35 and 6.36 show examples of 8- and 10-layer stack-ups, respectively. Only a few examples are shown, but by using the different stack-up strategies of the previous examples and capitalizing on the various finished board thicknesses available from most board manufacturers, it can be seen that the higher layer stack-ups can provide a multitude of routing possibilities.

Bypass capacitors and fan-out

Bypass capacitors serve two main functions, namely to short high-frequency noise to ground and to act as current reservoirs. Consequently there are three basic methods to

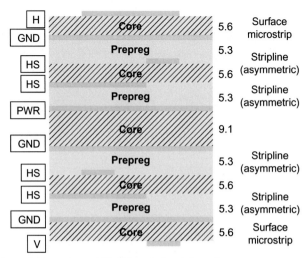

Figure 6.36 A 10-layer PCB stack-up. *PCB*, Printed circuit board.

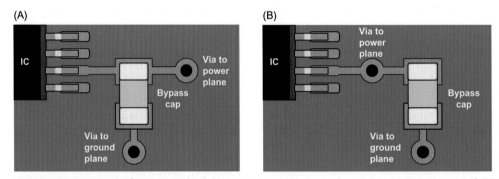

Figure 6.37 Power pin fan-out methodologies: (A) power pin to bypass capacitor to via and (B) power pin to via to bypass capacitor.

fan-out power pins. The first method usually occurs when the autorouter is used to create the fan-outs automatically. The autorouter creates individual fan-out vias for the IC's power and ground pins, and it creates individual fan-out vias for each pin on the bypass capacitor (resulting in no direct trace connection between the IC and its capacitor). In general this is acceptable. The second method is to deliberately route the power pin to the bypass capacitor before the fan-out via and to the power plane as Fig. 6.37A shows, and the third method is to route the power pin to the power plane first by placing the fan-out via between the power pin and the bypass capacitor as shown in Fig. 6.37B.

At first glance there may appear to be no difference electrically speaking, but the differences can be significant at high frequencies and fast rise times. Additionally, other issues such as the method of assembly (wave vs reflow soldering) and the available board real estate often influence orientation and routing of bypass capacitors. Sources in the literature do not all agree with which method is best, but more often Fig. 6.37A is recommended for analog circuits and Fig. 6.37B for digital circuits.

Trace width for current-carrying capability

When current flows through a conductor, it will heat up due to I^2R losses. Wider traces exhibit less resistance and therefore less heating. *To determine the minimum trace width* required to minimize heating, determine the maximum current a trace will carry and the thickness of the copper you will use on your board. Use Eq. (6.17) to calculate the minimum trace width,

$$w = \left(\frac{1}{1.4 \times h}\right) \times \left(\frac{I}{k \times \Delta T^{0.421}}\right)^{1.379} \tag{6.17}$$

where w is the minimum trace width (in mils), h is the thickness of the copper cladding (in oz/ft^2), I is the current load of the trace (in A), $k = 0.024$ is used for inner layers and $k = 0.048$ is used for outer (top or bottom) layers, and ΔT is the maximum permissible rise in temperature (°C) of the conductor above ambient temperature. You can also use the graph shown in Fig. 6.38, which was derived from curves in IPC-2221B.

Note: The glass transition temperature for FR4 is 125°C−135°C, so as ambient temperature increases, the allowed ΔT decreases and minimum trace width increases.

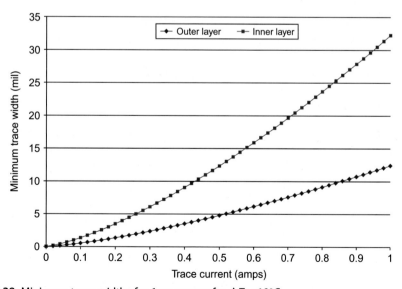

Figure 6.38 Minimum trace widths for 1 oz copper for $\Delta T = 10$°C.

Even if the board will be at room temperature, giving you a $\Delta T = 108°C$, it does not hurt to restrict the temperature rise intentionally just to be safe (e.g., specify $\Delta T = 20°C$ even if it could be much higher).

Note: Eq. (6.17) was derived from the IPC-2221B standard. Please refer to the standard for the actual equation and additional details.

You can usually use any of the standard technology files for most small signal applications. Fig. 6.38 shows the minimum trace widths for 1-oz inner and outer layer copper with $\Delta T = 10°C$. With 6-mil traces you can run up to about 300 mA on inner traces and about 600 mA on the outer traces, but for manufacturability and reliability considerations, they should be as wide as practical. Usually the applications of concern are power supply lines for large circuits and power supply boards in general.

Trace width for controlled impedance

Controlled impedance PCB can be significantly more expensive than standard process PCBs. Before going through the effort and expense of designing controlled impedance transmission lines on the PCB, it is a good idea to determine if they are needed by determining if traces are electrically long (and whether controlled impedance is required).

For digital systems, at a minimum, we want the PT, to be less than the rise time of the driver (i.e., PT < 1/2RT or RT > 2PT). From the discussion under Reflections another way to look at it is that the length of the trace, L_{trace}, should be less than one-half of the special extent (edge distance) of the rising edge (i.e., $L_{trace} < 1/2L_{SE}$ or $L_{SE} > 2L_{trace}$). To determine if this condition is met we need to determine RT and PT. RT (or fall time—FT) can be obtained from the data sheet for the device that is driving the trace and will have units of time (e.g., nanoseconds). Note that, since both the rise and the fall times must fall within limits, the smaller value should be used in the calculations. The RT and FT of several logic families are listed in Appendix C.

To calculate PT, we need to know the intrinsic propagation delay (t_{PD}), which has units of time/distance. The intrinsic propagation delay for various transmission line configurations is given in Tables 6.6 and 6.7. Bracketed numbers (IPC-2141A, 2004; IPC-2251, 2003, pp. 32, 33, 35, 36; Montrose, 1999, pp. 172, 177, Montrose, 2000, pp. 101, 103, 105, 107) refer to the numbered references at the end of the chapter.

Topologies, Z_0, T_{PD}, trace width, and trace separation design equations for various transmission lines

Units for h, w, and t can be mils, centimeters, inches, etc., as long as they are consistent.

Note that in general the values of intrinsic capacitance (C_0) and intrinsic inductance (L_0) are given by IPC-2251 (2003, p. 32)

$$C_0 = \frac{t_{PD}}{Z_0}$$

and

$$L_0 = \frac{Z_0^2 C_0}{1000},$$

where C_0 is in picofarads per inch, t_{PD} is in picoseconds per inch, Z_0 is in ohms, and L_0 is in nanohenrys per inch.

For values of C_0 and L_0 that are specific to particular transmission line topologies, please see the appropriate references as listed.

There are two families of transmission lines described here, microstrip and stripline, including their subfamilies, which are as follows:

Microstrip:
1. Surface microstrip,
2. Surface differential microstrip,
3. Embedded microstrip, and
4. Embedded differential microstrip.

Stripline:
1. Balanced (symmetric) stripline,
2. Unbalanced (asymmetric) stripline,
3. Broadside coupled differential stripline (symmetric), and
4. Edge coupled differential stripline (symmetric and asymmetric).

Next calculate the PT using Eq. (6.18),

$$PT = L_{trace} \times t_{PD} \tag{6.18}$$

where PT is a trace's one-way PT, L_{trace} is the length of the trace as measured on the PCB, and t_{PD} is the intrinsic propagation delay from Tables 6.6 and 6.7.

If it is determined that the trace is electrically too long, then either it needs to be shortened or the impedances need to be matched. To determine the maximum allowable trace length use Eq. (6.19).

Since we want $PT < 1/2RT$ and $PT = L_{trace} \times t_{PD}$, then

$$L_{trace} < \frac{RT}{2t_{PD}}. \tag{6.19}$$

For analog systems we determine the critical length of a trace with respect to the wavelength rather than the rising edge. The wavelength, λ, is determined using Eq. (6.20),

$$\lambda = \frac{v_P}{f}, \tag{6.20}$$

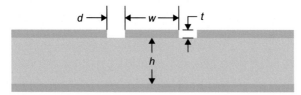

Figure 6.39 A coplanar transmission line.

where v_P is the intrinsic propagation velocity ($v_P = 1/t_{PD}$) and f is the frequency of the signal on the trace.

Various trace length limits are stated in the literature: anywhere from $L_{trace} < 1/6\lambda$ to $L_{trace} < 1/20\lambda$. IPC-2251 recommends $L_{trace} < 1/15\lambda$, where L_{trace} is the length of the trace as measured on the PCB, and λ is the wavelength of the highest frequency component of the signal (the shortest wavelength).

Tables 6.6 and 6.7 can be used to determine t_{PD} (ps/in.) for critical length calculations for both analog and digital circuits. If it is determined [using Eq. (6.19) or (6.20)] that a trace is electrically long, then source and load resistance and the transmission line impedance need to be controlled.

To design a controlled impedance transmission line on a PCB, use Tables 6.6 and 6.7 to determine the trace width, w (in.). The trace thickness, t (oz/ft^2), dielectric thickness, h (mils), and dielectric constant, ε_r (unitless), are determined by your board manufacturer.

There is another type of transmission that is not in Tables 6.6 and 6.7: The coplanar transmission line is shown in Fig. 6.39. Typically $d < h$, and w is relatively wide compared to the other configurations. The copper along each side of the trace is a topside return plane and is typically $>5w$ in both directions. The bottom side return plane is not present in some applications.

The equations are not well documented in the literature for this type of coplanar configuration except in the *Transmission Line Design Handbook*, by Wadell (1991). The equations are not presented here because they are rather unwieldy, and this type of transmission line is not widely used on FR4. If you need to use this type, please see the reference for a full description of its design, use, and limitations. An approximation of this type of configuration is sometimes attempted using guard traces and copper pours, but because of the relative dimensions (i.e., $h < d$ or $h \approx d$), an accidentally designed surface microstrip is usually the result.

Trace spacing for voltage withstanding

There are two reasons for controlling the spacing between traces: (1) to ensure adequate voltage-withstanding capability (insulation resistance) between high voltage lines and (2) to minimize cross talk between signal lines. Table 6.8, which is an abridged

Table 6.8 Minimum conductor spacing (Mil).

Voltage between conductors (V_{DC} or V_{p-p})	External traces			
	Internal traces	Bare	Soldermask only	Conformal coating
0−15	2	4	2	5
16−30	2	4	2	5
31−50	4	24	5	5
51−100	4	24	5	5

Source: After IPC-2221B. (2012). *Generic standard on printed board design.* Northbrook, IL: IPC-Association Connecting Electronics Industries.

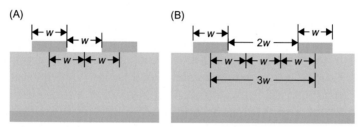

Figure 6.40 Trace spacing methods: (A) typical trace spacing and (B) 3w spacing to minimize cross talk.

version of a similar table given in IPC-2221B (Table 6-1), shows the required trace spacing for various voltage ranges on internal and external layers. The spacing on external traces depends on both the voltage and the external coating of the board.

Trace spacing to minimize cross talk (3w rule)

The default trace spacing used in PCB Editor's Constraint Manager is shown in Fig. 6.40A, in which the edge-to-edge spacing between traces is typically one conductor width. If a trace is susceptible to cross talk from adjacent traces then it should be kept a minimum of two trace widths (edge to edge) from the other traces, as shown in Fig. 6.40B. This is referred to in the references as the 3w rule because the center-to-center spacing is 3w, as indicated in Fig. 6.40B. At 3w, the traces are out of reach of about 70% of each other's magnetic field if the traces are controlled impedance transmission lines (or reasonable approximations). Keeping the traces 10w apart at the centers will keep the traces out of about 98% of each other's field (Montrose, 1999, p. 210).

To adjust the trace-to-trace spacing in PCB Editor select the Constraint Manager button on the tool bar to display the Constraint Manager as shown in Fig. 6.41. Select

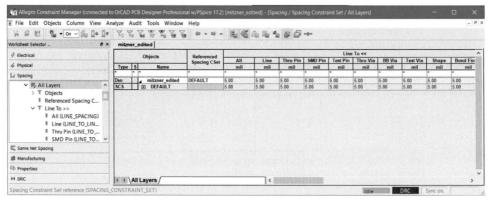

Figure 6.41 Setting the trace spacing in PCB Editor's Constraint Manager.

the Spacing tab and the Line icon as shown. Enter the edge-to-edge spacing in the Default row or click the " + " box and enter edge-to-edge values for specific nets.

Traces with acute and 90 degree angles

Routing high-frequency analog or high-speed digital traces with acute or 90 degree angles has long been discouraged, but not everyone agrees anymore as to how much of a problem it really is (see Bogatin, 2004, p. 317; Brooks, 2003, p. 383; Johnson & Graham, 1993, p. 174; Montrose, 2000, p. 220). The argument is that the trace width increases by a factor of 1.414 at the corner of the trace (as shown in Fig. 6.42) and causes a change in the characteristic impedance (due to an increase in capacitance) of the trace. As discussed above impedance mismatches cause reflections. The reflections in turn cause ringing in digital circuits with fast rise times and standing or traveling waves in high-frequency analog circuits. In theory then, 90 degree corners should be avoided—at least when routing controlled impedance transmission lines.

A look through the references shows that, when using high-end network analyzers and time domain reflectometers to measure the reflection from a 90 degree corner, the effects are evident (Bogatin, 2004, p. 318). However, the frequencies at which the reflections occur are high (into the upper gigahertz for traces greater than 50 mil wide and tetrahertz for traces less than 10 mil), and compared to other sources of impedance discontinuities (such as vias), the effects are insignificant.

Another thing to consider is that (with PCB Editor anyway) the corners are not as sharp as shown in Fig. 6.42. Recall from Chapter 1, Introduction to PCB design and computer-aided design, that pads and traces are drawn as flashes or draws, respectively, by photoplotters with apertures of a given diameter. While square apertures can be used (and are for square pads), PCB Editor uses round apertures for traces in the Gerber codes. A 90 degree corner produced by PCB Editor is shown in Fig. 6.43A.

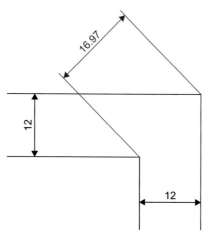

Figure 6.42 Trace geometry of a sharp 90 degree corner.

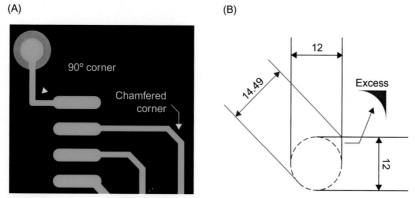

Figure 6.43 Geometry of a 90 degree corner in layout: (A) 90 degrees and chamfered corners and (B) excess copper at a 90 degree corner.

Although the corner may look fairly sharp, a close-up of the corner shown in Fig. 6.43B shows that the outer edge is rounded and only the inner corner is square. This results in a smaller increase in width than with a square corner.

The difference between a 12-mil trace and a 17-mil trace (at the sharp corner in Fig. 6.42) is about 11.5 Ω for a dielectric constant of 4.2 and a core thickness of 10 mil, and the difference between the 12-mil trace and a 14.5-mil trace (at the rounded corner in Fig. 6.43) is about 6.2 Ω. However, the effect of extra width (and change in impedance) is very small. The excess area shown in Fig. 6.43B is 7.73 mil^2 for a 12-mil trace. If the equivalent area is divided by 2 and each piece is placed on either side of a straight segment of a 12-mil trace (as indicated by the arrow in Fig. 6.44) it is clear that the excess area created by the 90 degree corner is insignificant

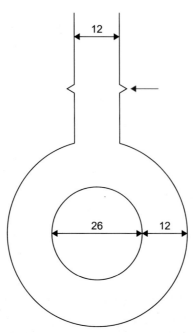

Figure 6.44 Excess area compared to a typical via (Bogatin, 2004, p. 315; Brooks, 2003, p. 383; Johnson and Graham, 1993, p. 174; Montrose, 1999, p. 220).

compared to other factors such as vias and land patterns. The literature (Brooks, 2003, p. 385) suggests that even acute angles of 135 degrees can be used up to about 1 GHz.

Using PSpice to simulate transmission lines

PSpice is used in Chapter 7, Making and editing Capture parts, to develop PSpice subcircuit models for creating new Capture parts with simulation capabilities. In this section PSpice is used to simulate transmission lines. When used in conjunction with the design equations from this chapter and the simulations in the PCB Design, Examples, PSpice can be of help in designing PCB-based transmission lines and in understanding transmission lines in general.

The circuit shown in Fig. 6.45 is used here to simulate a high-speed digital circuit (it was also used previously to generate the plots used to explain ringing). The circuit consists of a driver gate with an output impedance of 10 Ω, a 50-Ω transmission line, and a receiving gate within input resistance of 1 kΩ and input capacitance of 15 pF. The PSpice parts used in the circuit are

$V2$ is VPULSE (from the Source library), the driver.

$R1$ is R (from the Analog library), the source resistance (output impedance) of the driver.

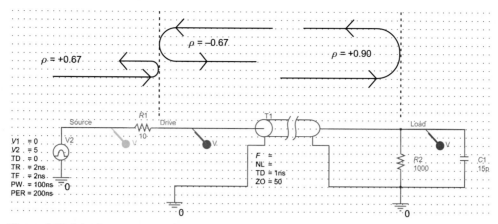

Figure 6.45 A basic transmission line simulation circuit.

$T1$ is T (from the Analog library), the transmission line.
$R2$ is R (from the Analog library), the load resistance.
$C1$ is C (from the Analog library), the load capacitance.

Fig. 6.45 also shows the various reflection coefficients at the impedance mismatches, where the reflection coefficient looking into $T1$ from $R1$ is

$$\rho = \frac{Z_0 - R1}{Z_0 + R1} = \frac{50 - 10}{50 + 10} = +0.667,$$

the reflection coefficient looking into $R1$ from $T1$ is

$$\rho = \frac{R1 - Z_0}{R1 + Z_0} = \frac{10 - 50}{10 + 50} = -0.667,$$

and the reflection coefficient looking into R2 from T1 is

$$\rho = \frac{R2 - Z_0}{R2 + Z_0} = \frac{1000 - 50}{1000 + 50} = +0.90.$$

To simulate a transmission line ($T1$) the characteristic impedance, Z_0, must be specified and either the transmission delay (TD) must be specified or the frequency (f) and number of wavelengths (NL) must be specified. Use TD for digital signals, and use f and NL for analog signals.

Simulating digital transmission lines

The value TD is the same as PT from the earlier discussion and is calculated by

$$TD = t_{PD} \times \text{Length}_{\text{trace}}$$

Table 6.9 TD calculations considering various trace lengths.

	Long	Critical	Safe
Length$_\text{trace}$ (in.)	30	7.3	3.5
k (approximate)	1/2	2	4
TD (ns)	4.1	1	0.24

TD, Transmission delay.

where t_PD is found from Tables 6.6 and 6.7. For the surface microstrip t_PD is

$$t_\text{PD} = 85\sqrt{0.475\varepsilon_\text{r} + 0.67}.$$

In laying out a digital circuit (for instance the digital design example in the PCB Design Examples in Chapter 9), it is important to know the critical trace length so that traces can be kept short enough and parts can be placed accordingly. The critical trace length is calculated by

$$\text{Length}_\text{trace} < \frac{\text{RT}}{k \times t_\text{PD}},$$

where k is a safety factor and is essentially the ratio PT/RT. The maximum trace length recommended is $k = 2$; larger values of k (shorter traces) are better.

For ALS family logic RT is approximately 2 ns. Using $\varepsilon_r = 4.2$ the critical trace length is then 7.3 in. Table 6.9 shows the effective k value and TD for various transmission line lengths.

Knowing RT (2 ns) and TD (from Table 6.9), we can use PSpice to simulate the various transmission line lengths. Set up the PSpice simulation as shown in Fig. 6.46. To display the `Simulation Settings` dialog box select `Edit Simulation Profile` from the `PSpice` menu. Note: If the step time is too large, the simulation may become unstable and you will not get good results. A maximum step size of 1/1000 of the total run time usually produces satisfactory results. Click `OK`.

To start the simulation, press the blue triangle on the menu bar, or press `F11` on the keyboard, or select `Run` from the `PSpice` menu. The results are shown in Figs. 6.47−6.49.

Simulating analog signals

For analog signals f is the frequency on the trace and NL is the length of the trace in relative wavelengths (e.g., NL would be 0.25 for a quarter wavelength trace at frequency f). To determine NL you need to know the wavelength, λ. You can calculate the wavelength using Eq. (6.21):

$$\lambda = \frac{v_\text{PD}}{f} = \frac{1}{f \times t_\text{PD}} \tag{6.21}$$

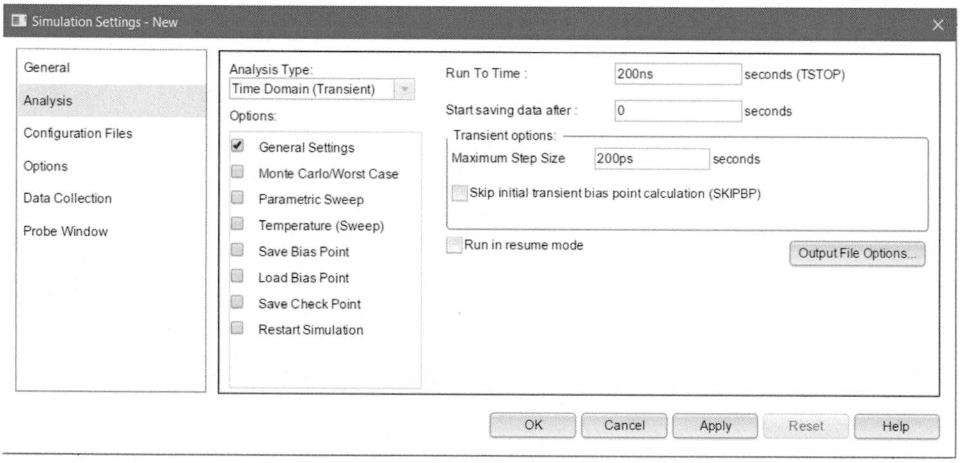

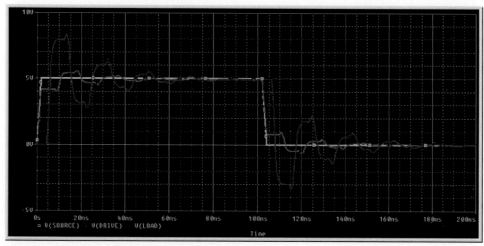

Figure 6.46 PSpice simulation settings.

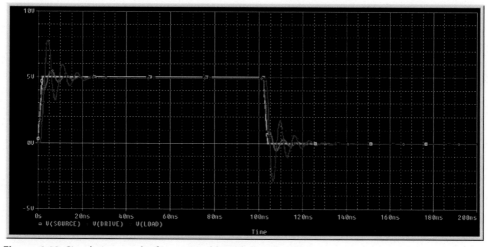

Figure 6.47 Simulation results for a long line ($k = 1/2$).

Figure 6.48 Simulation results for a critical length line ($k = 2$).

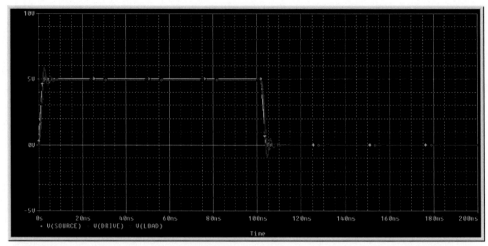

Figure 6.49 Simulation results for a short line ($k = 4$).

where v_{PD} is the propagation velocity (distance/time) of a wave through a dielectric; f is the frequency of the wave; t_{PD} is the PT (time/distance) as described above.

So for a 66-MHz signal traveling through the same surface microstrip from the above example ($\varepsilon_r = 4.2$), $\lambda = 110.8$ in. The critical length for traces carrying analog signals varies depending on which book you read but is often cited as being $\lambda/6$, $\lambda/15$, or $\lambda/20$ (or somewhere in between). As with the digital signals, the shorter the trace, the better.

References

Bogatin, E. (2004). *Signal integrity—Simplified*. Upper Saddle River, NJ: Pearson Educational.

Brooks, D. (2003). *Signal integrity issues and printed circuit board design*. Upper Saddle River, NJ: Pearson Educational.

IPC-2141A. (2004). *Controlled impedance circuit boards and high speed logic design*. Northbrook, IL: IPC-Association Connecting Electronics Industries.

IPC-2221B. (2012). *Generic standard on printed board design*. Northbrook, IL: IPC-Association Connecting Electronics Industries.

IPC-2251. (2003). *Design guide for the packaging of high speed electronic circuits*. Northbrook, IL: IPC-Association Connecting Electronics Industries.

Design Guide Manual, 1992, http://www.ipc.org (Superseded by IPC-2221, IPC-2222 and IPC-7351)

Johnson, H., & Graham, N. (1993). *High-speed digital design: A handbook of black magic*. Upper Saddle River, NJ: Prentice Hall.

Montrose, M. I. (1999). *EMC and the printed circuit board: Design, theory, and layout made simple* (2nd ed., p. 172)New York: IEEE Press.

Montrose, M. I. (2000). *Printed circuit board design techniques for EMC compliance: A handbook for designers* (2nd ed.). New York: IEEE Press.

Serway, R. A. (1992). *Physics for scientists and engineers with modern physics* (3rd ed.). Orlando, FL: Harcourt Brace Jovanovich.

Wadell, B. C. (1991). *Transmission line design handbook*. Norwood, MA: Artech House.

CHAPTER 7

Making and editing Capture parts

Contents

Capture provides many libraries of parts that you can use to build your schematic, perform simulations, and generate printed circuit board (PCB) layouts. However, you will need to make your own parts at some point. There are a couple of ways to work with the part libraries. You can build custom parts and save them to existing or new libraries, modify existing parts and save them to a library, or modify and save parts to just a specific project. Whether you decide to build custom parts or modify existing ones, you need to know about the libraries, the different ways of making parts, how different parts are packaged, what types of pins to use with different parts, and how to connect them properly for board layout and simulation purposes.

The Capture part libraries

The full version of Capture contains more than 200 libraries, in which the components are collected by the component type or series, such as 74LS for well-known TTL (transistor-transistor logic) series. The libraries are located in the "tools/capture/library" folder of the OrCAD® installation path; each library contains many ready-to-use parts. As mentioned in Chapter 2, Introduction to the printed circuit board design flow by example, two types of libraries are supplied with the software, both of which are located in the `Capture` folder. The parts located immediately inside the `"tools/capture/library"` folder are schematic parts that have no footprints or PSpice models associated with them, but the parts located in `"tools/capture/library/ PSpice"` folder have PSpice models and footprints associated with them. For the purpose of board layout, it does not matter from which library you select parts, because it is a simple matter of assigning (or changing) a footprint regardless of the library from which it came. Assigning PSpice simulation capabilities to a part is another matter altogether and is discussed later in this chapter.

The libraries are fairly well labeled and, in most cases, particular parts are relatively easy to find. However, there is some overlap between some of the libraries (especially the PSpice libraries), which can make searching for parts in general less than straight-forward. For example, MOSFET transistors can be found in the `discrete.olb` and the `fairchild.olb` libraries, the `infineon_x.olb` series libraries, and the `jpwrmos.olb` library, to name a few. Some of the libraries contain mixtures of parts (`anlg_dev.olb` contains amplifiers, multipliers, and multiplexers), while others contain only certain types of parts (`fairchild.olb` contains only FETs (field effect transistors)). So at times you may have to search manually through the libraries to find parts.

If you cannot find the part you are looking for in the actual parts libraries, you may be able to download parts and models from the Internet. Many semiconductor manufacturers generate their own parts and models and provide them free of charge from their websites. You can also download symbols and footprints from the UltraLibrarian.com online library that contains millions of verified parts.

If you cannot locate a ready-made part, you will need to make your own. It is highly recommended that you make your own folders and libraries to save your parts, rather than adding parts to the libraries supplied with the software. Instructions on how to create Capture parts and PSpice models are provided later. You can find additional information on the subject in the Capture and PSpice user's guides in the document folders and on the frequently asked questions section of the OrCAD website.

Types of packaging

One package can contain one or more parts. All the parts can be the same (homogeneous), or the parts can be different (heterogeneous), and either type of part can be passive or active.

Homogeneous parts

Fig. 7.1 shows examples of homogeneous packages (a single bipolar junction transistor and a digital integrated circuit (IC) with four identical NAND (not-AND) gates). In both cases, all the parts within a package are identical (even if there is only one).

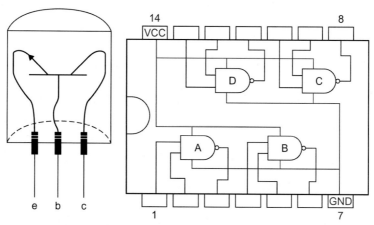

Figure 7.1 Homogeneous—one or more identical parts in a package.

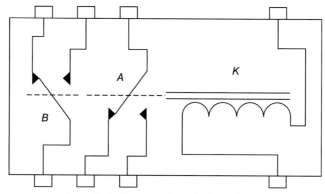

Figure 7.2 Heterogeneous relay—dissimilar parts in a single package.

All the gates in the IC are active and have to share the two power pins (7 and 14). Homogeneous parts can be placed independently in a schematic. Setting up the gates to share the power pins at the schematic level and the board level is addressed later.

Heterogeneous parts

Fig. 7.2 shows an example of a heterogeneous package. This package has three parts, which can be placed independently on the schematic: one relay coil (*K*) and two sets of contacts (*A* and *B*). Although power can be applied to the coil and to one or more of the contacts, it is a passive device and does not have power pins (strictly speaking).

Pins

When making parts in Capture that will be used to make a PCB design in PCB Editor or perform simulations in PSpice, one of the most important issues in making

Table 7.1 Pin types and attributes.

Pin types	Pin shapes	Visibility settings
Power/GND	Clock	A pin's number
Three state	Dot	A pin's name
Bidirectional	Dot—clock	The pin itself[a]
Open collector	Line	
Open emitter	Short	
Input	Zero length	
Output		
Passive		

[a]Applies to power pins only. All other pins are always visible.

new parts is properly handling the pins. All the other lines that make up a part are cosmetic, but how the pins are assigned determines if you get design rule check errors when checking your schematic, if you have a PCB laid out correctly, and if a design simulates correctly.

There are eight types of pins. Each pin has a pin number, a pin name, and an order number. Each of these pins also has specific characteristics that determine how it performs in the package with regard to what it looks like in the schematic, how it is "judged" during design rule checks, how routing is governed in PCB Editor, and the like. In addition, you can assign pin characteristics that determine what the pins look like once the part is placed into a schematic. Table 7.1 shows a summary of the pin types, shapes, and visibility options.

With regard to PCB layout the pin types are simply power pins and nonpower pins (any of the other types). Power pins require special handling because of their global nature. For example, it needs to be considered if any of the pins will be shared between parts within one package, if they will be visible on all the parts, and if they will be connected to a global net or a nonglobal net. Examples that explain how to deal with pins are given later.

Part editing tools

Before we begin making parts, we take a quick look at the part editing tools.

The Select tool and Settings
Select tool
The Select tool is used to select pins, text, or any of the graphical objects (lines, arcs, etc.). You can select objects by clicking on them or dragging a box across or around objects (see the Area Select description).

Snap to Grid

The Snap-to-Grid function is actually a coarse or fine setting and never actually fully disables the snap to functionality.

 Area Select method

Objects can be selected by fully enclosing the object with a selection box (using the Select tool) or just intersecting the object with the selection box. Using the fully enclose method is handy for selecting an object that is surrounded by many closely located objects.

 Drag connected object mode

When you move parts on your schematic, nets stay connected to the part and may cross over other nets or part pins may land on nets. Two modes of operation determine how these events change or allow connectivity. If the drag connect mode is off, , the default, then connectivity changes are prohibited. This helps prevent unwanted connections, but if you are placing new parts on the schematic, you need to specifically add new nets from existing nets to the new part to make connections. If the drag connect mode is active, ![icon] (menu `Options→Preferences→Miscellaneous→Wire Drag`), then connectivity changes are allowed. New or moved parts will automatically be connected to existing nets if the pins touch them. Red dots are used to indicate where connections will be made.

The Pin tools
Place Pin tool

The Place Pin tool is used to place pins on a part one at a time.

Place Pin Array tool

The Array tool is used to simultaneously place multiple pins on a part, such as a 16-bit output port on a microcontroller. The Array tool enables you to establish a base name common to all the pins then automatically number each pin chronologically.

The Graphics tools
◈ *Place IEEE Symbol*

The Place IEEE Symbol tool is used to place Institute of Electrical and Electronics Engineers (IEEE) standard symbols, such as arrows, math symbols, or pin states, on your part. The symbols are graphic only and do not add functionality to the part or its pins.

▔ *Place Line tool*

The Line tool is used to place a single orthogonal or diagonal line segment.

Place Polyline tool

The Polyline tool is used to place multisegment lines or closed and filled shapes. Polylines are orthogonal by default but can be made diagonal by holding down the Shift key on the keyboard while drawing the line.

Place Rectangle tool

The Place Rectangle tool is used to place closed rectangular parallelograms.

Place Ellipse tool

The Place Ellipse tool is used to place circles and ellipses.

Place Text tool

The Place Text tool is used to place text objects, which have font, color, and rotation settings.

Place Elliptical Arc tool

The Place Elliptical Arc tool gives you the possibility of placing the part of the circle or ellipse.

Place Bezier Curve tool

The Place Bezier Curve tool is used to place the approximation curve using a set of four points.

Place Picture tool

The Place Picture tool (Place → Picture... menu) allows one to import and place the bitmap picture.

The Zoom tools

Zoom In

The Zoom In tool zooms in by set increments. You can use the button, press the I key on your keyboard, or select Zoom In from the View menu.

Zoom Out

The Zoom Out tool zooms out by set increments. You can use the button, press the O key on your keyboard, or select Zoom Out from the View menu.

Zoom to Region

Use this tool to zoom to a particular region of your design by dragging a box around the area to which you want to zoom.

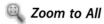 *Zoom to All*

Use this tool to see the entire design.

Methods of constructing Capture parts

Four methods are used to construct Capture parts. Three of the methods are completed from the Capture `Library Manager` and the fourth is initiated from the PSpice `Model Editor` and finished with the Capture `Library Manager`. PCB footprints can be assigned to any of the parts regardless of how it was constructed, and the footprints can be assigned when the parts are first constructed or assigned later when the parts are actually used in a design.

Methods from Capture `Library Manager`:

1. `Design menu` → `New part`.
2. `Design menu` → `New part from spreadsheet`.
3. `Tools menu` → `Generate part`.

   Method from PSpice `Model Editor`:

4. `File menu` → `Export to Capture part Library` (then use the Capture `Library Manager` to modify the part's appearance).

Table 7.2 shows which method to use by how the part will be used. The procedure for each method is described later. Later in the chapter, it will be shown how to make and download PSpice models, which can be converted to Capture parts using Methods 3 and 4.

Method 1. Constructing parts using the New Part option (Design menu)

Constructing parts using the `New Part` option in Capture is demonstrated through three design examples. The first two examples are used to design homogeneous parts; the first is a transformer (a single-part, passive device) and the second is an operational

Table 7.2 Methods of constructing parts.

Function/Purpose of the part	Method of construction
Schematic entry only	Method 1 (easier) or Method 2
Schematic entry and PCB layout	Method 1 (easier) or Method 2
Schematic entry and PSpice simulation using functional OrCAD schematic designs (can also assign footprints for PCB layout)	Method 3 (easier) or Method 4
Schematic entry and PSpice simulation using new or existing PSpice models (can also assign footprints for PCB layout)	Method 4

PCB, Printed circuit board.

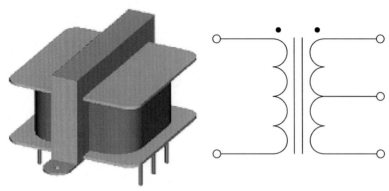

Figure 7.3 Single-part, homogeneous transformer and its schematic.

amplifier (a multipart, active device) with shared power pins. The third example is a heterogeneous, multipart, and passive device.

Design example for a passive, homogeneous part

The first design example is of a transformer with an iron core, a single primary, and a center-tapped secondary. The transformer and its schematic representation are shown in Fig. 7.3.

To begin, start Capture and from the session frame navigate to File → New → Library as shown in Fig. 7.4.

The window that opens is the Capture Library Manager. Select the Library icon (Fig. 7.5, left), then select New Part from the Design menu (Fig. 7.5, right), or right click and select New Part from the pop-up menu.

A New Part Properties dialog box will open as shown in Fig. 7.6. Enter a name for the part (e.g., Spri_CTsec) and T for the reference prefix. You can also enter a PCB footprint if desired (or known). You can change this later if you are not sure what the footprint is or will be. Leave the Parts per Pkg: as 1 and Package Type as Homogeneous. Select Part Numbering as Numeric and check Pin Number Visible if you want the pin numbers to be visible on the schematic. Click OK.

A part editing window will open with a dotted outline as shown in Fig. 7.7. The dotted box defines the boundary of the part. Pins are placed on the boundary (as described later) and text and graphics are placed inside the boundaries. You can also add pictures and IEEE symbols if desired.

Snap to Grid:

On   Off

The default grid is 0.1 × 0.1 in., with the upper left corner as coordinate (0, 0). By default pins and graphical objects are placed on the grid. The snap-to-grid setting can

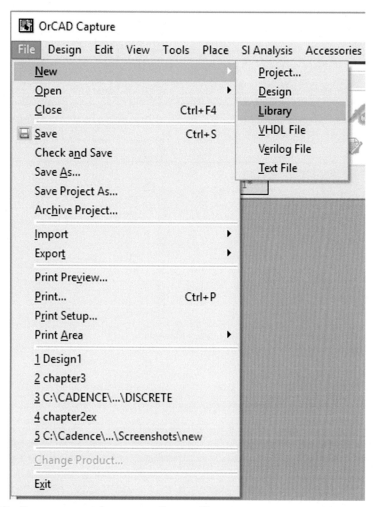

Figure 7.4 Starting a new part from a new Capture library.

be changed by toggling the `Snap-to-Grid` tool on, 🔳, or off, 🔳. The grid can be shown as dots or lines. To make changes to the grid settings, navigate to `Options →  Preferences` from the menu bar and select the `Grid Display` tab as shown in Fig. 7.8. Select the desired part and symbol grid properties on the right side of the dialog box.

The first step to make the transformer is to make the part outline 0.4 in. wide by 0.6 in. tall (for comparison, resistors from the `Analog` library are 0.2×0.2 in., not including the pin lengths). To resize the part outline, left click on the dotted border then click to hold one of the corner handles, and drag the corner to resize the box.

`Pin tool`

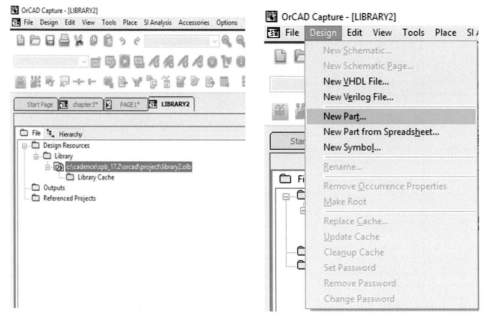

Figure 7.5 Beginning a new part from the Library Manager.

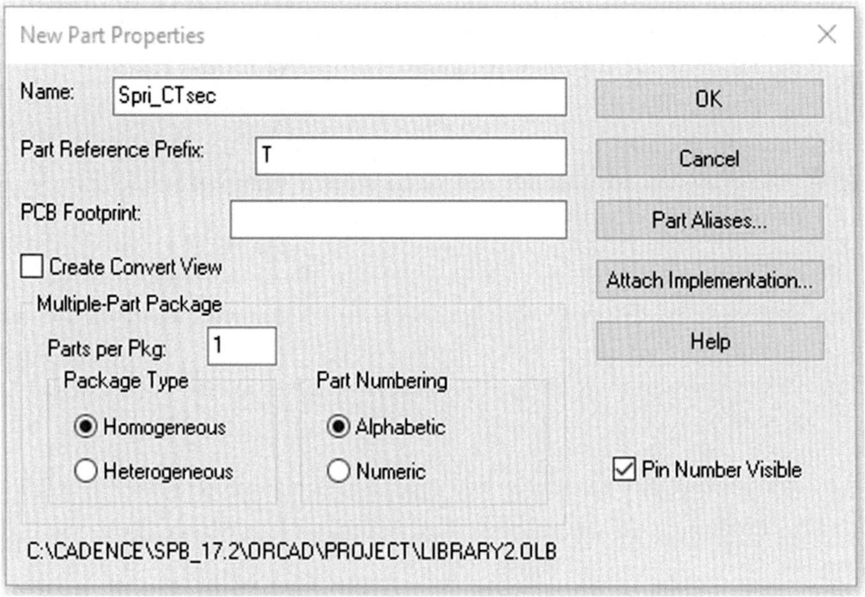

Figure 7.6 New Part Properties dialog box.

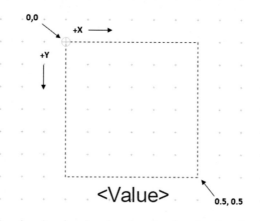

Figure 7.7 The new part shown in the part editing window.

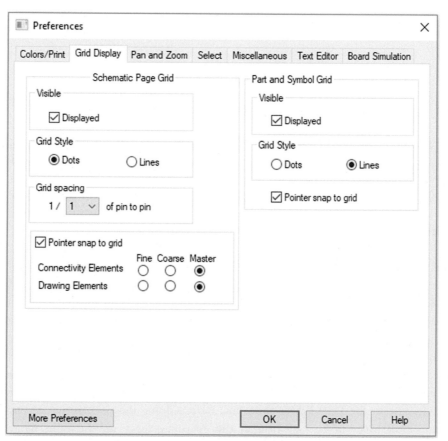

Figure 7.8 Grid settings in the Options → Preferences dialog box.

Place Pin   ⊠

▾ Pin Properties

Name: Pin1

Number: 1

Shape: Short   ▼

Type: Passive   ▼

Width: Scalar   ▼

Pin Visible   ☑   User Properties...

▾ Additional Options

Pin# Increment for Next Pin   1

Pin# Increment for Next Section

                               OK   Cancel   Help

Figure 7.9 The Place Pin dialog box.

The next step is to add the pins that make the transformer's five leads. To add pins to the part, toggle the pin tool, 🔧. The Place Pin dialog box shown in Fig. 7.9 will be displayed. Select Short for the pin shape and Passive for the pin type; click OK. Place the first two pins on the left side of the border at positions (0.0, 0.1) and (0.0, 0.5) for the primary winding leads and three pins on the right side at positions (0.4, 0.1), (0.4, 0.3), and (0.4, 0.5) for the secondary winding leads. To quit placing pins, press the Esc key on the keyboard or right click and select End Command from the pop-up menu.

The Place Pin tool will automatically increment the name and number of each pin as it is placed. To change a name or number of a pin, click on it and at the right hand side in the Pin Properties folder you will see the editable properties of the selected pin. Below you can see the "Edit Pins" button that allows you to edit the Pin Name, Pin Number, Pin Group, and other properties for the selected pin(s) or for all pins of the package in a single table (Fig. 7.10).

The pin type and pin shape cells have dropdown lists that allow you to select the desired pin properties. The other columns require you to enter the values manually. Naming pins is flexible, but the pin numbers you assign in a Capture part must match the pad name in the PCB Editor footprint. Before doing the next step you may turn off the check box Pin Name Visible in the Property Sheet on the right hand side, which will hide the pin names inside the part.

Edit All Pins								

Pin Number ☑   Pin Group ☑   Pin Ignore ☑   Order ☑   Pin Type ☑   Pin Shape ☑

Normal View: Pin Name	Section: Pin Number	Section: Pin Ignore	Order	Pin Group	Normal View: Pin Shape	Normal View: Pin Type	Normal View: Pin Visible
Pin1	1	☐	0		Short ▼	Passive ▼	☑
Pin2	2	☐	1		Short ▼	Passive ▼	☑
Pin3	3	☐	2		Short ▼	Passive ▼	☑
Pin4	4	☐	3		Short ▼	Passive ▼	☑
Pin5	5	☐	4		Short ▼	Passive ▼	☑

OK   Apply   Close   Help

Figure 7.10 Use the Browse Spreadsheet dialog box to modify pin parameters.

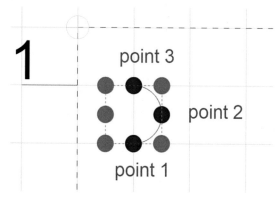

Figure 7.11 Making coils with the Place Elliptical Arc tool.

Grid button

Place Elliptical Arc tool

The next step is to *place graphics* to build the transformer coils. Turn the Snap to Grid off (toggle the grid button, , so that it turns red). Select the Place Elliptical Arc tool, . Arcs are defined by three points, as shown in Fig. 7.11. The first point you click defines the most upper or lower vertical point of the arc, in which the arc starts, the second point defines the most left or right horizontal point of the ellipse or circle that will define the arc shape, and the third point, which you should make with right-click and choosing End Mode, defines the end point of the arc on the ellipse. The arc is drawn counterclockwise from point 1 to point 3, so the location of point 3 depends on the direction you want the arc to face. To end the arc placement mode hit Esc, or right click over the point 3 and choose End Mode. When you finish the arc, it remains selected and you can drag the start or end point around the ellipse using the blue handles located on the arc ends. Also you can change the size of the ellipse by dragging the pink handles on the corners of the surrounding rectangle. After you have placed the first couple of arcs, you can copy and paste the rest in place. *To copy graphic objects*, toggle the Select tool, 🔍, and select the objects you want to copy. Left click once to select a single object or use Ctrl + left click to select multiple objects. Copy and Paste the objects as will be shown in Fig. 7.13 using typical Windows Copy/Paste techniques. You can also left click once to select an object, then hold down the Ctrl key, left click and hold the part and "drag a duplicate part" out of the first, release the mouse button to place the copy, then release the Ctrl key.

Select tool

Place Line tool

Place Ellipse tool

Next, place two parallel lines between the arcs to represent the core. Use the Place Line tool, 🔍, to draw them. Then make the dots for the dot−coil indicator by using the Place Ellipse tool, 🔍. Draw a small circle near one of the coils, then hit the Esc key to exit drawing mode. Change the circle to a filled dot. To change the fill style, use the Select tool to select the circle then you can see the Basic Attribute window at the right hand side as shown in Fig. 7.12. You can leave the Line Style as is but change the Fill Style to Solid. Hit the Esc key to clear selection.

Fig. 7.13 shows the transformer design so far. Note that the pin numbers are on the pin outside the boundary and pin names are inside the boundary with the graphics. You can use the Package Properties and Part Properties pane to change the visibility of the pin names and numbers, as shown in Fig. 7.14.

The final task is to save the new part shown in Fig. 7.15 (and the library if it is new). *To save the part*, close the part editing window and click Yes at the Save changes

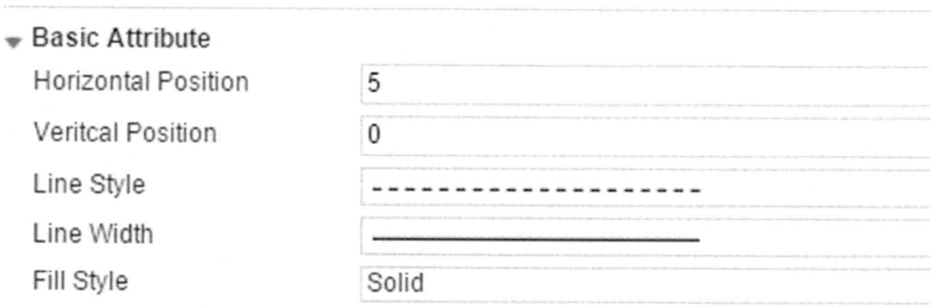

Basic Attribute
Horizontal Position
5
Veritcal Position
0
Line Style
Line Width
Fill Style
Solid

Figure 7.12 Use the Edit Filled Graphic dialog box to fill objects with patterns.

PIN# 1
PIN NAME
1
3
3
4
4
2
2
5
5
<Value>

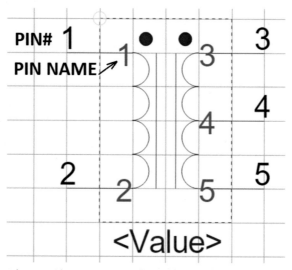

Figure 7.13 Completed part with pin names and numbers visible.

to partname Part? prompt. *To save the library*, select the Library icon in the Project Manager, right click, and select Save As... from the pop-up menu. Save the library with a new name in your UserLibrary folder in the Capture/Library path.

Design example for an active, multipart, and homogeneous component

The second example shows how to construct the dual op-amp component shown in Fig. 7.16. The parts are identical (homogeneous) and share the power pins.

Begin as you did with the transformer, by selecting New Library from the File menu in the Capture session frame. Select the Library icon in the Library Manager, right click, and select New Part from the pop-up menu. In the New Part Properties dialog box (Fig. 7.17), enter a name for the op-amp, make U the part reference prefix, and change the parts per package to 2. Leave the package type as homogeneous, assign

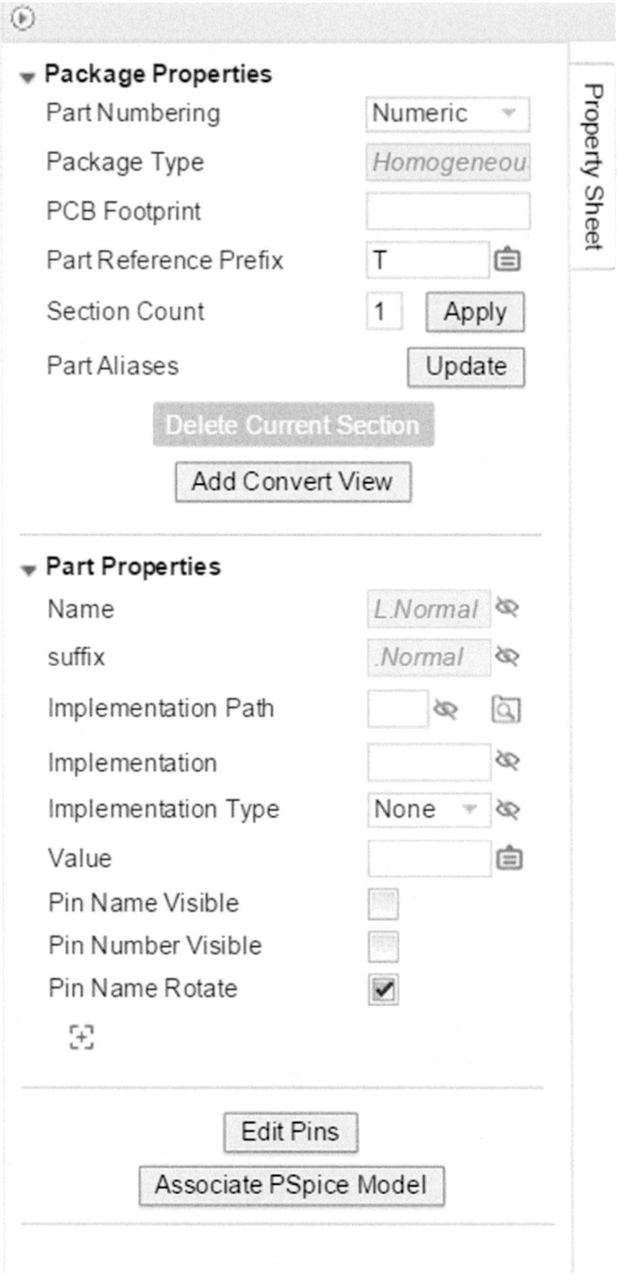

Figure 7.14 Package Properties and Part Properties pane.

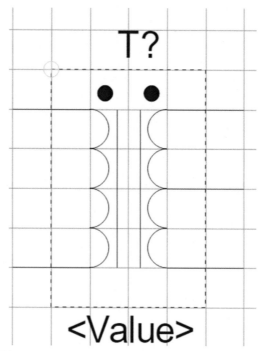

Figure 7.15 Completed homogeneous part (pin names and numbers not visible).

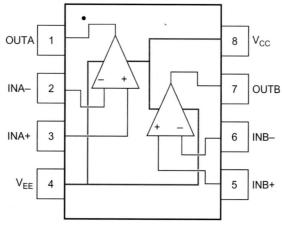

Figure 7.16 Data-sheet diagram for a dual op-amp component.

alphabetic part numbering, and make the pin numbers visible. Click OK when you are finished.

Use the Place Pin tool to begin placing pins as shown in Fig. 7.18. Label and place pins for part A per the data sheet (Fig. 7.16). You can either correctly name and

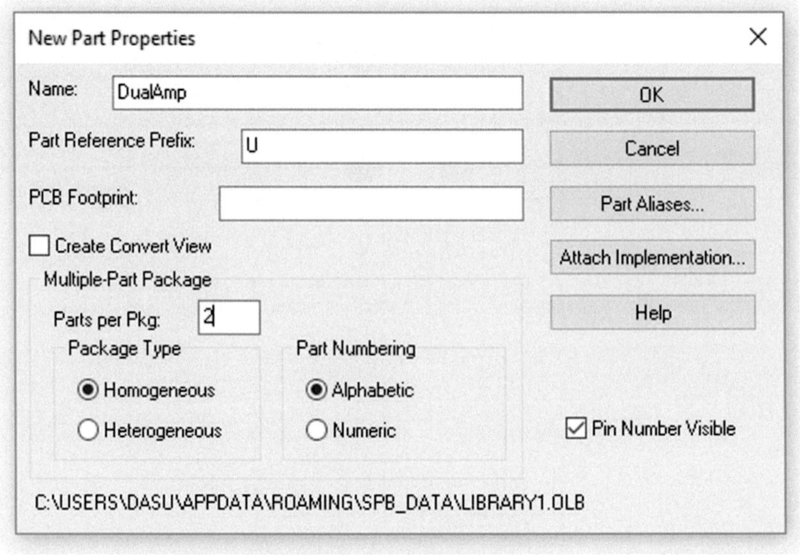

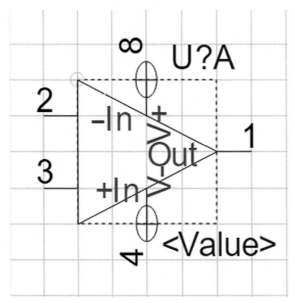

Figure 7.17 New Part Properties dialog box for a dual op amp.

Figure 7.18 Dual op amp, part A.

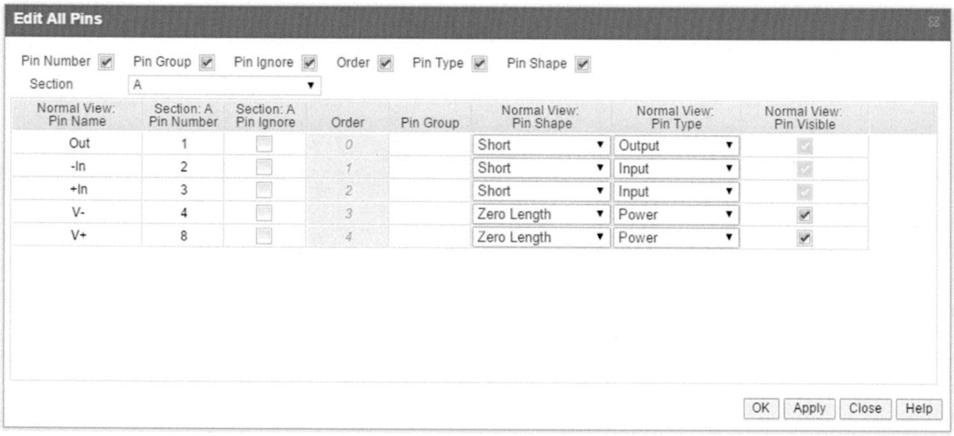

Figure 7.19 Pin characteristics for part A of the dual op amp.

number each pin as you place it or place all the pins, then go back, and fix the names and numbers after they have been placed. Whichever method you use, set the pin characteristics as shown in Fig. 7.19. When you select the pin you see its properties at the right hand side, in the Property Sheet. Finally, add short graphic lines between the op-amp body and the power pins using the Place Line tool,  . You may need to turn off the Snap to Grid to do this. Remember to turn the Snap to Grid back on when you are finished with the graphics.

In the Part Editing window, make the part border 0.4 × 0.4 in. Then make the triangular body of the op-amp using the Place Line tool (see Fig. 7.18).

As you can see from Fig. 7.18, the op amp looks cluttered because the pin names run into each other and are not optimally spaced with respect to each of the pins. You can fix this by making the pin names *not* visible and adding your own text. First, *turn the pin names off* by setting the Pin Name Visible to False in the Property Sheet. Use the Place Text tool or the Place Line tool to add − and + signs in place of the − In and + In pin names. Add text near the power pins to indicate, which is the positive supply pin and which is the negative supply pin.

For the time being, that completes part A. There are a couple of ways to get to part B. Use Ctrl + N or Ctrl + B to switch forward or backward between the parts of the package, or use the list of parts in the lower left corner of the screen. Change the properties of the pins in part B in accordance to Fig. 7.21.

Assigning power pin visibility

If you compare Figs. 7.19−7.21, you see that parts A and B share pins 4 and 8. Shared power pins require special handling to make the parts look and function properly between Capture, PSpice, and PCB Editor. The type and visibility of shared pins can

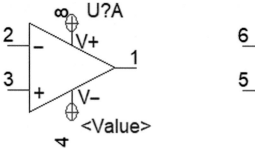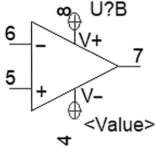

Figure 7.20 Package view of dual op-amp part.

Normal View: Pin Name	Section: B Pin Number	Section: B Pin Ignore	Order	Pin Group	Normal View: Pin Shape	Normal View: Pin Type	Normal View: Pin Visible
Out	7	☐	0		Short ▾	Output ▾	☑
-In	6	☐	1		Short ▾	Input ▾	☑
+In	5	☐	2		Short ▾	Input ▾	☑
V-	4	☐	3		Zero Length ▾	Power ▾	☑
V+	8	☐	4		Zero Length ▾	Power ▾	☑

Edit All Pins

Pin Number ☑   Pin Group ☑   Pin Ignore ☑   Order ☑   Pin Type ☑   Pin Shape ☑
Section   B   ▾

OK   Apply   Close   Help

Figure 7.21 Pin characteristics for part B of the dual op amp.

be power and visible, power and nonvisible, or nonpower and visible. Digital parts usually share nonvisible power pins, while active analog parts typically share either visible power pins or nonpower-type pins (such as input or passive pins), which are always visible. The type and visibility are initially established at the part level in the library, but you can change these parameters at the schematic level after you have placed the part into your design.

Several methods can be used for changing the part properties on the schematic, and each method has slightly different effects. If you add up all the combinations of the possible pin settings with the various methods of setting them, you will find over 200 possible combinations! Some of the combinations can cause severe errors that prevent you from creating netlists, others cause various types of warnings, and still others cause no errors or warnings but fail to route correctly on the board. We do not discuss

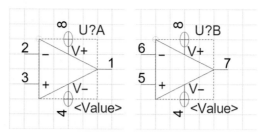

Figure 7.22 The finished op amps.

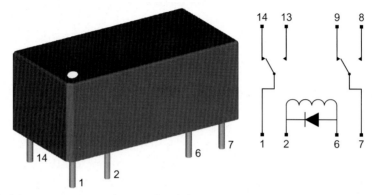

Figure 7.23 A heterogeneous, multipart relay: (left) relay package, (right) relay schematic.

all 200 combinations here, but we look at various scenarios in the PCB design examples. For now, to keep things simple and to follow typical convention, we make the shared power pins visible power pins since this is an analog part.

To set a power pin's properties, select pin 8. In the `Pin Properties` section of the `Property Sheet` pane on the right you will see all available properties of the pin. Make sure that the `Pin Visible` box is checked, an option that is available only for power pins (since all other types of pins are always visible). If you use the `Part Properties` section, you can set the visibility of only the pin's names or numbers, not the visibility of the pin itself. The completed op amps are shown in Fig. 7.22. Save the parts as described previously.

When you use a part with shared pins in a schematic then create a PCB Editor netlist, you may get the warning, `WARNING [MNL0016] Duplicate pin number "4" on` "`LM324`." If the part was constructed properly and the duplicated pins have the type "`Power`" you can ignore the warning.

Design example for a passive, heterogeneous part

In this design example we make a passive, heterogeneous part, namely, a double-pole, double-throw relay, as shown in Fig. 7.23. The relay consists of three parts: two sets of identical double-throw contacts and an operating coil with its clamping diode.

Begin by navigating to File → New → Library from the Capture session frame as described in the preceding examples. Select the Library icon and select New Part from either the Design menu or the pop–up menu after right clicking on the icon. Fill out the New Part Properties dialog box as shown in Fig. 7.24. Make sure you select 3 parts per package and select the heterogeneous package type. Click OK.

Begin by constructing the relay coil and clamping diode as part K?-1. Bring up K?-1 in an editing window. Use Ctrl + N on your keyboard to toggle through the parts. Resize the border so that it is 0.3 in. wide by 0.6 in. tall. Using Fig. 7.25 as a guide, use the Place Pin tool to place two pins on the left side of the part border. Define the top pin as number 2 with name C1 and the bottom pin as number 6 with name C2. Make both pins short, passive, and make the names nonvisible. Use the Place Polyline and Place Elliptical Arc tools to add the coil, diode, and " + " graphics to the part as described in the previous examples. To draw the triangle you can draw a closed poly-line and change its Fill Style in the properties pane to Solid. Hold Shift to draw the slanting lines.

The next step is to add the first set of contacts. Toggle to part K?-2 using Ctrl + N. Resize the border to 0.4 in. wide by 0.6 in. tall. Using Fig. 7.26 (A) as a guide, add two pins to the right side of the border and one on the left. Set the pin parameters as shown in Fig. 7.26 (B) and make the pin names nonvisible. Use the Place Line tool to add the pole and the lines leading from the pin to the throws (the contacts).

Use the Place Polyline tool to draw the throws so that you can make them solid filled.

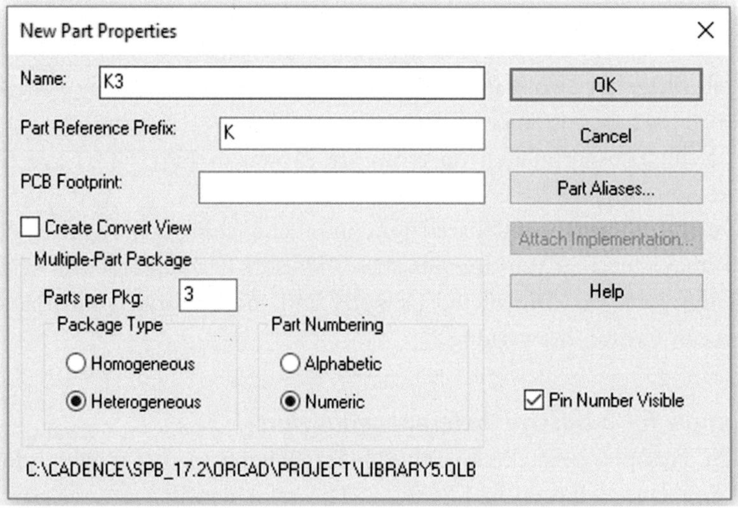

Figure 7.24 New Part Properties dialog box for the heterogeneous relay.

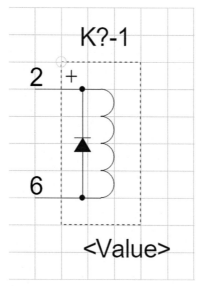

Figure 7.25 Relay coil and diode.

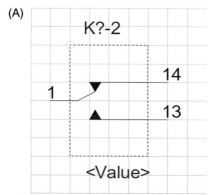

Figure 7.26 Schematic (A) and pin settings (B) for the first set of relay contacts.

Edit Pins

Pin Number ☑   Pin Group ☑   Pin Ignore ☑   Order ☑   Pin Type ☑   Pin Shape ☑

Section   2   ▼

Normal View: Pin Name	Section: 2 Pin Num...	Section: 2 Order	Section: 2 Pin Group	Pin Ignore	Normal View: Pin Shape	Normal View: Pin Type	Normal View: Pin Visible
NC1	14	0		No	Short	Passive	Yes
NO1	13	1		No	Short	Passive	Yes
P1	1	2		No	Short	Passive	Yes

OK   Apply   Close   Help

Once you have the pole and throws for contact assembly K?-2 finished, copy and paste the graphics into the K?-3 part. To copy all the pieces at one time, hold down the Ctrl key on your keyboard and select each line and polyline one at a time, or use Ctrl+A. Once you have them all selected, right click and select Copy from the pop-up menu (or select Copy from the Edit menu or use Ctrl+C on your keyboard). Go to the next part, K?-3, by typing Ctrl+N on your keyboard then paste the copied graphics into the empty part outline (right click and select Paste from the pop-up menu, select Paste from the Edit menu, or use Ctrl+V on your keyboard). Add three pins using the Place Pin tool as you did for the previous parts, or rename and renumber them, if you have already copied them. Set the pin parameters as shown in Fig. 7.27. The completed, multipart relay is shown in Fig. 7.28.

When you place the relay in a Capture schematic, use the Part: dropdown list as shown in Fig. 7.29 to place the desired part of the multipart package.

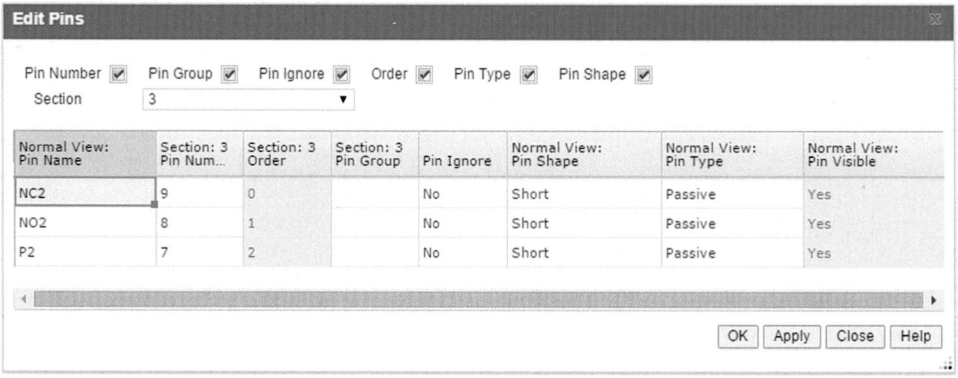

Edit Pins

Pin Number ☑	Pin Group ☑	Pin Ignore ☑	Order ☑	Pin Type ☑	Pin Shape ☑
Section	3				

Normal View: Pin Name	Section: 3 Pin Num...	Section: 3 Order	Section: 3 Pin Group	Pin Ignore	Normal View: Pin Shape	Normal View: Pin Type	Normal View: Pin Visible
NC2	9	0		No	Short	Passive	Yes
NO2	8	1		No	Short	Passive	Yes
P2	7	2		No	Short	Passive	Yes

OK   Apply   Close   Help

Figure 7.27 Pin parameters for the second set of contacts.

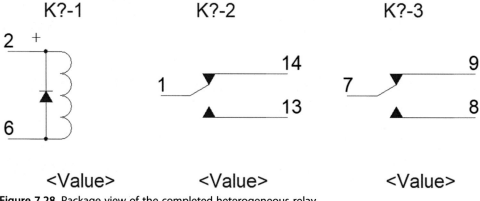

Figure 7.28 Package view of the completed heterogeneous relay.

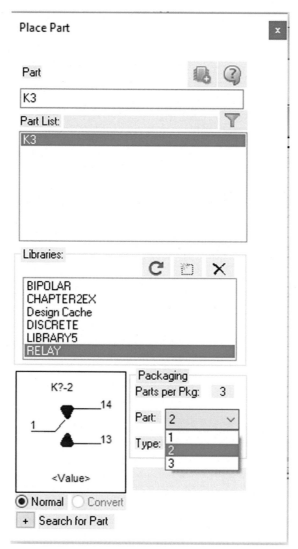

Figure 7.29 Selecting a part from a multipart package in a Capture schematic.

Method 2. Constructing parts with Capture using the Design Spreadsheet

Use this method to automatically generate a single or multipart package. It is similar to the first method, except that you define all the pins up front, using the spreadsheet shown in Fig. 7.30. New Part from Spreadsheet generates heterogeneous packages by default, but you can use it to make homogeneous parts. The parts will be defined as heterogeneous but will be homogeneous in effect if there are no differences between the parts.

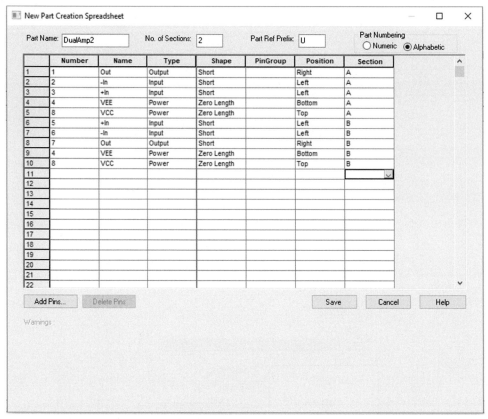

Figure 7.30 The New Part Creation Spreadsheet.

To construct a part using the design spreadsheet, open an existing Capture library to which you want to add a part, or start a new library as described previously. In the Library Manager, select the Library icon and select the New Part from Spreadsheet... option, which you can do either from the Design menu or from the pop-up menu by right clicking on the Library icon.

The spreadsheet will initially be blank. Add the pin numbers and parameter settings as needed per the part requirements. The pin settings shown in Fig. 7.30 would generate a part identical to the dual op amp designed in a previous example (see Fig. 7.16).

Notice that pins 4 and 8 were added twice, once for each part, to define the shared power pins.

After the pin parameters have been entered into the spreadsheet, click the Save button. The information box shown in Fig. 7.31 will be displayed, letting you know that there were warnings (because of the duplicate pins). You can view the warnings or click Continue to view the parts.

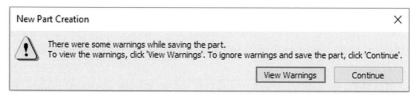

Figure 7.31 New Part Creation warnings.

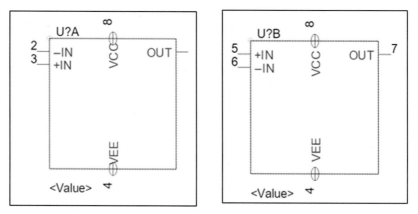

Figure 7.32 Package view of new parts generated using the spreadsheet.

Fig. 7.32 shows the package view of the parts generated by the spreadsheet. By default, the "parts" are initially represented by simple boxes. All you have left to do at this point is to modify the graphics to make the op amps cosmetically correct. In the homogeneous op amp from the previous example, all parts were automatically made identical when you made a graphic in one part. Since this is a heterogeneous part by default (which you cannot change), you have to modify the graphics on every part individually. Delete the boxes that lie on top of the dotted outline and add the graphics as described in the previous op-amp example.

Method 3. Generating or copying the schematic symbols from other sources

Use this method to create a new Capture part (or parts) from a PSpice model library (based on the PSpice model description), or from a Capture project (to create the appropriate hierarchical block symbol based on hierarchical pins in the schematic), or from a pin description or netlist file from the external source, such as Altera or Verilog/VHDL. The part(s) can then be used for schematic entry, PSpice simulations, and PCB layout (if you assign a footprint to it). To use this method, a PSpice model library, a functional schematic design, or a pin netlist file needs to preexist. In the section that follows, several methods for developing or obtaining the PSpice libraries and functional schematic designs are described. For the time being we will go ahead under the assumption that a PSpice library is available so that the part creation process is kept separate from the library/ schematic creation process.

Begin by starting a new Capture library or open an existing library from the Capture session frame's File menu. Select the Library icon and then select Generate Part... from the Tools menu. The Generate Part dialog box will pop up, as shown in Fig. 7.33.

Select PSpice Model Library (or Capture Schematic/Design if appropriate) from the Netlist/source file type: list, then select the library (*.lib etc.) or design

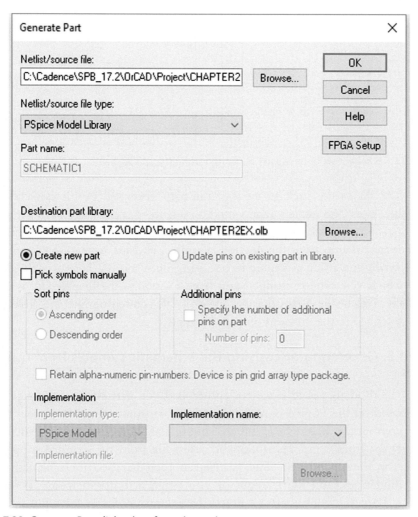

Figure 7.33 Generate Part dialog box from the tools menu.

(*.dsn etc.) from the Netlist/source file: box using Browse button. Make sure that the Destination part library: path and name are what you want and that the library has an .OLB extension. Make sure that the Create new part option is checked and click OK.

After Capture has completed importing the library/design file you selected, you should see the list of parts in the Library Manager. If the Capture library (.olb) you are working with was originally empty, the new parts will be located directly under the Library icon in the Design Resources folder. But if the library already had some parts in it, the new parts and copies of the previously existing parts will be added to both the Library icon and the Library folder under the Design Resources folder (see Fig. 7.34, in which, as an example, the breakout library was added to the analog_p library). The Library folder will also contain a duplicate of the Library icon. This allows you to copy, add, or delete parts to or from the new library. Once you are satisfied with which parts are contained under the Library icon under the

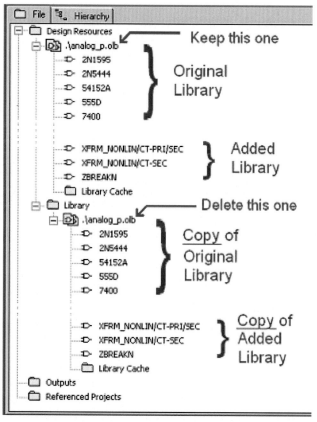

Figure 7.34 Structure of Library Manager after copying a library into an existing library.

Design Resources folder, you can delete the duplicate Library icon and parts in the Library folder, keeping just the library in the Design Resources folder.

Once you finish adding parts to the library, you can edit the added part(s), pins, and links to footprints using the Capture Part Editor.

Notice that there is an easy way to copy the needed parts from the existing schematic DSN project or from existing OLB library to the new library in which you can then rename and modify these parts as you need. To do that, open the existing Capture project or library, then add the target library to the Project Manager by right click on Library subfolder and choosing Add File, or just by creating the new library using menu File→New→Library. Then select the needed parts in the Design Cache subfolder of the opened Capture design, or in the opened library parts list, right click and Copy, and then select the target library, right click and Paste. All selected parts will be copied to the new library. Now you can rename any part using the right click and Rename, and modify it with right click and Edit Part. Only a collection of "Parts" or a collection of "Symbols" can be copied and pasted, a "mixed collection" of "Parts" and "Symbols" won't be copied.

Finally, save and close the new library. You can now use the library and its parts in new designs for schematic entry, simulations, and PCB layout.

Method 4. Generating parts with the PSpice Model Editor

You can use the PSpice Model Editor to make Capture parts that can be used in schematic designs, circuit simulations, and PCB layouts. The difference between using the PSpice Model Editor and the Generate Part method (Method 3) is that you can work directly on and with the PSpice models with the Model Editor.

Start the PSpice Model Editor. Go to All Programs → Cadence Release 17.2-2016 → Model Editor. If there are multiple licenses available that could start the Model Editor, you will need to select a license option, such as PSpice A/D, and the proper editor, such as Capture. You will begin with a blank Model Editor session window. *To open an existing PSpice library* (*name*.LIB), select Open from the File menu and navigate to /tools/PSpice/library. Select one of the libraries (e.g., bipolar.lib) and click Open; then select one of the models from the model list. Fig. 7.35 shows the Q2N696 BJT transistor model from the bipolar library. The Models List window pane shows all of the simulation models contained in the library, and it tells you whether it is a primitive model or a subcircuit model. The text window under the Models List displays the "code" that describes the selected model.

You can construct new models by using existing model listings as examples and/or the model descriptions found in the PSpice reference guide. The following sections describe how to download primitive models (models beginning with .model) from the Internet and how to create your own subcircuit models (models beginning with .subcircuit). The models can then be added to a PSpice library from which

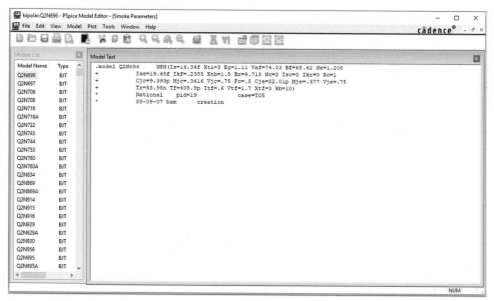

Figure 7.35 A PSpice library as viewed from the Model Editor.

you can generate Capture part libraries as described. Creating custom primitive models is not described here.

Generating a Capture Part Library from a PSpice Model Library

Beginning with a PSpice library (.lib) open in the Model Editor (similar to Fig. 7.35), select Export to Part Library... from the Model Editor's File menu. You will be presented with the Create Parts for Library dialog box shown in Fig. 7.36. In the Enter Input Model Library: text box, use the Browse... button to find the PSpice library (*.LIB) for which you want to make parts. Use the second Browse... button to specify the location for the new Capture library (*.OLB). Remember that the PSpice models and the libraries that contain them are usually stored in the /tools/PSpice/library path, while the Capture parts that use the models are stored in the /tools/capture/library/PSpice path. Click OK once you have the input and output libraries and paths specified.

Note: A Capture part library (bipolar.olb) already exists in the /tools/capture/library/PSpice folder for the bipolar PSpice Model Library (bipolar.lib). The bipolar library is used here for demonstration purposes. If you perform the following procedure on an existing library, save the new library to a temporary or user folder so that you do not overwrite the existing library.

When the Model Editor has finished, it will display an information box similar to the one in Fig. 7.37. If the Model Editor is successful at generating the Capture part library, the last line will say 0 Error messages, 0 Warning messages. Click OK to close the information box.

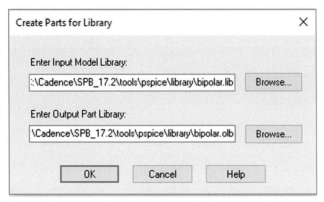

Figure 7.36 The PSpice Create Parts for Library dialog box for Capture parts.

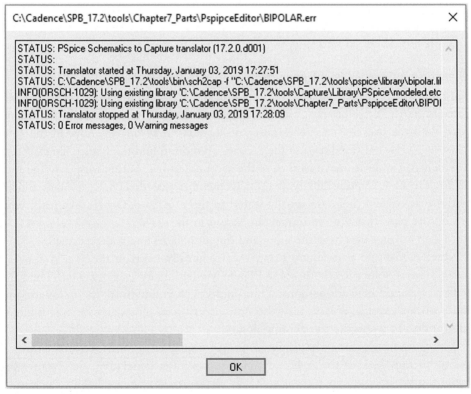

Figure 7.37 New part creation information box.

The PSpice `Model Editor` generates Capture parts with correctly named and numbered pins, but the parts are generic boxes because the PSpice `Model Editor` describes only how the parts function, not what they look like. You need to use the Capture `Part Editor` to modify the graphical appearance of the parts, as described in the earlier examples. To view the new part library, start Capture and select `Open` → `Library...` from the `File` menu in the Capture session frame.

Making and/or obtaining PSpice libraries for making New Capture parts

Before you send a final board design to be manufactured, at some point in the design process you will want to simulate your design. A detailed explanation of PSpice model development and simulation process is outside of the scope of this text and many references are available on the subject. But in the interest of completeness, a brief explanation of how to add PSpice models to your Capture parts is discussed here.

PSpice contains many libraries, but manufacturers continually design new parts, which may not be included with your version of PSpice, so eventually you will want to be able to develop your own Capture parts that have simulation capabilities (and ultimately will be used in a board design).

A PSpice library contains primitive models (such as resistors and diodes) and subcircuit models (such as logic gates and op amps). Primitive models are basically limited to behavioral descriptions of single devices. Subcircuit models can describe the behavior of a single device, multiple devices, or complete circuit designs. Unless specifically noted, *model* in this text refers to both primitive and subcircuit models in general.

With regard to PSpice model libraries the models that the libraries contain begin as simple text files that have a .mod extension. A model is added to a library by importing it into the PSpice (`.lib`) library. Once the model is imported into the library, the original `.mod` file is no longer needed and can be deleted or archived in a separate folder.

Downloading libraries and/or models from the Internet

The easiest way to make a new PSpice model library is to download it from the manufacturer's website (when available). You can often download complete libraries, but sometimes OEMs (original equipment manufacturers) provide only individual models that you can add to your own libraries.

Here is an example. Suppose you were to use an RS2A fast recovery diode (Diodes Incorporated) in your design and need its PSpice model. By going to the Diodes Incorporated website (www.diodes.com/products/spicemodels/index.php), you can obtain the model for the RS2A diode as shown in Fig. 7.38. Copy the text as is from the Internet Explorer window and paste it into any simple text editor such as Notepad. Save the text file with a .mod file extension (e.g., `RS2A.mod`) to a convenient folder that is set up for your collection of model files.

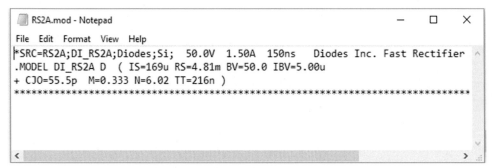

Figure 7.38 Spice diode model downloaded from the web.

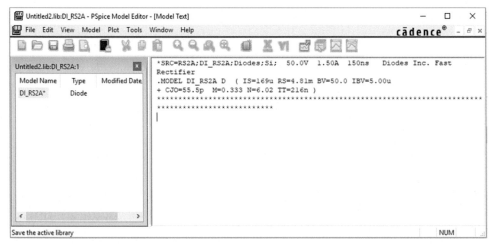

Figure 7.39 New RS2A PSpice model in the Model Editor.

Next start a new PSpice library for Diodes Incorporated parts. From the PSpice Model Editor File menu choose New. You will be presented with a blank Models List window pane and an unsaved library (Untitled1.lib). From the Model menu, choose Import. From the Open File dialog box, navigate to and select the RS2A.mod file. Click Open. The Models List window should now contain the new model as shown in Fig. 7.39. Save the new library in a user folder with a name such as DiodesInc.lib (select Save As... from the File menu). You now have a PSpice library (with only one part in this case) for which you can make a new Capture part.

Making a PSpice Model from a Capture project
In the event that a PSpice model (or even a generic spice model) cannot be located, you can make your own *.mod file. There are two ways to do this, depending on the type of model you are trying to make. If you need to make a primitive model for a

device (e.g., a diode or P-channel MOSFET transistor), you need to compose a .mod file using a text editor then import the model into a library as just described. To make an accurate model, you need to be familiar with model parameters for the part and the "code" that PSpice understands. This is not described here, but you can read about the details in the *PSpice Reference Guide*. As a starting point you can also look at examples from the PSpice Breakout library (breakout.lib).

If you need to make a model for a nonprimitive part (an IC or a transformer, for example), you can compose a subcircuit model without having extensive knowledge of model parameters or PSpice code. You simply "draw" the circuit using Capture, have Capture write the subcircuit model for you, then save it as a PSpice library. Once you have the .LIB file, you can use Method 3 or 4 to make a Capture part with the model attached to it and attach a PCB Editor footprint if so desired.

An example of how to make a PSpice model and subsequent Capture part is given here for a transformer with a single primary winding and a center-tap secondary, similar to the one in Fig. 7.3 (except that the transformer in this example has a PSpice model—a PSpiceTemplate property—associated with it). Using this procedure you can specify the inductance and DC resistance of the windings and the coupling between the windings for a specific part as described in a data sheet.

The basic process is as follows:

1. Use Capture to draw a circuit that you can simulate. The circuit will consist of inductors, resistors, and coupling coefficients.
2. After the transformer "circuit" is simulated to verify it behaves correctly, simulation sources are deleted, and hierarchical ports are added to the schematic, which will become the leads (pins) of the transformer.
3. Use Capture to create a PSpice library netlist file (.lib) for the circuit. Method 3 or Method 4 is used to generate a Capture part from the .lib file. In this example, we use Method 4 so that we can look at and modify the model prior to making a Capture part for it.

To make a .subcircuit model, open Capture and choose New → Project... from the session frame's File menu. From the New Project dialog box, choose PSpice Analog or Mixed A/D as shown in Fig. 7.40.

Select the location of the new project using the Browse... button at the bottom of the dialog box. If you plan on making more models in the future, it is a good idea to create a new folder just for model development. Once you have your models fully developed and tested, you can copy the finished libraries into the normal Capture and PSpice library folders.

After you click OK, the Create PSpice Project dialog box (Fig. 7.41) will be displayed. Check the Create based upon an existing project radio button and select either the empty.opj or the simple.opj project template. Different templates may be displayed depending on which version you have. For what we are going to do in this example, it really does not matter which template you start with. Click OK.

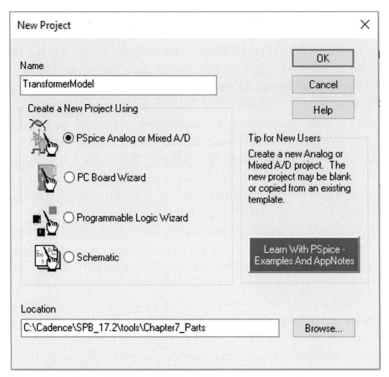

Figure 7.40 New Project dialog box for a PSpice project.

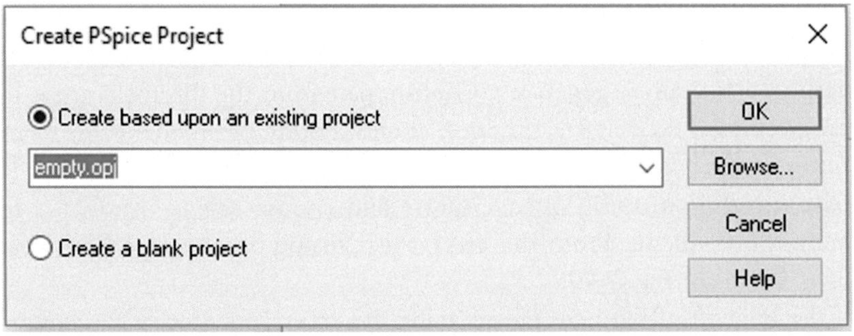

Figure 7.41 The Create PSpice Project dialog box.

In the `Project Manager` window, expand the `Design` (.dsn) icon if it is not already expanded, and double click the `SCHEMATIC1` folder (see Fig. 7.42). The name of the design (`DesignName.dsn`) will become the default name of the PSpice library (`DesignName.lib`), and the name of the root folder (`FolderName`) will become the name of the PSpice part model (`.subcircuit FolderName…`), so you want to change the name of the folder from `SCHEMATIC1` to the name you wish for your part.

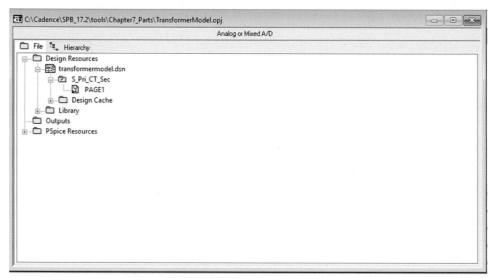

Figure 7.42 Project Manager view of design.

To rename the schematic folder, select the folder by left clicking once, then right click, and select Rename from the pop-up menu. Change the name of the schematic folder to the name that you want the part to have (e.g., S_Pri_CT_Sec for single-winding primary, center-tap secondary).

Double click on the PAGE1 icon to display the schematic page (see Fig. 7.42). Delete any parts or text provided by the template by dragging a box around (or across) the parts to highlight them, then hit the Delete key on your keyboard.

Place part button

Place four resistors from the Analog library (which have PSpice models associated with them) on the schematic page (see Fig. 7.45 later). The resistors are used to simulate the DC resistance of the windings and a dummy load resistor (RL). You can get the resistors from the Place Part dropdown list located on the toolbar at the top of the window frame or by selecting the Place Part button, , on the toolbar at the right of the schematic page. If you use the Place Part button, you will be presented with the Place Part dialog box shown in Fig. 7.43. Select ANALOG from the Libraries: list then scroll down the Part List: and select part R. Notice that the parts in the Analog library have PSpice models and footprints associated with them, as indicated by the PSpice and PCB Editor icons located under the part preview box in the lower right corner of the dialog box. Click OK. Click on the schematic page in four places to place four resistors (see Fig. 7.45 later for reference).

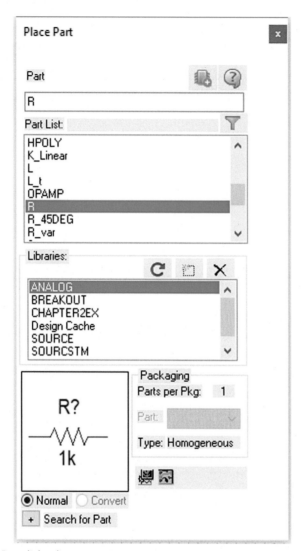

Figure 7.43 Place Part dialog box.

Repeat this procedure to place three inductors (part L) on the schematic page. One inductor serves as the primary winding, and the other two serve as the secondary windings. The inductors will be used to define the inductance (the turns ratios) of the primary and secondary windings.

Next, place one K_Linear part from the Analog library on the schematic page. K_Linear defines the coupling coefficient of the windings.

Place a VSIN part on the schematic page so that we can test the operation of the transformer. VSIN is located in the SOURCE library.

Place Ground tool

Finally, place three zero (0) ground references on the schematic page. Click the Place Ground tool, , then select 0/SOURCE from the Place Ground dialog box, as shown in Fig. 7.44. If you don't see SOURCE in the Libraries list, push Add Library... button and select /PSpice/source.olb to add it. For schematic entry and PCB Editor, you can use any ground symbol, but for PSpice simulations you have to have at least one 0 reference ground.

Position and rename the components, wire the circuit, and change the values of the components as shown in Fig. 7.45. To change the reference designators (e.g., R1 or Rp1) or the component values (e.g., 10 Ω), double click the property you want to change. In the Display Properties dialog box, enter the appropriate value and click OK.

You can also modify a part's properties by double clicking the part (or click once to select it then right click and select Edit Properties... from the pop-up menu). A Property Editor spreadsheet will pop up, as shown in Fig. 7.46. If the properties are listed across in rows instead of vertically in columns, you can change the view by

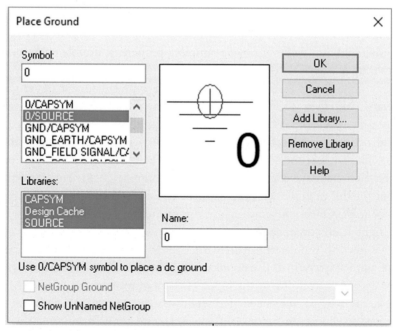

Figure 7.44 Place Ground dialog box.

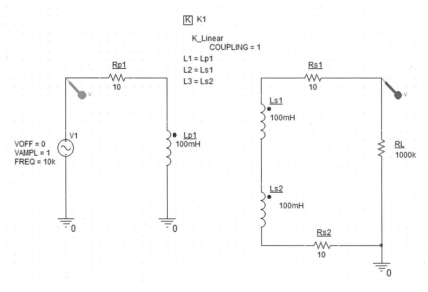

Figure 7.45 Schematic of the linear step-up transformer.

selecting the upper left-hand (corner) cell to highlight the entire spreadsheet, then right click and select `Pivot` from the pop-up menu. You can also specify how many items are listed by using the `Filter by:` dropdown list just above the spreadsheet cells. To display all pertinent properties, select the `<Current properties>` option located at the top of the list.

To modify and display the `K_Linear` coupling properties, double click the coupling part to bring up the spreadsheet. Make sure that either the `Current properties` or the `OrCAD-PSpice` filter option is selected. In the L1 cell, type `Lp1` (or whatever you named your primary coil) then click the `Display` button, which is located just above the spreadsheet cells, to enable displaying `Name` and `Value` of this property. Enter `Ls1` in the L2 cell and `Ls2` in the L3 cell and click the `Display` button for both of these parameters too. The L1, L2, and so on cells establish which coils are coupled as part of the transformer. You can have additional or separate coupling coefficients for different sets of coils, but for this example we have all three equally coupled together using the single linear coupler. Once you finish, close the spreadsheet by clicking the `X` button in the upper right-hand corner of the spreadsheet.

Now we need to simulate the part with PSpice. You can perform several types of simulations with PSpice. In this example, we perform a time-domain analysis so that we can see the AC waveforms at the input of the transformer and at the RL.

To test the circuit we need to set up a simulation profile. *To set up a simulation profile,* choose `Edit` (or `New`) `Simulation Profile` from the `PSpice` menu. In the `Simulation Settings` dialog box (see Fig. 7.47), select the `Analysis` tab.

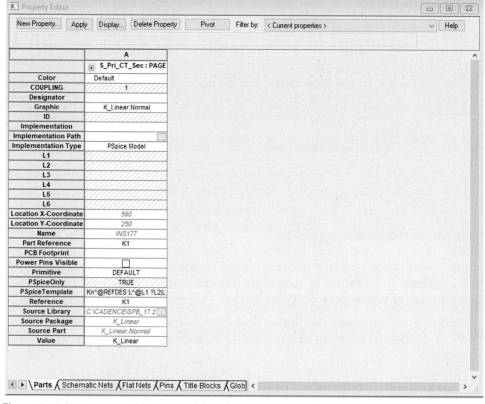

	A
	⊞ S_Pri_CT_Sec : PAGE
Color	Default
COUPLING	1
Designator	
Graphic	K_Linear.Normal
ID	
Implementation	
Implementation Path	
Implementation Type	PSpice Model
L1	
L2	
L3	
L4	
L5	
L6	
Location X-Coordinate	590
Location Y-Coordinate	250
Name	INS177
Part Reference	K1
PCB Footprint	
Power Pins Visible	☐
Primitive	DEFAULT
PSpiceOnly	TRUE
PSpiceTemplate	Kn^@REFDES L^@L1 ?L2|L
Reference	K1
Source Library	C:\CADENCE\SPB_17.2
Source Package	K_Linear
Source Part	K_Linear.Normal
Value	K_Linear

Figure 7.46 Part Property Editor spreadsheet (vertical view).

In the `Analysis type:` dropdown list, select `Time Domain (Transient)`. Set the rest of the parameters as shown in Fig. 7.47, then click `OK`.

Voltage Probe button

Run PSpice button

Place voltage markers on the circuit to specify which voltages to display in the PSpice probe window. Click the `Voltage/Level Marker probe` button, , on the toolbar and place a probe on the wire coming from VSIN (green marker) and one (red marker) on the wire going from the secondary winding to the RL.

Start the simulation by clicking the `Run PSpice` button, . PSpice runs the simulation and displays the results in a probe window, as shown in Fig. 7.48. The voltage curves show that the transformer functions as a 1:2 step-up transformer, since the

Figure 7.47 Simulation Settings dialog box.

output (red marker curve) is twice as high as the input (green marker curve). Additional tests (e.g., frequency response) could be performed to validate the circuit model further, but these are not discussed here.

Since the circuit model has been validated, we now prepare to make a PSpice model of the circuit. Begin by deleting the VSIN source, the RL, and all the ground references.

Place Port tool

Add ports to the schematic, which will serve as the leads of the transformer. Click the Place Port tool, . Select the PORTBOTH-L port from the Place Hierarchical Port dialog box, as shown in Fig. 7.49. All the port symbols behave identically in Capture, PSpice, and PCB Editor. The only difference is the appearance on the schematic. Since the transformer is a passive device, we use the symbol that indicates that an applied signal can go in either direction. Click OK and place five of the ports on the schematic page (two for the primary and three for the secondary).

Reposition, connect, and label the ports as shown in Fig. 7.50.

Display the Project Manager window by minimizing the schematic page or *ProjectName*.opj from the Window menu. Select the Design icon then select Create Netlist from the Tools menu.

In the Create Netlist dialog box (Fig. 7.51), select the PSpice tab. Check the Create SubCircuit Format Netlist box and the Descend radio button. A default name will be given to the netlist. Modify the path and name as desired. Make sure the name ends with the .LIB file extension. Click OK.

You now have a PSpice library file with one model (the transformer) in it. You can use Method 3 or 4 to generate a Capture part library from this model or add it to

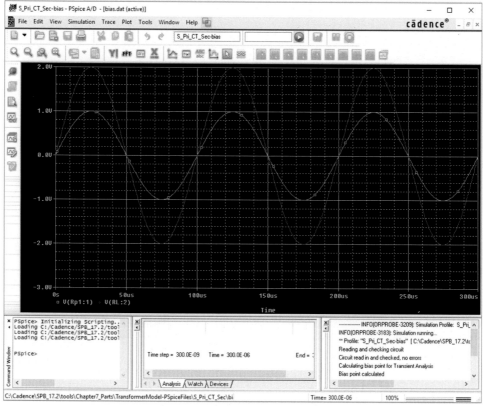

Figure 7.48 PSpice probe window for the center-tap transformer.

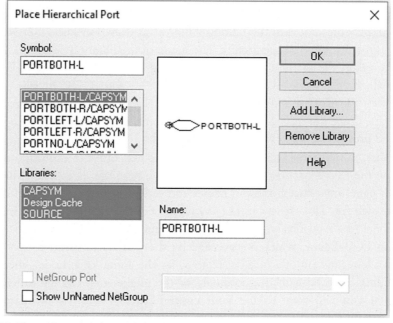

Figure 7.49 Place Hierarchical Port dialog box.

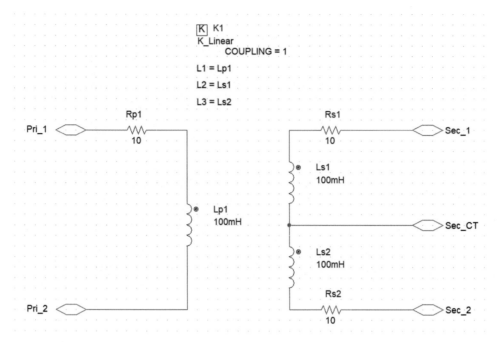

Figure 7.50 Hierarchical ports in the circuit design.

an existing part library. To complete the example we use Method 4 so that we can take a look at the PSpice model generated by Capture.

Start the PSpice Model Editor again. From the File menu, select Open, and navigate to the transformer library that you made previously. Click on the S_Pri_CT_Sec model to display the model in the editing window, as shown in Fig. 7.52. Notice that the model type is .SUBCKT (a subcircuit). The name of the library is whatever you specified as the netlist file in the Create Netlist dialog box, and the name of the part is the schematic folder name in the Capture design (see Fig. 7.42). At this point you could modify the part to add specific requirements that will carry forward into the Capture part. Once the model editing (if any) is completed, generate the Capture part as described previously, using Method 3 or Method 4.

Note that, as shown in Fig. 7.52, the pin names and order in the Capture part (as indicated in the Pin Properties pane) and the PSpice template (as indicated in the Part Properties pane) must match the pin names and order in the PSpice model exactly or simulations will fail. The Implementation name must also match the model name in the PSpice model file. And finally, the part's pin numbers must match the pin (padstack) numbers in PCB Editor, which is governed by the part's data sheet. If the part's pin number is a letter instead of a number (e.g., A, for the anode of a diode), then the pad in PCB Editor must be named accordingly or the engineering change order will fail. Now, when you will want to use your custom part S_Pri_CT_Sec with its custom

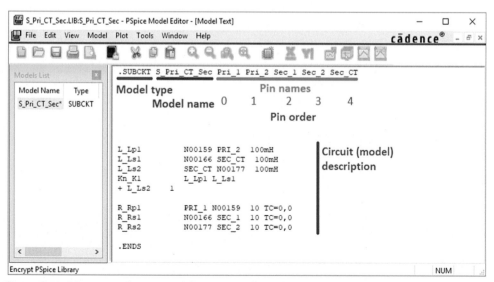

Figure 7.51 Create Netlist dialog box.

Figure 7.52 PSpice transformer model generated from a Capture project.

PSpice model in simulation, don't forget to add your appropriate custom LIB library to a PSpice simulation profile (Configuration Files/Library, use the upper Browse button to locate your LIB file and use the Add as Global button).

Adding PSpice templates (models) to preexisting Capture parts

Rather than using Method 3 or 4 to make a new part from the transformer PSpice library, you might wonder why we did not just add the PSpice model to the transformer already created in the first example (for which Method 1 was used). In older versions of OrCAD, this was somewhat of a challenging task (e.g., you have to know what "X^@REFDES %A %B %Y %VCC %GND @MODEL PARAMS: \n1IO_LEVEL5@IO_LEVELMNTYMXDLY5@MNTYMXDLY" means). Fortunately, it is a simpler matter with the newer versions.

In this example, we see how to add an existing PSpice model to an existing Capture part. We add a basic capacitor model to one of Capture's capacitor parts that has no model associated with it.

A basic capacitor (part C or CAP) from Capture's discrete.olb library has no PCB Editor footprint or PSpice model associated with it. We add a PSpice model to the part now. The location of the basic PSpice capacitor model that we use is in the PSpice Breakout library (breakout.lib).

To add a PSpice model to an existing Capture part, start Capture and select File → Open → Library and select the library with the part to which you want to add a PSpice model (use discrete.olb for this example). Find the capacitor (C or CAP, for example) in the Capture Library Part Manager and click the part's icon to select it. Right click and select Associate PSpice Model... from the pop-up menu.

At the Model Import Wizard dialog box (Fig. 7.53), use the Browse... button in the upper right-hand corner to find the PSpice model you want to associate with the Capture part. Locate and choose the library file "tools/PSpice/library/breakout.lib" in the Cadence installation folder. The wizard automatically searches through the PSpice library and lists all models in the Matching Models window that have the same number of pins as the Capture part you selected in the Part Library Manager. Select the CBREAK model and click Next.

The wizard then displays the Pin Matching tool shown in Fig. 7.54. This is where you connect the PSpice model pin to the Capture part pin. Click Finish when the pins are matched. The information box shown in Fig. 7.55 should be displayed, indicating that the capacitor now has a PSpice model attached to it.

Constructing Capture symbols

Four types of symbols are used in Capture: (1) power/ground symbols, (2) off-page connectors, (3) hierarchical ports, and (4) title blocks.

To make a new power symbol, open the capsym.olb library. Select the Library icon under the Library folder. From the Design menu, choose New Symbol. In the New

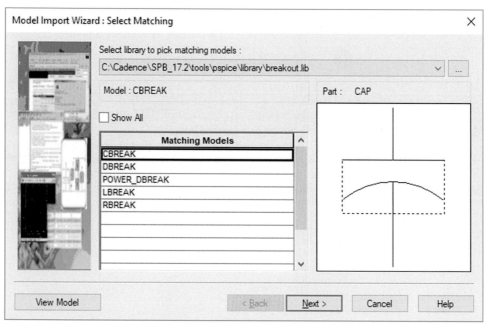

Figure 7.53 Matching parts with the Model Import Wizard.

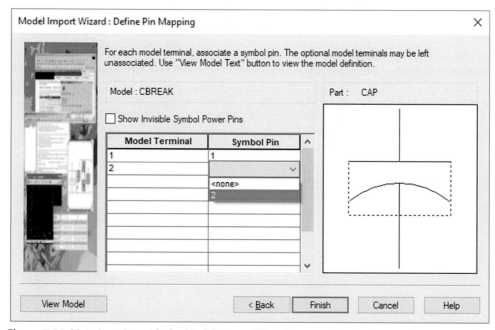

Figure 7.54 Mapping pins with the Model Import Wizard.

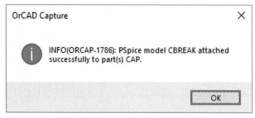

Figure 7.55 PSpice model successfully added to a Capture part.

(A)

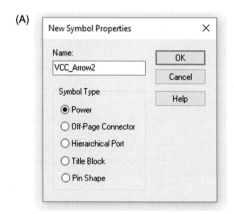

(B)

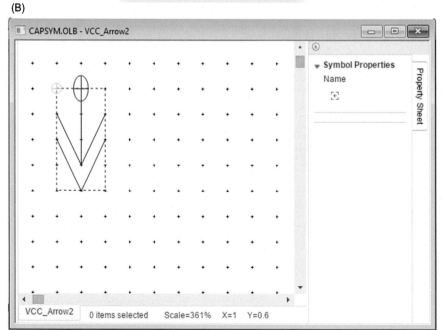

Figure 7.56 Making a new power symbol.

Symbol Properties dialog box (Fig. 7.56), enter the name, and select the appropriate radio button from the Symbol Type group box. Click OK. A part editing window and drawing toolbar will be displayed. Use the graphics tools to make the symbol. Close the editing window and save the symbol. Save and close the library.

The construction methods for power symbols, off-page connectors, and hierarchical ports are similar (although each has its own functionality). The title block is different from the other three. The two types of title blocks are default and optional. Every new project has a default title block, which you can specify for each project by going to Options → Design Template... on the schematic page menu. You can also add optional title blocks by going to Place → Title Block..., also on the schematic page menu. The title blocks are located in the capsym library. You can construct your own title blocks and save them with the others or make your own library of power symbols, title blocks, and the like.

To make a new title block, open or make a new Capture library, select the Library icon, and from the Design menu, select New Symbol.... At the New Symbol Properties dialog box (Fig. 7.56), select Title Block and click OK. Just as with the power symbols, you will be presented with a part editing window. Use the graphics tools to construct the box (Fig. 7.57 shows an example of a parts list) and the Text tool to add titles and headers. To enter text that can be modified on schematic pages click on the [+] sign of Symbol Properties section in the Property Sheet pane. The name field is required but you can either leave the value field empty or enter a default value. Click the green check icon to add the property. To make an object field visible click the "eye" icon

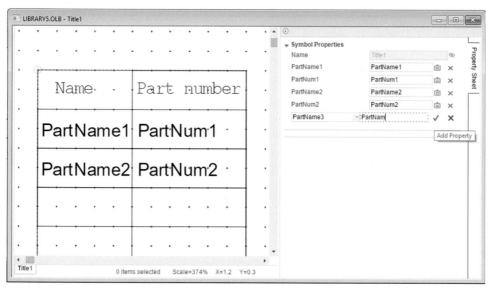

Figure 7.57 Example of a new title block used as a parts list.

on the right of the property and choose value only. The property value will be displayed, and you can move it to any position inside your symbol. After you have saved the block, you can add it to any schematic page.

That completes the chapter on making Capture parts and symbols.

CHAPTER 8

Making and editing footprints

Contents

Footprints provide a means of physically attached components to your printed circuit board (PCB), and they provide electrical connectivity as defined by the netlist generated in Capture. Footprints in PCB Editor are known by several names, including *symbols*, *components*, and *packages*. The PCB Editor library has over 400 symbols, but

Complete PCB Design Using OrCAD® Capture and PCB Editor
DOI: https://doi.org/10.1016/B978-0-12-817684-9.00008-4

there will be occasions when you will need to make your own. This purpose of this chapter is to

1. introduce PCB Editor's symbol libraries,
2. discuss the composition of a footprint,
3. provide a detailed explanation of padstacks,
4. demonstrate how to create a padstack using the Padstack Editor,
5. provide footprint design examples, and
6. provide design examples for mechanical symbols and flash symbols.

Introduction to PCB Editor's symbols library

The standard PCB Editor footprint library is located in the special folder of the OrCAD® installation path, for example, "C:\Cadence\SPB_17.2\share\pcb\pcb_lib". In that folder, there are two more folders: devices and symbols. The devices folder contains text files that define logic information for certain component types, which are not discussed here. The symbols folder contains eight types of files. Table 8.1 lists the files and explanations of their functions.

Some files have the same file name but different file extensions, for example, the cap300.dra file and cap300.psm file. In general, you work with the .dra files with PCB Editor, then PCB Editor creates the other files that it actually uses in a board design.

For each type of file, PCB Editor operates in a particular mode. When you open an existing drawing (.dra) file, PCB knows which mode to be in as it reads the file and knows the corresponding symbol file (bsm, psm, etc.). When you start a new drawing (.dra) file, you select the mode by choosing a drawing type from a list. When

Table 8.1 Library files and their extensions and functions.

File extension	File type
name.bsm	A *mechanical symbol* file (derived from the .dra file with the same name)
name.dra	A PCB Editor drawing file
name.fsm	A *flash symbol* file (derived from the .dra file with the same name)
name.log	A text file containing the date and time the symbol or padstack name was last modified or saved
name.osm	A *drawing format* file (derived from the .dra file with the same name)
name.pad	A *padstack definition* file generated/used by Padstack Editor
name.psm	A *package symbol* file (derived from the .dra file with the same name)
name.ssm	A *custom pad shape file* (derived from the .dra file with the same name)

you finish working with the .dra file and save it, PCB Editor automatically creates the correct symbol file. This is demonstrated in the examples that follow.

Symbol types

Package symbols (*name*.dra and *name*.psm) are the component footprints used on your board design that are assigned in Capture. After finishing the footprint drawing the .psm file is generated when you save the drawing.

Mechanical symbols (*name*.dra and *name*.bsm) are objects, such as mechanical pins, which are not connected to a net, mounting holes on boards, or predefined board outlines. The mechanical pins don't have the pin number, and there is no need to include such pins to the Schematic.

Flash symbols (*name*.dra and *name*.fsm) are used to define thermal reliefs, which are used to connect padstacks to power and ground planes when the planes are defined as negative plane layers. Positive plane layers do not need flash symbols, because PCB Editor automatically creates thermal reliefs based on trace width and spacing settings as defined in the Constraint Manager (more on that in PCB Design Examples in Chapter 9). Also flash symbols can be used to create the array of openings in the pastemask stencil.

Drawing formats (*name*.dra and *name*.osm) are predefined drawing templates that follow standard paper sizes. The six drawing formats range from drawing size A (8.5×11 in.) to size E (44×34 in.). You can add these to your board designs to give them a professional look and follow industry drawing standards.

Custom pad shapes (*name*.dra and *name*.ssm) are specially designed shapes used to create padstacks that have pad shapes other than round or rectangular/square pads.

Padstack files (*name*.pad) define surface-mount and through-hole padstacks and include conductor information, such as inner and outer pad shape and size, and thermal relief sizes and physical information, such as drill diameter and soldermask and pastemask shapes and sizes. Padstack definitions are handled using an application called Padstack Editor, which can be launched from within PCB Editor (Tools menu) or launched separately from the Windows Start menu. Padstack Editor is demonstrated in the design examples that follow.

Composition of a footprint

A footprint symbol is made up of pins (padstacks), graphics (for outlines etc.), and text. As an example, Fig. 8.1 shows the through-hole resistor symbol res400.dra found in the symbols library. To view it, open PCB Editor, navigate to the symbols folder, and select the .dra file type and then select the res400.dra. Table 8.2 describes what the objects are and to what class and subclass each object belongs. As described earlier the pin (padstack) is actually defined by the Padstack Editor and simply inserted into the symbol drawing along with the graphics and text objects.

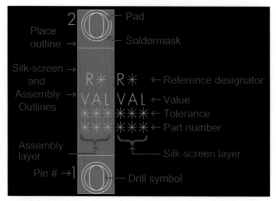

Figure 8.1 Anatomy of a footprint symbol.

Table 8.2 Types of symbol objects and their classes.

	Object type	Class	Subclass
Pin (padstack)			
Pad	Conductor	Pin	Top
			Bottom
Soldermask	Nonconductor	Pin	Soldermask_Top
			Soldermask_Bottom
Drill symbol	Drill symbol figure	Manufacturing	Ncdrill_Figure
Pin number	Text	Package geometry	Pin_Number
Graphics			
Place outline	Filled rectangle (frectangle)	Package geometry	Place_Bound_Top
Silk-screen outline	Rectangle	Package geometry	Silkscreen_Top
Assembly outline	Line segment	Package geometry	Assembly_Top
Text			
Reference designator	Text	Ref Des	Silkscreen_Top
			Assembly_Top
Value	Text	Component value	Silkscreen_Top
			Assembly_Top
Tolerance	Text	Tolerance	Silkscreen_Top
			Assembly_Top
Part number	Text	User part number	Silkscreen_Top
			Assembly_Top

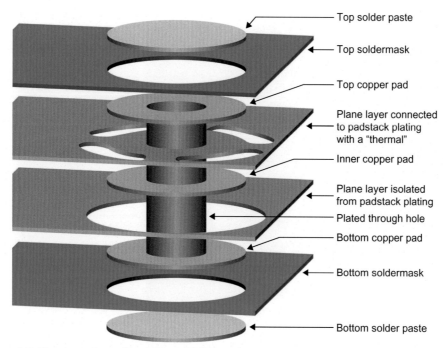

Top solder paste
Top soldermask
Top copper pad
Plane layer connected to padstack plating with a "thermal"
Inner copper pad
Plane layer isolated from padstack plating
Plated through hole
Bottom copper pad
Bottom soldermask
Bottom solder paste

Figure 8.2 Elements of a padstack from the PCB perspective. *PCB*, Printed circuit board.

Padstacks

Fig. 8.2 shows an example of a multilayer padstack from a PCB's perspective. Padstacks define every aspect of how a component's pins will be fastened to the PCB and how traces will be connected to them. Padstack definitions specify areas for copper pads on outer and inner routing layers, thermal reliefs and clearance areas in plane layers, openings in soldermasks, and solder paste (optional). From the PCB designer's perspective the drill hole may also be considered part of the padstack definition, but the copper used to plate through holes and vias is not. This is because the plating thickness is controlled by the board manufacturer and is typically insignificant relative to the drill and lead diameters. More about the lead-to-hole diameter is discussed in a later example and Chapter 5, Introduction to design for manufacturing.

Fig. 8.3 shows how a padstack is actually displayed in PCB Editor. When a particular layer is made active (using the Options pane), the pad on that layer is brought to the front of the view, while pads on inactive layers are pushed behind. Padstacks have clearance areas defined, which remove copper from plane layers and are visible when the plane layers are visible. The pin number of the padstack is displayed if the Pin_Number subclass box is checked in the Package Geometry class in the Color Dialog panel. Drill holes are also displayed if the Display / Plated Holes option is enabled in the Design Parameter Editor as shown in Fig. 8.4.

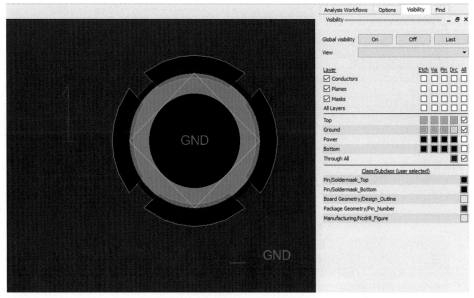

Figure 8.3 A padstack as seen in PCB Editor.

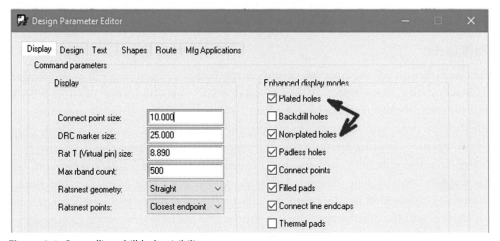

Figure 8.4 Controlling drill hole visibility.

Graphical objects

Graphical objects (lines, arcs, etc.) are placed on footprints and board layouts to show information not defined by padstacks (such as the outline of a part). The objects may be visible on the final board or visible only in the design files. Objects that are visible on the board include silk-screen markings (component outlines, for example), while objects visible only in the design files include things such as board outlines, assembly outlines, and boundaries (e.g., keep-out areas where traces and components are not allowed).

Several types of objects can be used in PCB Editor, but usually only three or four types are used when making footprints (i.e., detail objects, place boundary outlines, and occasionally copper areas). Detail objects are used to indicate silk–screen markings that will be visible on the board and on assembly layers to provide information during assembly. Place boundary outlines are used by the design rule checker (DRC) utility to maintain required distance between parts while laying out the board. On routing layers, copper areas can be used as heat spreaders or mini-ground planes for components that require them. Examples of how to use these objects are given in the PCB Design Examples (Chapter 9).

Text

Like graphics, text objects can be visible on a PCB or visible only within the design files. Text objects on PCBs are often part of the silk screen and may be used for such purposes as component reference designators (part of the footprint silk screen) and displaying board serial numbers or design revisions (part of the board silk screen). Text that is visible only in the design files might be placed on one of the assembly or documentation layers to show board dimensions or special manufacturing instructions.

Minimum footprint requirements

As shown in Fig. 8.1, many types of text and graphics objects can be used when you make a new footprint. Not all the items shown in the figure are required, but four minimum objects are required on a footprint design. Packages must have the following elements: (1) at least one pin, (2) at least one reference designator, (3) a component outline, and (4) a place-bound rectangle. As shown in Table 8.2, reference designators and component outlines are placed on the silk-screen or assembly layers (usually both), and the place-bound rectangle is on the `Place_Bound_Top` (or `Bottom`) layer.

Optional footprint objects

The following is a list of other elements you can add to a package symbol during the symbol building process:
- Device type (text for the component device type).
- Component value (text for the component value).
- Tolerance (text for the component tolerance).
- Component height (text for the physical height of the component).
- User part number (text for the package part number).
- Route keep-out shapes, identifying areas where etch is not allowed.
- Via keep-out shapes, identifying via keep-out areas.
- Etch (etch lines, arcs, rectangles, shapes, or text added to the symbol).
- Vias.

For further reading on design requirements, please see Chapter 4, Introduction to industry standards, for a list of industry standards and Appendix B for more information on packages and footprints.

Good sources for package information can be found at the following websites:

- IPC website (www.ipc.org)
- www.PCBLibraries.com/downloads—Land Pattern Calculator based on the IPC-7351B standard
- http://www.onsemi.com/PowerSolutions/supportDoc.do?type = drawing
- http://www.ti.com/support-packaging/packaging-information.html
- www.diodes.com/datasheets/ap02002.pdf

Introduction to the Padstack Editor

The two basic kinds of padstacks are through-hole padstacks (for leaded components and vias) and surface-mount pads. Through holes are made using padstacks that allow connections from any one layer to any other layer (as in Fig. 8.2). Surface-mount pads are isolated from all layers except for the top layer and therefore do not use drilled and plated holes. Surface-mount pads may be on the top or bottom layer (or both in the case of edge connector footprints). Connecting a surface-mount pad to another layer is accomplished using a special padstack called a *fan-out via* (also called a *stringer pad*). A fan-out via is usually not part of the footprint; it is added to a PCB when placing parts and routing traces. You will see how to do this in the PCB Design Examples (Chapter 9).

Through-hole padstacks are often named by their shape and size. The typical pad geometries are null, circle, square, oblong, rectangle, rounded rectangle, chamfered rectangle, octagon, donut, n-sided polygon, and user-defined shape. For circular through-hole padstacks, one naming convention used is *padXcirYd*.pad, where *pad* indicates padstack, *X* is the pad outer diameter (OD), *cir* is the circular shape (*sq* for a square top pad or *rec* for rectangular), and *Y* is the pad inner diameter (ID)—the drill size (see Fig. 8.5). For example, if the outer dimension of a round pad is 62 mil (0.062 in.) and the drill hole is 25 mil, the pad name would be pad62cir25d.pad.

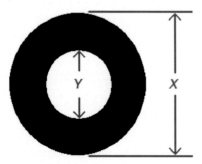 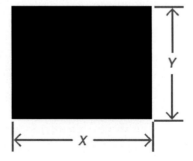

Figure 8.5 Pad dimensions.

Making and editing footprints

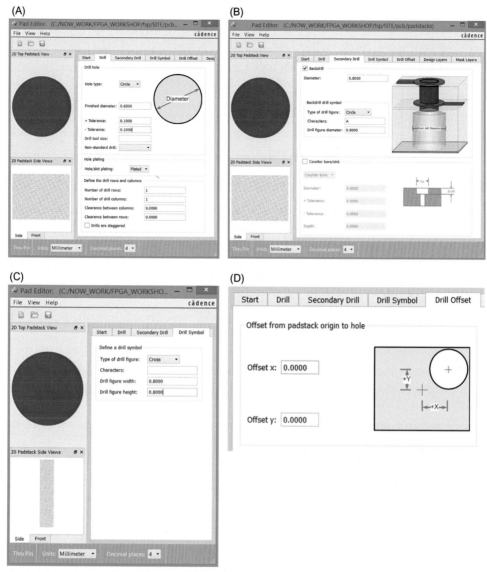

Figure 8.6 (A) Drill tab of the Padstack Editor. (B) Secondary Drill tab. (C) Drill Symbol tab. (D) Drill Offset tab.

For surface-mount pads a typical naming convention is *smdXrecY*.pad (or *smdX_Y*. pad), where *smd* indicates the padstack, which is for a surface-mount device, *X* is the pad width, *rec* (or _) indicates a rectangular pad, and *Y* is the pad height.

Through-hole and surface-mount pads are designed and modified using the Padstack Editor (see Fig. 8.6). There are two ways to launch the Padstack Editor.

Padstack Editor can be launched from within PCB Editor or separately from the Windows Start menu. In the PCB Design Examples (Chapter 9), it will usually be launched from within PCB Editor. In the footprint design examples that follow, Padstack Editor will be launched from the Windows Start menu, where detailed instructions on how to use it are provided in the examples. Padstack Editor has several tabs, arranged from left to right in the sequence of steps required to create the new padstack.

Start tab

Here you can define the Type (usage) and Shape (geometry) of new pad.

Drill tab

The Drill tab (Fig. 8.6A) is where the drill diameter and related parameters are set.

Hole type—the shape of the hole, typically Circle, but you can also choose Square for some special technology reasons.

Finished diameter—the diameter of the hole after plating. Usually this is more important value than the drill tool diameter, because for the assembly reasons we need to know the finished diameter.

$\pm$ Tolerance—the tolerance of the hole size, typically ± 0.1 mm, but for the press-fit holes it can be set as ± 0.05 mm.

Drill tool size—the optional field to set up the tool size. If not specified, the PCB manufacturer will choose the tool diameter according to the technology used, based on needed finished diameter value. Some component assembly techniques require a specific hole size for correct assembly.

Nonstandard drill—you can optionally choose a specific technology to make the hole, such as laser, plasma, and punch.

Hole plating—here you should choose if the hole is plated, or nonplated. Note that Padstacks of Thru Hole Type are always plated according to the IPC-2581B standard.

Define the drill rows and columns—if you need the array of the holes located inside the pad.

Units—this field in the left-bottom corner of the window allows you to choose the measurement units. It's recommended to use millimeter or mils.

When you enter the hole parameters, you can see it on the left, in 2D Top Padstack View and 2D Padstack Side View windows (the Side View window, in its turn, allows to see the Side and Front cross section of the hole).

Secondary Drill tab

This allows you to define the secondary drill parameters. This operation is needed if you want to make additional drilling over the hole, for example, the *backdrilling* for high-speed signals (remove the extra part of the plated hole in thick PCB, to reduce the reflections and to improve the Signal Integrity), or the *counterbore* for the mounting holes.

Diameter—the tool diameter of backdrilling, which is usually 0.3. . .0.4 mm bigger than the finished plated hole diameter.

Type of drill figure—the graphics symbol of this hole on the drawing.

Characters—additional text inside of the drill figure.

Drill figure diameter—the size of the drill figure.

Drill Symbol tab

In this tab, you can define the drill figure for the drawings. This definition is not fixed and can be overridden later in the PCB.

Type of drill figure—the shape of the figure.

Characters—the additional text inside the figure.

Drill figure width/height—the size of the figure—typically equal to the hole size.

Drill Offset tab

If you need to shift the hole from the center of pad, you can use the `Drill Offset` tab.

Design Layers tab

In Fig. 8.7A the layer stack-up definition consists of a beginning layer (`TOP`), a default internal layer, an ending layer (`BOTTOM`) and the associated soldermasks. In Fig. 8.7B the layer stack-up definition consists of just a beginning layer (`TOP`).

Begin Layer/Default Internal/End Layer—here you should choose the pad shape and size for the beginning, inner, and end layers. `Default Internal`—any internal layer of the multilayer PCB. Note that it's no need to create in advance all types of blind or buried vias from scratch. Instead, in OrCAD PCB Editor, you can just select some existing plated through hole (PTH) or blind via as a template and simply create from it the new blind/buried via, with new name and new set of start/end layers.

`...`

`Browse... button`

Regular Pad—choose any of standard shapes. If your pad should have nonstandard shape, you can choose the `Shape Symbol`, where the polygon of needed special shape can be drawn. To see the list of available shape symbols, press the `Browse` "..." button, `...` , on the right of shape symbol name. Then, if you don't see the appropriate shape symbol in the list, push the "`Create new Shape Symbol`" button, set its name, and the PCB Editor window will open. Then you should draw the single polygon of needed size and shape located in the 0,0 point in the `ETCH/TOP` layer (which should be used for any shape symbol). Then save the shape symbol file to the proper path (e.g. */share/pcb/pcb_lib/ symbols*) and close the PCB Editor. Now you will see the created `Shape Symbol` in the list, you can choose it to place to the needed layer of your padstack.

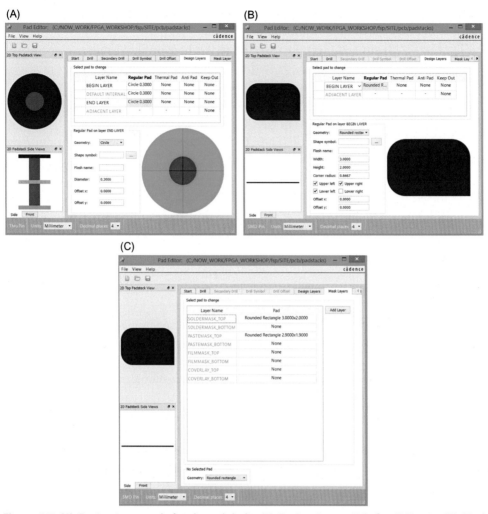

Figure 8.7 (A) Design Layers tab for through-hole. (B) Design Layers Tab for SMD pin. (C) Mask Layers tab for SMD pad. *SMD*, Surface-mounted device.

You can see the shape and size of the pad in the right bottom corner of the Padstack Editor window (see Fig. 8.7).

Thermal Pad/Anti Pad/Keep Out—these fields allow to set up the spacing of the pad in the negative plane layers for thermal connection and for nonconnected pads, as well as the route keep-out area for mechanical hole or under the surface-mounted device (SMD) pad in adjacent layer, for example, if you need to decrease the SMD pad capacitance and improve the signal integrity. Actually, you can leave these fields blank, because it's better to set the spacing parameters later, globally for all pads and vias in the PCB project.

Mask Layers tab

Here you can set the openings in the soldermask and pastemask layers, as well as the coverlays for rigid-flex PCBs and any other types of masks. You can create your own types of mask layers. The standard types of masks are listed below.

Soldermask TOP/BOTTOM—you should make the opening in the soldermask a little bit larger than the pad size, usually by 2 mil (0.05 mm) each direction. Ask your PCB supplier for the correct value.

Pastemask TOP/BOTTOM—the size of the opening in solder paste stencil depends on the assembly house. Normally it is a little bit smaller that the pad size, about 1 mil (0.025 mm) each side. For the large pads, it's recommended to split the pastemask opening to several smaller openings, which will cover not more than 50% of the pad area. For this purpose you can create the Flash Symbol file (similarly to Shape Symbol described earlier) and use it.

Filmmask_TOP/BOTTOM—the user mask which you can use for such tasks as capped via masks or plugged/tented vias.

Coverlay_TOP/BOTTOM—this mask is used to create the openings in the flexible part of PCB, in polyimide coverlay material. Usually the size of the opening is smaller than the pad size by 2 mil (0.05 mm) each side.

Options tab

In the Options tab, you can manage two parameters.

Suppress unconnected internal pads—to improve the manufacturability and signal integrity by removing the pads in the inner layers, if they are not connected to the traces or shapes. Contact to your PCB supplier to understand if this option will be acceptable.

Lock layers span—use this option while creating the blind via if you don't want to allow to insert the new layer between the start and end layers of this blind via.

When you have filled all needed values and parameters, you can check everything in the Summary tab and then save the padstack file to the proper folder. It is recommended to create the special folder for the user created pads, shape symbols, and footprints. Then in PCB Editor User Preferences, you should add this folder path to the padpath and psmpath preferences, to let PCB Editor know where the user padstacks and footprints are located.

Footprint design examples

Three footprint design examples are given. Before starting any footprint design, it is best to have the correct padstacks predefined, as it makes the design process go more smoothly. In each example, Padstack Editor is used to locate existing padstacks or design new ones.

The first design example demonstrates how to design a through-hole footprint from scratch. The second design example demonstrates how to modify an existing, surface-mount footprint to create a new one. The last example demonstrates how to use the Symbol Wizard to design a high pin-count pin-grid array (PGA) footprint.

Example 1. Design of a through-hole device from scratch

In this design example the footprint for a 1/4-W, 20%, carbon film resistor will be made. For this resistor the lead diameter is 25 mil, the body length is 250 mil, and the body width is 100 mil. An example is shown in Fig. 8.8. The component is similar to the res400 symbol that exists in the symbols library, but we design our own to learn the process, then we can compare the two.

Designing the through-hole padstack

Using the procedures described in Chapter 5, Introduction to design for manufacturing, the calculated hole diameter should be between 33 and 56 mil, and the pad should be between 51 and 74 mil for a Class A board design (using the IPC method). Several padstacks in the symbols library meet this requirement, but one that does not exist will be chosen so that the design process can be demonstrated. So, if we decide that, in addition to these requirements, we want an annular ring width of 10 mil, then a padstack with a drill hole diameter of 42 mil and a pad diameter of 62 mil suffices. The typical naming convention in PCB Editor for a pad with these dimensions is pad62cir42d, which does not exist in the native symbols library. To create this

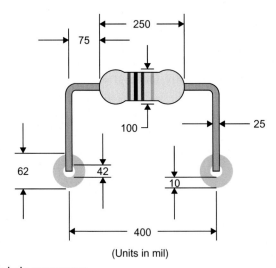

(Units in mil)

Figure 8.8 A through-hole component.

padstack, launch Padstack Editor by selecting `Start → All Programs → Cadence Release 17.2-2016 → Padstack Editor`, or just find the `Padstack Editor` in the list of applications (depending on operating system version that you use). A blank, unnamed padstack will be opened.

In the `Start` tab, choose the type of new padstack—`Thru Pin` and select the pad geometry—`Circle`. Switch to the `Drill` tab and enter the remaining values as shown in Fig. 8.9. You can choose any of the available drill symbol figures in the `Drill Symbol` tab. The size of the drill figure is not critical but is typically about the size of the drill diameter or a little smaller. Note that the drill symbol will not be visible initially when you ultimately place the component on the board. The drill symbol becomes visible only after you place a drill table in the design.

Next select the `Design Layers` tab. Enter the values shown in Fig. 8.10A. To enter values for a particular layer, select the layer by left clicking the row in one of the three

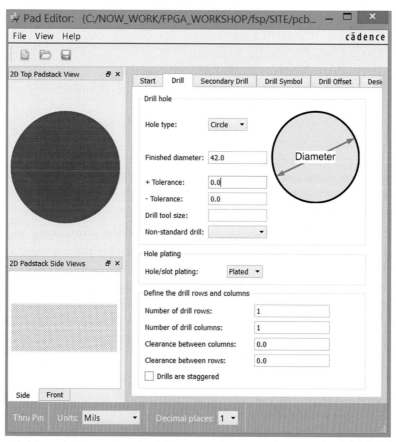

Figure 8.9 Drill parameters for a new padstack.

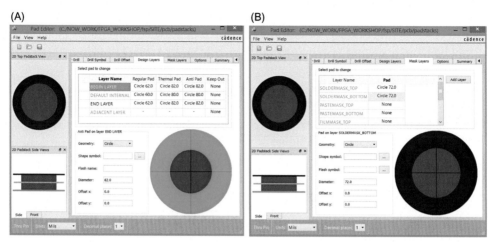

Figure 8.10 (A) Etch layers parameters for a new padstack. (B) Mask parameters for a new padstack.

columns to the right. If you left click the row in the name column, the name will be highlighted so that you can change its name.

- *Regular Pad*. The values on the internal layers can be identical to the outer layers but often they are slightly smaller (anywhere from 0 to 20 mil difference in diameter depending on the size).
- *Thermal Pad*. The native padstacks in the symbols library typically use circles that are the same size as the Anti Pad. But if you use negative artwork layers in your design, you need to use Flash symbols for the thermal pads. Flash symbols are described later, so for the time being leave the thermal pad as a circle. The procedure for defining padstacks that have no thermal pads (i.e., completely connected to a plane) is described later.
- *Anti Pad*. The clearance between a pad (or hole) and the surrounding copper (e.g., on a plane layer) should be similar to or larger than the trace spacing constraints you will likely use in your design. A typical clearance diameter is 10−20 mil larger than the pad diameter. Note that, if you make an inner pad much smaller than the outer layer pads, the Anti Pad on the inner layer should still be larger than the largest pad in the padstack; otherwise undesirable capacitive coupling can occur between the larger pad and an adjacent plane layer (see the discussion in Chapter 5: Introduction to design for manufacturing). This parameter is important only if you use negative artwork layers in your design.
- *Keep Out*. There is no need to set the route keep-out area for through-hole padstack, so we can leave it blank.

Before moving on to the next step, it would be a good idea to save your work to this point. When saving padstacks that you plan to use in active designs, you need to save them to the PCB Editor *symbols* folder or your working folder. Otherwise, if the design is just for practice, you can save it anywhere you like. When you save it for the first time, select `Save As...` from the `File` menu. After that you can just select `Save` from the `File` menu to save changes.

Select the `Mask Layers` tab. Set the `Circle` openings size in `SOLDERMASK_TOP` and `SOLDERMASK_BOTTOM` layers. Soldermask openings are usually a little larger than the outer pads so that registration tolerance errors will not cover any of the pad. The exact amount really depends on the capabilities of your board manufacturer, but in general the soldermask opening is typically 4–8 mil (0.1–0.2 mm) larger than the outer pad.

When you complete the entries shown in Figs. 8.9 and 8.10 save your design.

Designing a through-hole footprint symbol

When a new symbol drawing is first opened, the drawing area is quite large by default. So before opening a new drawing, it is a good idea to have an idea of what the footprint dimensions will be, so that it is known how large the drawing area should be. From the preceding description the body length is 250 mil, and the width is 100 mil. The footprint needs to be larger than that to account for the lead extensions, padstack dimensions and placement, and any other details that you might want to add to the drawing.

Using the procedure described in Chapter 5, Introduction to design for manufacturing, the padstacks are placed 400 mil apart. The diameter of each pad is 62 mil, so the drawing area should be another 200 mil or so wider. Silk-screen and assembly details and text are also added, so at the very least the drawing area should be 1000 mil square. The next thing to consider is where the origin of the part will be. Through-hole components often have their origins on pin 1. Surface-mount components typically have their origins at the body center. Since this will be a through-hole component, we put the origin on pin 1. Starting at the lower left corner the (x, y) coordinate will be $(-1000, -1000)$, and we make the working area 3 in. wide by 2 in. tall.

To begin making a new footprint, open `Start → All Programs → Cadence Release 17.2-2016 → PCB Editor`. PCB Editor will open the last design you worked on by default. Start a new footprint drawing by selecting `File → New...` from the menu bar. The `New Drawing` dialog box will be displayed (see Fig. 8.11). Enter a name in the `Drawing Name:` box for the new footprint and click the `Browse...` button to choose a location for the new footprint. If you have a project folder setup for your design, you can save it there. Otherwise save it in the *symbols* library. Select `Package symbol` from the `Drawing Type:` list (we will use the package `Symbol Wizard` in a later example) and click `OK`. Note that, at the beginning of the chapter, it was told that PCB Editor will be in a

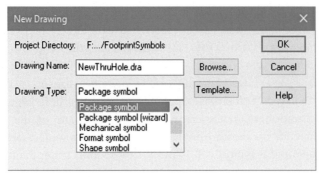

Figure 8.11 `New Drawing` package symbol dialog box.

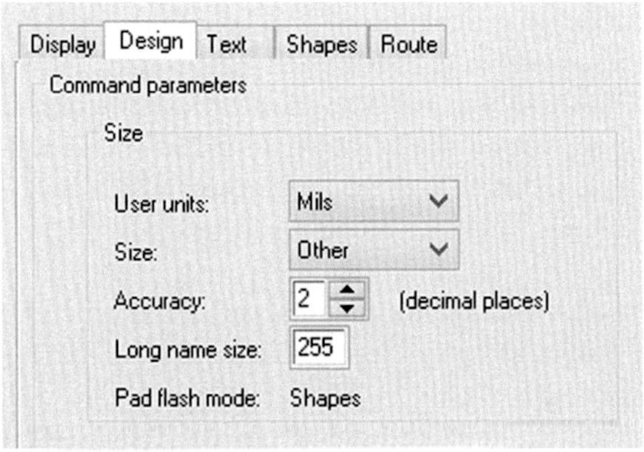

Figure 8.12 Setting up the drawing environment.

specific mode depending on what type of drawing you are making; this is where the mode is set.

When the new design is opened, a PCB Editor window will be displayed that looks just like a board design window.

The first thing that needs to be done is to set up a drawing area that is practical for this design. Select `Setup → Design Parameters...` from the menu bar. At the `Design Parameter Editor` dialog box, select the `Design` tab.

Select `Other` from the `Size:` list and enter the `Extents` values as shown in Fig. 8.12. Click `Apply` (but not `OK` yet). Select the `Display` tab. Enable the grid (if it is not already enabled) by checking the `Grids on` box and click the `Setup grids...` button to display the `Define Grid` dialog box shown in Fig. 8.13. For the time being set spacing for all of the layers to 25 mil. Entering values in the `All Etch` area will automatically change all the `Top` and `Bottom` values so that you do not have to set them individually.

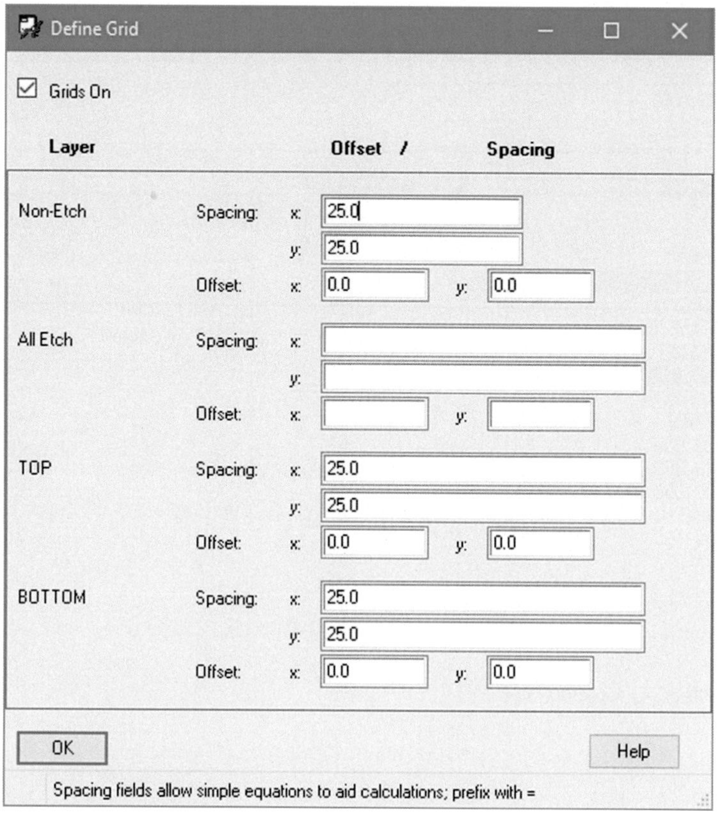

Figure 8.13 Setting the grid spacing.

Click OK to dismiss the grid dialog box then click OK to dismiss the Design Parameters dialog box and update the work area.

The next step is to place the padstacks. Select Layout → Pins from the menu bar and in the Options pane press "…" after the Padstack: box to display the Select a padstack: dialog box as shown in Fig. 8.14.

Scroll down until you find the padstack that was just made. If it is not displayed, make sure the Database box is checked. If you did not make the new padstack, you can just pick one that is close in size; pad60cir42d is a good choice. Click OK.

A padstack will be attached to your mouse pointer. Place the first padstack by clicking and releasing your left mouse button at location (0, 0). Another padstack will be attached to your mouse pointer. Place the second padstack at location (400, 0) then right click and select Done from the pop-up menu.

Zoom in so that the two padstacks take up about 75% of your display (use your mouse wheel if you have one or use the Zoom Points button, 🔍 , on the toolbar).

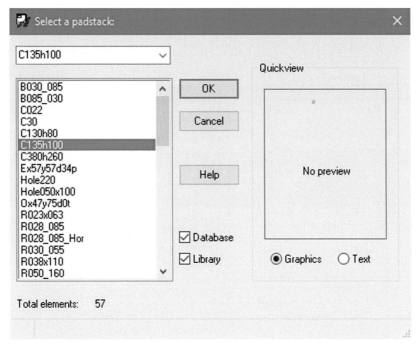

Figure 8.14 Selecting a padstack.

Zoom Points button

Add line tool

Next draw the body outline on the top silk-screen layer. To draw the silk screen, select the Add Line tool, , select the Package Geometry class and Silkscreen_Top subclass from the Options pane, and select a Line width of 10 mil or so, as shown in Fig. 8.15.

Note: By default, lines (including rectangles) have a width of zero. When PCB Editor processes the artwork lines, those with a width of zero are ignored unless you specify a default width on the artwork control form. In addition, you can change the width of only lines, not rectangles. So if you want wide silk-screen lines, you need to use lines.

Begin the body outline by clicking and releasing the left mouse button at coordinate (75, −50). With the Add Line tool still active change the coordinates display from Absolute (A) to Relative (R) by toggling the small square XYMode button located at the lower middle of the screen (Fig. 8.16). Now move the pointer 100 mil in the Y direction (up) from the starting point (watch the mouse coordinates indicator at the

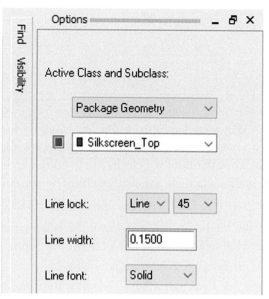

Figure 8.15 Setting silk-screen options.

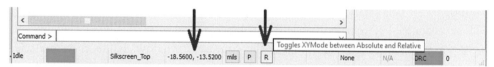

Figure 8.16 Coordinates tools.

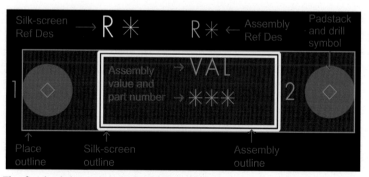

Figure 8.17 The finished through-hole footprint symbol.

bottom, the left arrow in Fig. 8.16) and left click and release to place a vertex. Move the mouse 250 mil to the right (X direction) and place another vertex. Move the mouse 100 mil down and place another vertex. Move the cursor back to the starting point, left click to place the last vertex, then right click, and select Done from the pop-up menu. The completed silk screen will be shown in white in Fig. 8.17.

Color button

Copy button

Note: To *change the color of objects* in the design, select the Color button, ⊞, on the toolbar. Select the desired class from the left window, select the desired color from the color palette, and left click the colored square for the item you want to change in the subclass list at the right. Click the Apply button and then OK.

Next we add the body outline for the assembly layer by copying the silk-screen outline. *To copy an object*, select the Copy button, ▯, on the toolbar. Left click the silk-screen outline. Move the mouse up and away from the existing outline and left click to place the new outline. Right click and select Done from the pop-up menu to dismiss the Copy tool. The copy is on the silk-screen layer, so we change it to the assembly layer.

To change the layer of an object, select Edit → Change from the menu bar. Move the mouse over to the Options tab to display the pane, if is not already displayed. Select the Assembly_Top subclass under the Package Geometry class and change the line width to 1 (see Fig. 8.18). Left click on the copied body outline to make the changes. The outline will be changed but remain highlighted. Notice too that PCB Editor will tell you: Changed 1 items out of 1 items found, in the command window. Right click and select Done to complete the change.

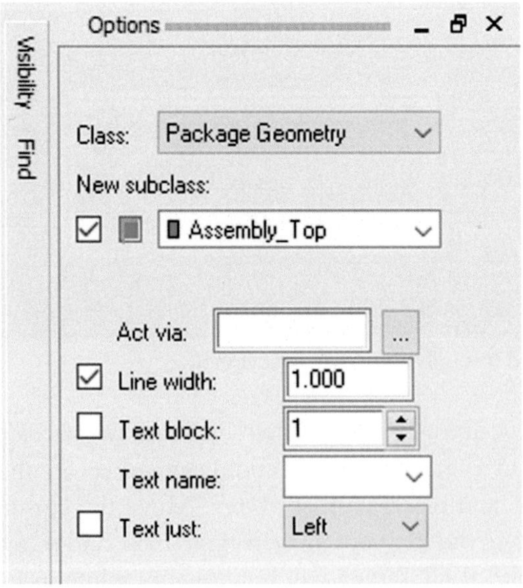

Figure 8.18 Changing outline properties.

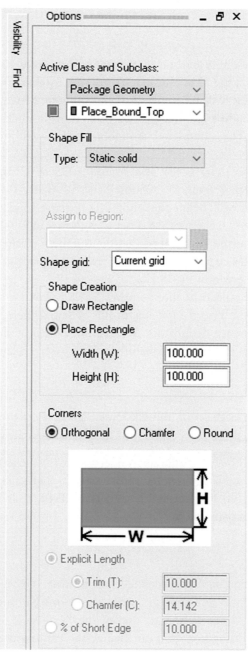

Figure 8.19 Place outline boundary options.

Move button

Move the new assembly outline to the same position as the silk-screen outline. *To move an object*, select the Move button, , on the toolbar and left click the assembly outline; the outline will be attached to the pointer. Move the assembly outline over the silk-screen outline and left click to place it, right click, and select Done from the pop-up menu. The completed assembly outline will be shown in blue in Fig. 8.17.

The next object to draw is the *place boundary outline*. Change the Non Etch grid spacing to 5 mil (Setup → Grids...). Choose Setup → Areas → Package Boundary from the menu bar, which will begin the outline as a static solid shape in the Package Geometry class and the Place_Bound_Top subclass as shown in Fig. 8.19.

Draw a box around the footprint so that the outline is as close to the padstack and detail objects as possible, while keeping them inside the place outline. The place outline will be shown in green in Fig. 8.17.

Now we add text objects to the footprint. Table 8.2 and Fig. 8.1 list some of the text objects that can be placed in a design, but as stated there, only one is required (i.e., at least one reference designator). See Fig. 8.17 while following the steps here. We begin by placing a reference designator on the silk-screen layer.

RefDes button

To place a reference designator on the silk-screen layer, select the Label Refdes button, , from the toolbar, or use Layout → Labels → RefDes menu. From the Options pane, select the Silkscreen_Top subclass (under the Ref Des class) and text block 2 for 31-mil high text (select a larger number for larger text). Left click above the place outline (off to the left a bit) to place the text marker. Type R* and then right click and select Done from the pop-up menu.

Controlling text size

The text block selection applies predefined text formats. To view or modify the formats, select Setup → Design Parameters from the menu bar and select the Text tab. Click the Setup Text Sizes button to display the Text Setup dialog box (shown in Fig. 8.20). From the dialog box, you can change the existing text formats or add your own custom text blocks.

Add Text button

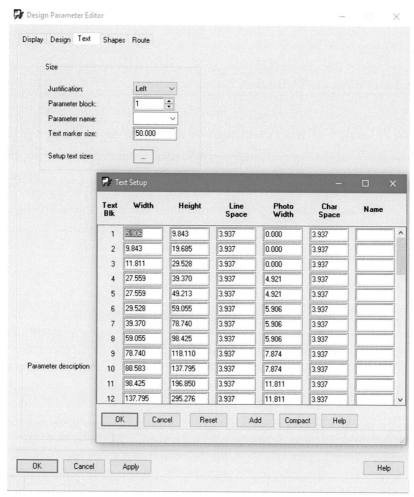

Figure 8.20 Text block options.

Next we place a couple of text objects on the assembly layer. Repeat the preceding process except select the Assembly_Top subclass and select text block 1. Left click above the place outline and to the right of the silk-screen reference designator. Again type R* and then right click and select Done from the pop-up menu. To add text object for the component value, select the Add Text button, 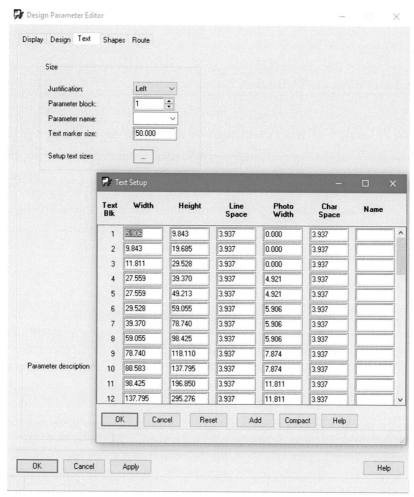, from the toolbar, select the Component value class and the Assembly_Top subclass from the Options pane, and again select title block 1 if it is not selected. Left click below the reference designator on the assembly layer, type VAL, then right click, and select Done. Repeat this process to enter a part number label. Again use the Add Text tool but select the User Part Number class and the Assembly_Top subclass. Select title block 1 and type *, right click, and select Done.

The finished footprint is shown in Fig. 8.17. Save the footprint by selecting `File →` `Save` from the menu bar. PCB Editor will tell you `Symbol 'name.psm' created` in the command window.

That completes the first design example. You can compare this to the `res400` symbol included in the native PCB Editor *symbols* library.

Example 2. Design of surface-mount device from an existing symbol

The native PCB Editor library contains one surface-mount capacitor footprint symbol and one surface-mount resistor symbol for general use. These capacitor and resistor footprints are roughly the 2010 and 1805 sizes, respectively. Many others are on the market. We could start from scratch and build our own; but since they are all similar, we start with an existing one and just modify it. In the next demonstration, we start with the existing smdcap footprint and modify it to make a size 1206 footprint.

Modifying an existing symbol: determining design requirements

Start PCB Editor and open the `smdcap.dra` located in the *symbols* library. Table 8.3 compares the `smdcap` symbol to the requirements of a 1206 chip capacitor as defined by IPC. Note that, since the origin is the body center, the outline coordinates will be ± 1/2 the listed values.

The closest equivalent padstack as defined in Table 8.3 for a size 1206 capacitor is `smd44rec72`. We use that as a starting point for a `smd45rec71` padstack. Start Padstack Editor, open `smd44rec72` and `File - Save As...` `smd45rec71` in the *symbols* library. Since this is an smd padstack, you can leave the `Start` and `Drill` tabs alone, but select the `Design Layers/Mask Layers` tabs, and change top and soldermask dimensions as shown in Fig. 8.21A and B. Save your changes (`File → Save` from menu).

Table 8.3 Surface-mount capacitor properties.

Parameter	smdCap		1206	
	X	Y	X	Y
PAD size	87	50	71	45
PAD C/C	195		118	
Body (L/W)	245	90	126	63
Place outline (L/W)	250	90	185	91
Padstack name	smd50_87		smd45_71	

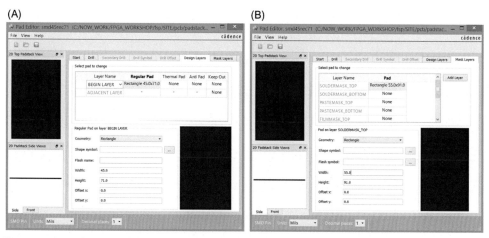

Figure 8.21 Surface-mount pad properties. (A) Design Layers and (B) Mask Layers.

Replacing a padstack definition

Next we will change existing padstacks in the footprint drawing to the new smd45rec71 padstack.

Select Tools → Padstack → Replace from the menu bar. Select pin 1. At the Options pane, click the "..." button on the right of New: box to look for the new pad (Fig. 8.22).

At the Select a Padstack dialog box, select the new one and click OK.

Show Element button

At the Options pane, select a pin number or leave as * for all pins. Click the Replace button. The pads should be visibly different. To verify that the new pads are correct, select the Show Element button, , and left-click on a pad. The information window will give you details about the padstack.

Next we need to move the padstacks. Set the grid to 1-mil-by-1-mil resolution (Setup → Grids...) and make sure XYMode is set to Absolute. Since pad center-to-center spacing is 118 mil, the left pad should be located at (−59, 0) and the right pad located at (59, 0). Use the Move tool and change the positions of the pads (turn off all other layers if it helps). Set the Find filter to Pins only if pins is unchecked. You can use P button in the status bar to pick the new location.

Next we need to modify the outline objects. Use Fig. 8.23 as a guide (which is based on IPC-7351B guidelines). When working with the outline objects, it can be helpful to change the layer colors and turn off all but the outline on which you are working. Otherwise you will need to use the right mouse button click and Reject menu to choose the required object for editing.

Shape Select tool

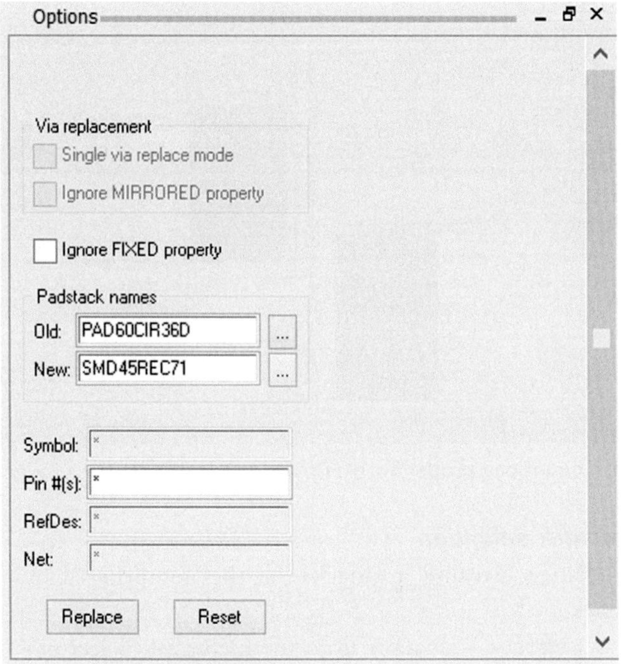

Options _____ − ⊟ ✕

Via replacement
☐ Single via replace mode
☐ Ignore MIRRORED property

☐ Ignore FIXED property

Padstack names
Old: `PAD60CIR36D`   ...
New: `SMD45REC71`   ...

Symbol: `*`
Pin #(s): `*`
RefDes: `*`
Net: `*`

[Replace]   [Reset]

Figure 8.22 Replacing a padstack definition.

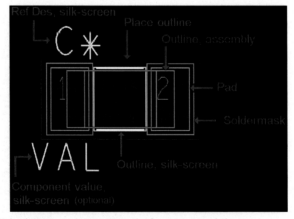

Figure 8.23 The finished SMD capacitor. *SMD*, Surface-mounted device.

We begin with the silk-screen outline. Turn off all layers except `Package Geometry/ Silkscreen_Top`. To resize the outline, select the `Shape Select` tool, 🔲, left click the object to select it. The line becomes dashed or highlighted, and square handles are displayed. Move your pointer over the outline, and it turns to a double-ended arrow.

Use the double arrow to grab a line and slide it to the proper location. Do this on all the sides to produce the proper outline size.

As stated already you cannot change the width of the lines that make up rectangles. In some footprints, silk-screen outlines are made with lines and others (such as this one) are made with rectangles. Since the rectangles have an actual width of 0, they will be ignored during artwork generation unless you specify a default value for all zero width lines in the `Undefined line width:` entry in the `Film Options` area of the `Artwork Control Form` dialog box (opened from the `Export → Gerber Parameters` menu). In this example, thicker Lines were added to the rectangle to emphasize the silk-screen detail on the component sides. To do so, select the `Add Line`, change the line width to 10 mil in the `Options` pane, and draw lines on the component sides as shown in the figure.

Next manipulate the assembly outline using the procedure just described. The assembly detail shown in Fig. 8.23 is drawn so that it is the actual size of the component and additional lines are drawn to show the end caps and their relationship to the pads.

Don't forget to change the size of the boundary shape in `Package Geometry/ Place_Bound_Top` and/or `Bottom` subclass—this will be used during the placement process as the component boundary.

Finally, move the text objects as appropriate for the new symbol size. For viewability Fig. 8.23 does not show the assembly text objects.

Save the drawing and make sure that `Symbol 'smdcap1206.psm' created` is displayed in the command window, which completes the design example.

Example 3. PGA or BGA package design using the Symbol Wizard

Using the preceding procedures allows you to construct the exact footprint symbol you want, but for large or complicated footprints, there is an easier approach. To create the complex multipin packages, the OrCAD Library Builder option allows you to create IPC-7351B compatible packages, schematic symbols and 3D STEP models based on a set of predefined scalable templates. If you don't have a license of OrCAD Library Builder, you still have a good basic `Symbol Wizard` in the OrCAD PCB Editor.

The `Symbol Wizard` performs many of the tasks automatically and can be used to create simple or complex footprint symbols. The resultant symbols contain all the necessary elements for a valid footprint symbol, but it may not meet your specific symbol requirements with regard to all the possible text and detail options. You can then use the procedures described in the second example to modify the automatically generated footprint symbol to meet your specific requirements. In this example the `Symbol Wizard` is used to construct a PGA footprint (Fig. 8.24) but you can easily create with it many other types of footprints such as DIP, SOIC, PLCC/QFP, discrete, SIP or ZIP.

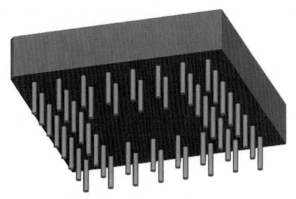

Figure 8.24 Bottom view of a PGA package. *PGA*, Pin-grid array.

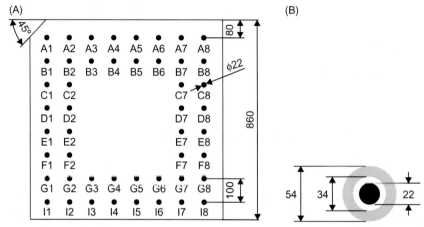

Figure 8.25 PGA design parameters.

The `Symbol Wizard` is used in this example to construct a footprint for the PGA. The first step is to obtain a data sheet for the part. The dimensions for a generic 48-pin, 8 × 8 PGA are shown in Fig. 8.25, left. Each pin has a 22-mil diameter and is spaced 100 mil from the others. A padstack with a 32-mil hole and 50-mil-diameter pads suffices for the 22-mil pin (Fig. 8.25, right). Before starting the footprint design, you need to have the correct padstack present in the symbols library. A `pad50cir32d` padstack is included with the symbols library and is used for this design.

To begin the design process, open PCB Editor and select `File → New...` from the menu. At the `New Drawing` dialog box (Fig. 8.26), enter a name for the drawing (e.g., `PGA_48pin`) and select `Package symbol (wizard)` from the `Drawing Type:` list. Use the `Browse...` button to select a directory (the *symbols* library is default). Click `OK` to start the wizard.

Figure 8.26 New PGA wizard. *PGA*, Pin-grid array.

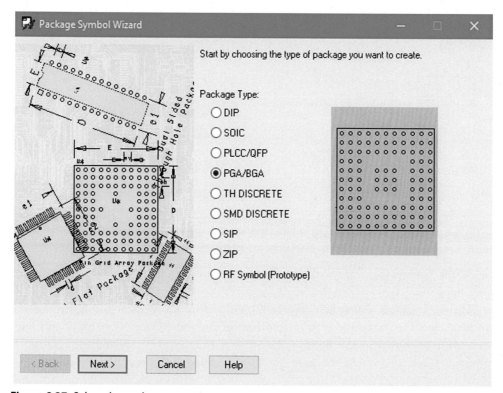

Figure 8.27 Select the package type.

At the Package Symbol Wizard dialog box (Fig. 8.27), select the PGA/BGA radio button, and click Next>.

At the Template dialog box (Fig. 8.28), leave the Default Cadence supplied template radio button selected and click the Load Template button, then click Next>.

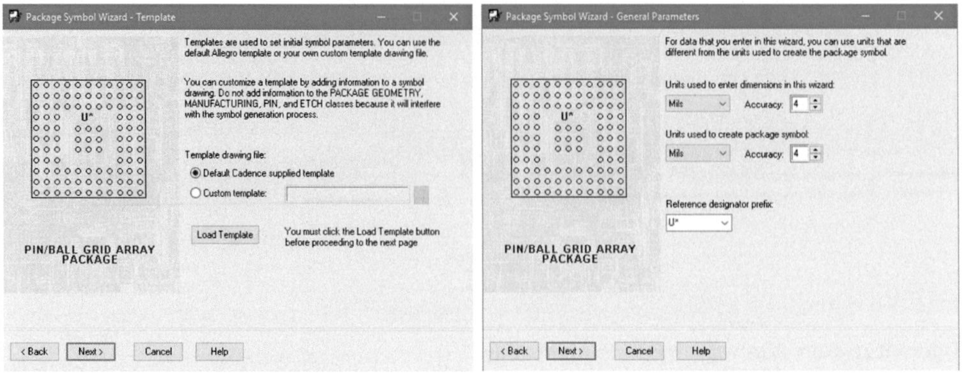

Figure 8.28 Select a drawing template.

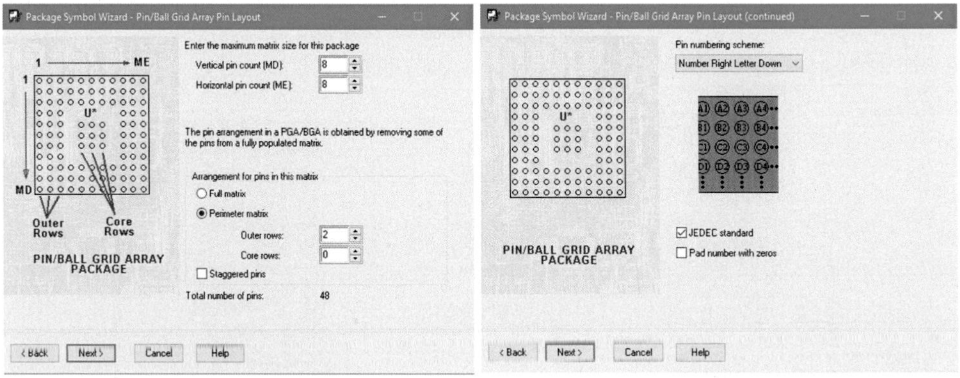

Figure 8.29 Setting up the pin layout.

At the General Parameters dialog box, leave the default settings (mils) and click Next>.

At the Pin Layout dialog box (Fig. 8.29), enter 8 for the vertical and horizontal pin counts. Select the Perimeter matrix radio button and enter 2 for Outer rows: and 0 for Core rows: as shown in the figure. The Total number of pins: should indicate 48, as required. Click Next>. At the next Pin Layout dialog box, leave the default settings (Number right and letter down and JEDEC standard). Click Next>.

At the next Array Parameters dialog box (Fig. 8.30), leave the Lead pitch at 100 mil, but change the Package width and length to 860 mil, and click Next>.

At the Padstacks dialog box (Fig. 8.31), click the Browse... button(s) and select pad50cir32d for the default padstack and pad50sq32d for the pin 1 padstack. Click Next>.

At the Symbol Compilation dialog box, leave the default settings and click Next>. At the Summary dialog box, click the Finish button to complete the design.

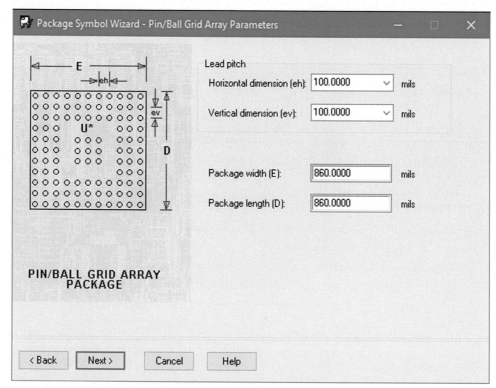

Figure 8.30 Setting up the array parameters.

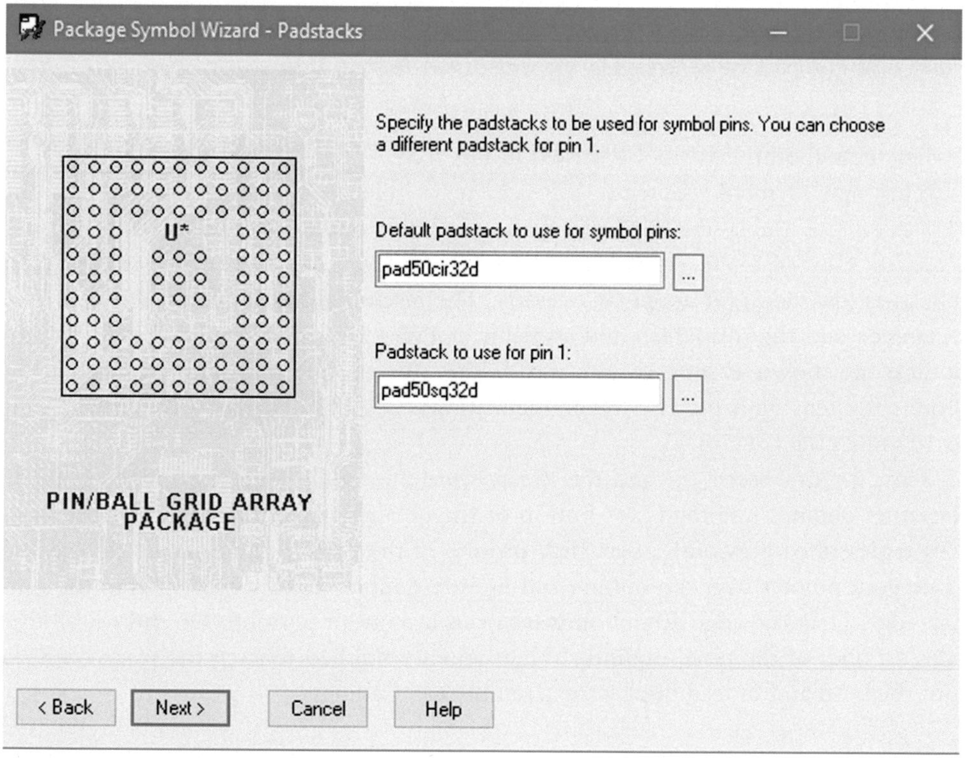

Figure 8.31 Select default pin types.

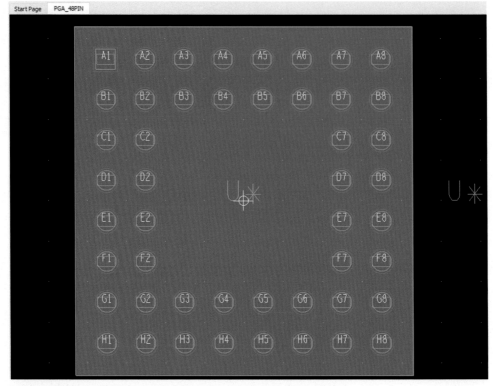

Figure 8.32 PGA package generated by the wizard. *PGA*, Pin-grid array.

The new footprint symbol is shown in Fig. 8.32.

The new symbol contains Ref Des text objects on the Assembly_Top and Silkscreen_Top sublasses (Ref Des class) and four outline details on the Package Geometry class: Assembly_Top, Silkscreen_Top, Place_Bound_Top, and Dfa_Bound_Top subclasses (dfa stands for *design for assembly*). The place outlines are constructed of filled rectangles, and the silk-screen and assembly outlines are constructed of lines. All four outlines are shown as squares with zero width. We modify the outline objects to be more consistent with the IPC recommendations. Begin by changing the grid spacing to 10 mil (Setup → Grids...).

First, the Dfa_Bound_Top and the Place_Bound_Top outlines are expanded. Turn off the other outlines (use the Color button or the Options pane). Select the Select Shape tool as described previously. Left click on one of the place outline shapes to select it. Place your pointer over the outline and use the double-ended arrow to grab the out-line edge. Drag the edge 40 mil outward from its current position. Do this for all four sides on both of the place outlines. When you are finished resizing the place outlines, turn them off and instead display the assembly top outline.

We now add a beveled corner to the assembly outline at the corner at pin 1, as shown in Fig. 8.33. Begin by zooming into the upper left corner of the symbol. Select the `Add Line` tool, at the `Options` pane select `Package Geometry/Assembly_Top`, and select `Line` and `45` as the `Line Lock:` options. Draw a diagonal line in the upper left corner as shown at (1) in Fig. 8.34.

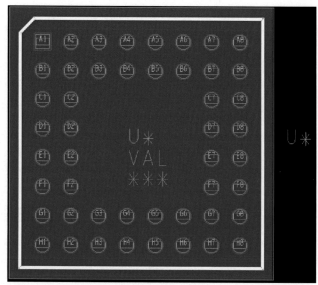

Figure 8.33 Final PGA design. *PGA*, Pin-grid array.

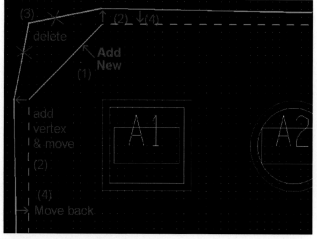

Figure 8.34 Modifying the outline object.

Next we need to trim off the excess part of the outline. To do so, vertices need to be added so that they can be used as the trim points. To add a vertex to a line, select Edit → Vertex from the menu bar. Left click at the intersection of the diagonal line you just drew and the existing outline [points at (2) in the figure] and move the vertex out 20 mil or so to create an angle. Do this at each end of the new line.

Next delete the line segments [shown at the Xs near (3) in Fig. 8.34]. To delete part of a line segment disable any current tool (right click and select Done from the pop-up menu—if it is available; if it is not, no tools are selected). Check that the current application mode is *General Edit*. Hover your pointer over one of the line segments so that just the segment is dashed. Left click the segment to permanently select it. Select the Delete tool from the toolbar, and only the segment will be deleted. If you select the Delete tool first then select the line, the entire outline will be deleted. Finally, move the remaining line segments [the (4)s in the figure] back to the ends of the diagonal line (use the Vertex tool not the Move tool). Perform the same steps on the silk-screen outline and change the silk-screen outline width to 5 mil.

The only text objects added by the wizard were the Ref Des objects on the Assembly_Top and Silkscreen_Top subclasses. You can add your own as described in the preceding examples (e.g., value and part number on the assembly layer).

Select File → Save to save the drawing and update the .psm file. The completed footprint is shown in Fig. 8.33. To make the BGA instead of PGA you just need to replace the through hole pads to SMD pads.

Flash symbols for thermal reliefs

As described in Chapter 2, Introduction to the printed circuit board design flow by example, thermal reliefs are used to provide resistance to heat flow from a PTH to the copper plane to which it is connected during solder operations. Fig. 8.35 (left)

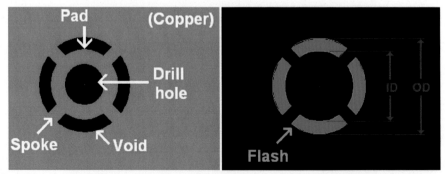

Figure 8.35 Thermal relief and its flash symbol: (left) copper plane, (right) the negative image flash symbol.

shows how this is accomplished. A pad, belonging to a plated padstack, is connected to the larger plane through spokes. Between the spokes (and between the pad and the plane) are void areas where the copper is removed. The spokes provide electrical conductivity while the voids provide a barrier to heat flow. Flash symbols are used to define the voids on negative plane layers. Fig. 8.35 (right) shows the flash symbol in the negative image view and is described further.

Thermal relief connections between PTHs and copper areas on positive planes are automatically generated by PCB Editor so flash symbols need not be defined for positive layers. The ID (inside diameter) of the thermal relief is defined by the pad diameter from the `Padstack` definition in the Padstack Editor, while the OD (outside diameter) is defined by the ID plus twice the `Shape to Pin` definition in the Constraint Manager, either from the applicable `Spacing` Constraint Set where Nets are different, or from the `Same Net Spacing` Constraint Set domain where the Nets are the same. The spokes are generated in the board design by PCB Editor using trace width specifications defined in the Constraint Manager as is shown in the PCB Design Examples (Chapter 9).

Note that PCB Editor allows to define the plane layers as either positive or negative. Many old PCB projects used negative planes because they were efficient when computers were not as fast as today.

The realization of thermal reliefs on negative planes is another story. Recall that, with negative plane layers, what you see is what is removed, so thermal reliefs are generated by creating flash symbols (shown in Fig. 8.35, right), which define the void areas. Once a thermal flash symbol is attached to a padstack, PCB Editor interprets it and creates the positive view that you see with the "what you see is what you get" view presented during the board design. When manufacturing artwork is created during the last phase of a board design, the negative planes and flash symbols are properly processed as negative Gerber images.

Most of the padstacks in the symbols library have no flashes assigned to them. During the board design process, flashes need to be assigned to the padstacks as the need arises. When you construct your own footprints and padstacks, you can follow the same practice or assign flashes right away. Another useful application of the flash symbol is to define the array of openings in the solder paste (stencil) layer, when you want to split the big surface-mount technology pad to several small patterns to improve the soldering quality. Either way you need to know how to construct flash symbols.

A flash symbol is just another type of drawing you construct using PCB Editor. Later, a flash symbol is made for the padstack designed in the first example.

Before we begin with the drawing, we "engineer" the thermal relief. Recall from the first example that the padstack we designed (`pad62cir42d`) had the characteristics summarized in Table 8.4.

Typically the flash's ID (ID in Fig. 8.35, right) is the same as the regular pad for that layer, and flash's OD (OD in Fig. 8.35, right) is the same as the antipad diameter

Table 8.4 Pad dimensions.

Parameter	Mils
Drill-hole diameter	42
Outer-pad diameter	62
Outer-antipad diameter	82
Outer-pad to antipad clearance	10
Outer annular-ring width	10
Inner-pad diameter	60
Inner-antipad diameter	80
Inner annular-ring width	9
Inner-pad to antipad clearance	10

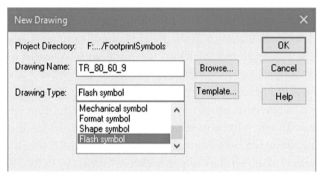

Figure 8.36 Starting a new flash symbol drawing.

plus or minus a couple of mils. The spoke width is calculated per IPC-2222A (pp. 21–22) and depends on the pad diameter and the number of spokes. The spoke width is calculated as

$$W = 60\% \left(\frac{P}{n}\right)$$

where W is the spoke width, P is the pad diameter, and n is the number of spokes.

If we assume that only internal layers have negative planes, then only the internal layers need a thermal flash. If it turns out that this is not the case, flashes can be added later.

Then for the pad62cir42d padstack, the flash ID is 60 mil, the OD is 80 mil, and the spoke is 9 mil wide.

To construct a flash symbol, open PCB Editor. Start a new flash symbol drawing by selecting File → New... from the menu. At the New Drawing dialog box, enter a name for the symbol (e.g., TR_80_60_9 for thermal relief with OD = 80 mil, ID = 60 mil, and spoke width = 9 mil), then select Flash symbol from the Drawing Type: list, as shown in Fig. 8.36. Click OK.

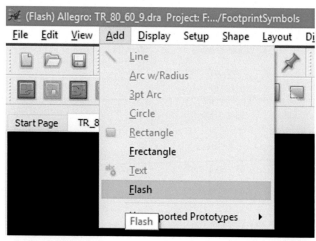

Figure 8.37 Add a flash symbol to a drawing.

Turn on the grids and check to make sure that the origin coordinates (0, 0) are at the center of the drawing area. If not adjust the drawing area by giving the Left X and Lower Y Extents negative values (Setup → Design Parameters from menu, Design tab).

Next, select Add → Flash from the menu bar, as shown in Fig. 8.37.

As shown in Fig. 8.38, enter the design values for the pad diameters and spoke definitions into the Thermal Pad... dialog box, and click OK.

The completed symbol should look like Fig. 8.35 (right). Save the symbol (to *symbols* path) and make sure you see Symbol 'tr_80_60_9.fsm' created in the command window.

Now we can add this symbol to the pad62cir42d padstack definition. Use Padstack Editor to open pad62cir42d.pad again. Select the DEFAULT INTERNAL layer in the Design Layers tab. Click the Flash symbol: Browse... button, [...], and from the Library Shape Symbol Browser dialog box (Fig. 8.39), select the Tr_80_60_9 flash symbol and click OK. Fig. 8.40 shows the flash symbol assigned to the padstack. Check that the Anti Pad in the DEFAULT INTERNAL layer is already defined as a circle with diameter = 80, so that the pad shape for negative layer is defined for both connected, and not connected, pads on the negative plane(s). You can save the new definition with the same name or rename it to something like pad62cir42f to indicate that it has a flash symbol assigned to it.

You can now use this padstack definition in the footprint design you did in the first footprint design example. To do so, open the footprint drawing and change the padstack definition to pad62cir42f instead of pad62cir42d. Use the procedure in the second design example to change the padstack (Tools → Padstack → Replace...). This completes the flash symbol design example.

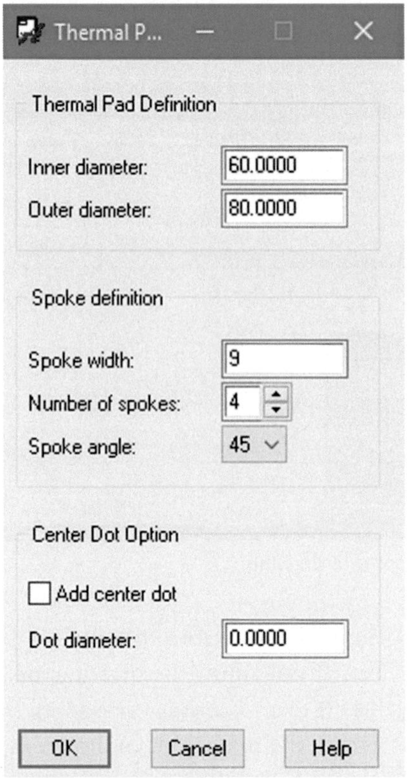

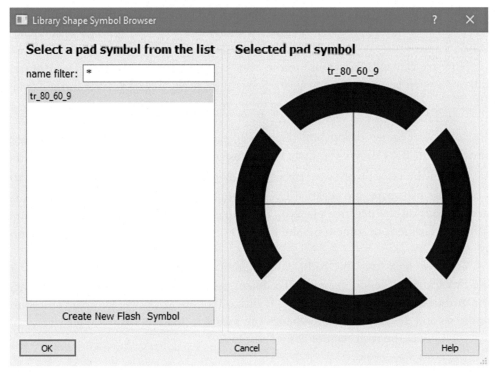

Figure 8.38 Setting up the flash symbol parameters.

Figure 8.39 Selecting the flash symbol.

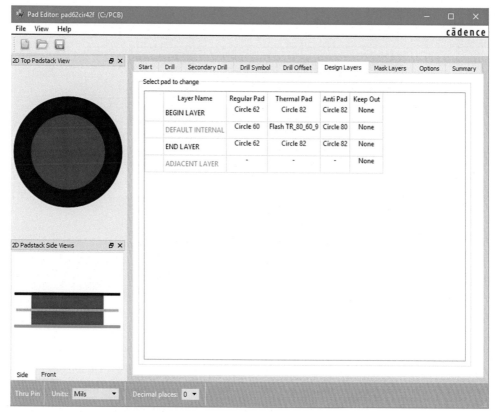

Figure 8.40 The new flash symbol assigned to a padstack definition.

Mechanical symbols

Mechanical symbols represent physical objects you can place in your board design, which are not part of the netlist and therefore have no connectivity. They can be things such as *mounting holes* (mechanical pins without pin numbers), nonconnecting copper etch objects, design outlines, dimensions, and areas (e.g., keep-in and keep-out). Several mechanical symbols are located in the *symbols* library for you to look at for ideas. The mechanical symbols consists of two files which have the same file name and .dra/.bsm file extensions.

One important mechanical symbol that you may need to have is the mounting hole.

Mounting holes

Mounting holes can be used for attaching the PCB to mounting hardware (such as stand-offs) or for attaching hardware to the PCB (such as *heat sinks*). The four basic

Table 8.5 Basic hole types.

	Plated	Nonplated
With pads		
Without pads		

hole types are shown in Table 8.5; in them holes can be made with or without lands (pads), with or without plating, or in any combination of the two. Mounting holes that are plated can be attached to a net or isolated from all nets.

When mounting holes are attached to a net that is assigned to a plane layer, the mounting hole can be connected to the plane through thermal reliefs or with full contact, just like any other PTH. When mounting holes are attached to a net, they should be included on the schematic using a Capture part with an assigned footprint, which consists of a connection-type padstack. When mounting holes are not attached to a net, they are not included in the schematic and are added to the board design in PCB Editor. Some designers do not like to have connected mounting holes on the schematic, and it is possible to add them to the board in PCB Editor only. But it is the author's opinion that this is a messy way to do it, since you have to override DRC errors to do it, and since it is connected to a net, it really is part of the circuit and should be on the schematic (if only for documentation purposes).

The PCB Editor contains several mounting-hole symbols, which have the name mtgXXX, where XXX indicates the hole diameter in mils. These are nonplated holes with no lands (without pads). The IPC standards call this an *unsupported hole*.

The mounting hole mtg125.dra is shown in Fig. 8.41. The drawing consists of a Pad125 padstack and a route keep-out area. The Pad125 padstack is a mechanical pin rather than a connect pin.

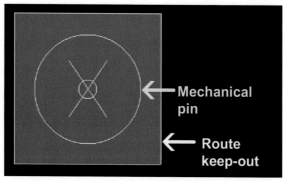

Figure 8.41 A mechanical pin.

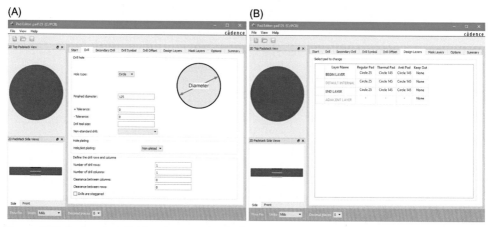

Figure 8.42 Padstack definition for a mechanical pin.

The mechanical pin properties are shown in Padstack Editor in Fig. 8.42. To view the pin launch, go to the Padstack Editor from PCB Editor by selecting `Tools →`
`Padstack → Modify Design Padstack...` from the PCB Editor menu bar. Fig. 8.42 (left) shows the drill land plating settings in the `Drill` tab and Fig. 8.42 (right) shows the layers setting on the `Design Layers` tab.

The drill diameter is 125 mil, and the pads are 25 mil. This means that the pads will be drilled out during the manufacturing process. You will also notice that this is a nonplated hole, which means that this hole will be drilled toward the end of the manufacturing process, separately from the plated holes. A clearance (`Anti Pad`) is included and is 20 mil larger than the drill diameter, so that any copper planes are always kept away from the hole (this allows for fabrication tolerances).

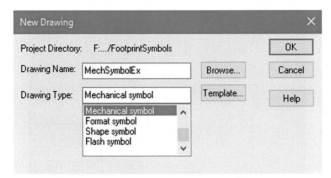

Figure 8.43 Starting a new mechanical symbol drawing.

Figure 8.44 Options for adding a mechanical pin.

Creating mechanical symbol drawings

To start a new mechanical symbol drawing, open PCB Editor and select File → New... from the menu bar. At the New Drawing dialog box, enter a name for the drawing and select Mechanical symbol from the list as shown in Fig. 8.43. Click OK.

You will be presented with a typical PCB Editor drawing window. Add lines, text, and dimensions as needed to complete your drawing.

Add Pin button

To add mechanical pins, select Layout → Pins from the menu bar or select the Add Pin button, ⊕, on the toolbar. Click the Padstack: Browse... button. At the Select a Padstack: dialog box, you can select any padstack or mounting hole you like. When you select a padstack and click OK in the dialog box then move your cursor to the drawing area, you will have the selected pin (padstack) attached to your pointer. Left click in the drawing to place the pin. Right click and select Done to quit. You can

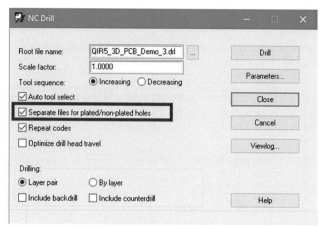

Figure 8.45 Specifying combined or separate NC drill files. NC drill, Numeric control drill.

place an array or row of pins by entering X: and Y: Qty values in the Options pane. In that case, left click in the drawing area where you want the first pin located and PCB Editor will automatically place the remaining pins for you.

You can use the Padstack Editor to make the types of mounting holes shown in Table 8.5. A reference table of drill and screw sizes is provided in Appendix D as an aid to designing mounting holes for standard screw sizes.

For nonplated holes, select that option on the Drill tab. If you do not want pads on the hole, make the pad diameters smaller than the drill size. Remember though to make the antipads at least 20 mil larger than the drill diameter.

If you want to make a mounting hole that can be connected to a plane or net, the mounting hole will simply be a plated padstack. You can use them or design your own as described previously.

When a board design has both plated and nonplated holes, you can combine the drill information for both holes types into one NC drill file or have PCB Editor generate separate drill files. A second set of drill instructions is needed because nonplated holes are drilled after all plating processes are complete, whereas PTHs are drilled before the plating process; therefore a completely separate set of instructions is required.

The standard, plated drill file has the naming convention *name*.drl file, and non-plated drill files have a *name*-np.drl naming convention. You specify combined or separate drill files by selecting Manufacture → NC → NC Drill... from the menu bar. At the NC Drill dialog box (shown in Fig. 8.45), uncheck the Separate Files option to generate a combined drill file for all types or select the option to make separate files.

Placing mechanical symbols on a board design

Examples of how to place a mounting hole (a mechanical symbol) on a board are given in Chapter 10, Artwork development and board fabrication, and Example 2 of

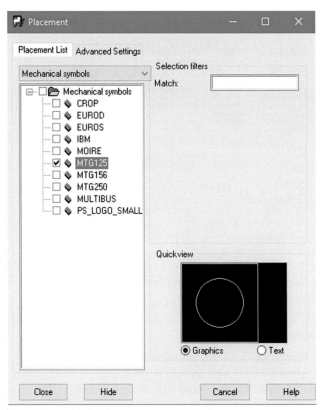

Figure 8.46 Placing a mechanical symbol on a board design.

the Chapter 9, PCB Design Examples. Open your board design and select Place → Mechanical Symbols... from the menu bar to display the Placement dialog box, as shown in Fig. 8.46. Make sure that both the Library and Database options are checked in the Advanced Settings tab and press the " + " icon to open the list of available mechanical symbols in the library. Check the box next to the symbol you want and left click on the board area at the desired location to place the mechanical symbol. Repeat the process for each symbol you want placed on your board.

Blind, buried, and microvias

Blind and buried vias are not typically designed into footprints but are added to a PCB during the layout and routing process. Example of PCB design using blind vias is given in the Chapter 9, PCB Design Examples (Example 3), but is discussed briefly here since it is related to padstack design.

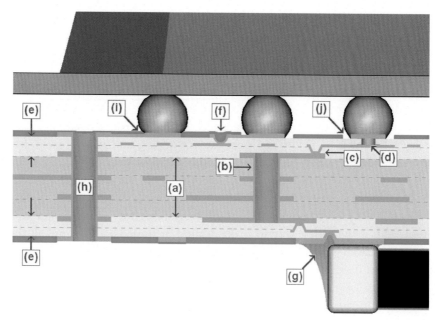

Figure 8.47 Via technologies and their applications.

Fig. 8.47 shows how microvias and blind and buried vias can help route high-density ball grid arrays (BGAs). Using these via types, a BGA can be placed on the top of a board and another SMD can be placed directly opposite the BGA on the bottom of the board. To accomplish this, typical through-hole vias cannot be used. Fan-out and routing of the BGA are accomplished using blind and buried vias. Blind vias are visible on only one side of the PCB and connect traces on the one outer layer to inner traces only, while buried vias are not visible from either side of the board and connect traces only between inner layers.

Blind and buried vias are realized on built-up PCBs. A built-up PCB often has a standard layer stack-up core [section (a) in Fig. 8.47] and additional layers are sequentially added to the board [the two (e) sections in Fig. 8.47]. PTH vias in the standard core become buried vias [(b) in Fig. 8.47]. As the outer layers are built up on top of the base core, buried microvias [(c) in Fig. 8.47] and blind microvias [(d) in Fig. 8.47] can be embedded into the outer layers. During the build-up process, resistors or capacitors can also be buried in the layers. After all the layers have been built up, additional microvias [(f) and (g) in Fig. 8.47] and standard PTHs [(h) in Fig. 8.47, tented on the top end] can be added to the entire assembly.

Fig. 8.47 shows three types of microvias: (c) and (g) are laser-drilled, plated vias; (d) is a laser-drilled, paste-filled via; and (f) is a plasma-etched, plated via. To learn more about designing built-up boards and microvias, see *Coombs' Printed Circuits Handbook* (Coombs, 2001).

Blind and buried vias can be constructed in three ways. One way is to use Padstack Editor to design the via padstack directly. In general, you use the Padstack Editor the same as was described for the normal through-hole padstacks. If you know exactly what the layer stack-up definition will be in your board design, you can add those layers to a padstack definition in Padstack Editor and remove other layers. In Padstack Editor, select BBVia or Microvia on the Start tab. On the Design Layers tab, add and name layers to the stack-up as needed. *To add layers to the stack-up*, right click on the layer that will be below the one you are adding and select Insert from the pop-up menu. Give the new layer the name used in your board design and select the appropriate pad sizes. For all other layers, select Null for the pad values. An example of a blind via design is shown in Fig. 8.48.

The second way is to set up vias interactively from the board design using the BBvia tool. This is the easiest way to make a blind or buried via, because all the layer definitions are already set up in the board design (using the Layout Cross Section dialog box) and will be known to the BBvia tool. The BBvia tool is launched by selecting Setup → Define B/B Vias... from the menu, as shown in Fig. 8.49. As indicated in the figure, two options are available.

If you select Define B/B vias..., you get the dialog box shown in Fig. 8.50. This method creates one padstack that goes from the Start Layer to the End Layer and includes all the layers between them.

If you select Auto Define B/B vias..., then you get the dialog box shown in Fig. 8.51. You can select which layers will be used in setting up the via. This method actually makes several vias. Essentially, it makes as many blind and buried vias as necessary to connect each of the adjacent planes (two layers at a time) and to connect the

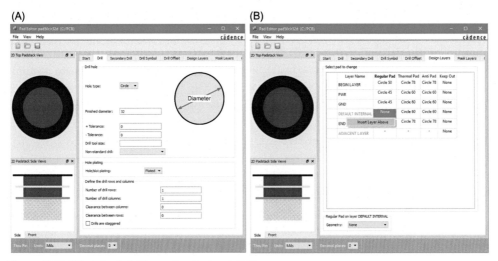

Figure 8.48 Blind via definition using Padstack Editor.

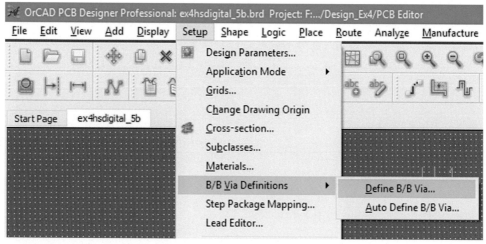

Figure 8.49 Starting the BBvia tool.

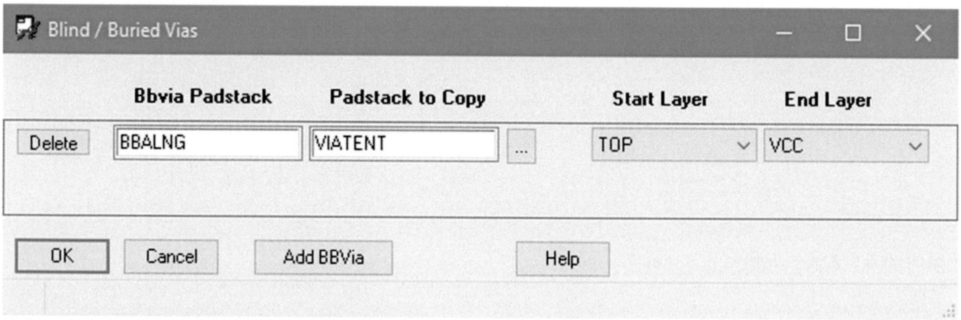

Figure 8.50 Define B/B vias... option.

start and end layers to each other and to each of the internal layers. Which vias get used in the board design is controlled by the Constraint Manager in PCB Editor as demonstrated in the PCB Design Examples (Chapter 9).

OrCAD Library Builder

The OrCAD Library Builder is a tool that allows to generate semiautomatically the schematic symbol, footprint, and 3D model in accordance to IPC-7351B standard. It offers a lot of templates for many typical packages, such as small outline integrated circuit, thin quad flat package, quad flat no leads package, and BGA, which can be scaled to the needed number of pins, pitch, and size of the package. Also it allows to extract the pin table from the data sheet in Adobe PDF format. In order to use it, you need the appropriate license.

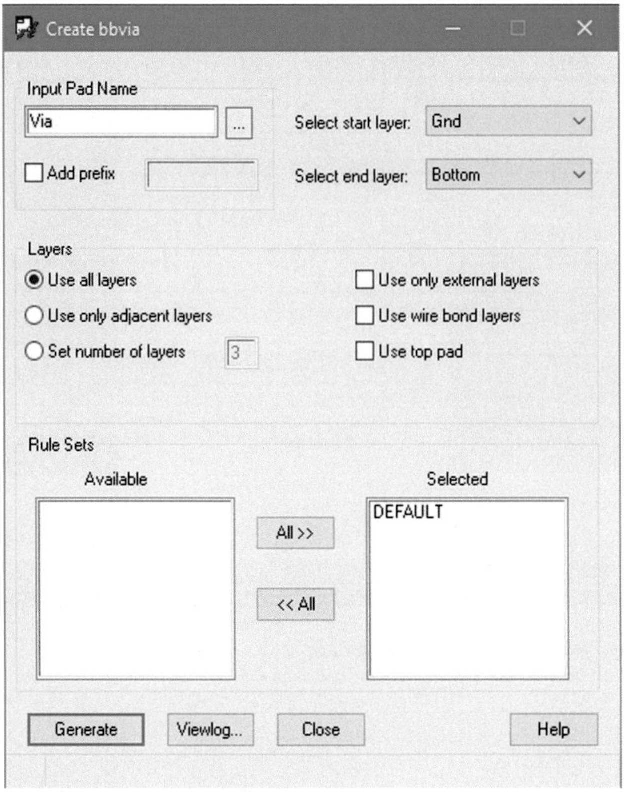

Figure 8.51 Auto Define B/B vias... option.

3D canvas

The 3D viewer embedded to OrCAD PCB Editor supports the use of 3D models of the components in STEP format (files `*.stp` or `*.step`). It displays the bare PCB with all holes, traces, and pads, the PCB with components and mechanical items, including the enclosure, if its 3D model is attached to the project as a mechanical symbol. In order to attach the STEP model to the footprint or mechanical symbol, you can use `Setup → Step Package Mapping...` menu during creation of the footprint or mechanical symbol, or directly in the PCB project. The `*.STEP` files should be located in the appropriate library path (`STEPPATH`) which is set in PCB Editor `User Preferences`. Also it's possible to view in 3D only the selected group of objects, for example, single net routing.

You can run the 3D viewer via the `Display → 3D Canvas` menu option or the `3D` icon. It's possible to tune the set of the objects which you want to see, by clicking the black triangle located near the `3D` icon (see Fig. 8.52).

The 3D canvas is shown at Fig. 8.53. Zooming is controlled by rolling the mouse wheel, panning is controlled by pressing the mouse wheel (middle-click) then

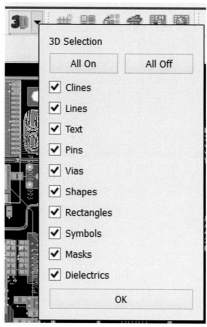

Figure 8.52 Selection of 3D objects to view.

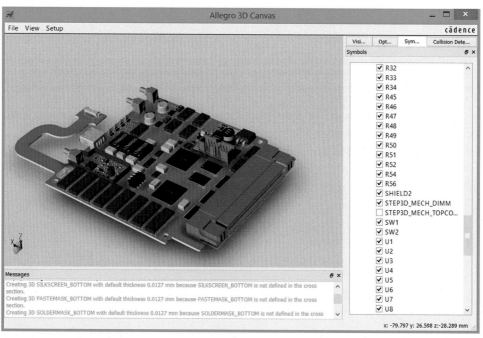

Figure 8.53 3D canvas.

dragging the mouse, rotating is controlled by holding the Shift key while pressing the mouse wheel then dragging the mouse. The File menu contains the following choices: Export, Output, and Close. The View menu allows to change the view direction, and to control the visibility of different panes of the viewer. The Setup menu allows to set the Preferences, such as colors and cutting plane. The Control panel on the right contains several tabs that allow to control the canvas: Visibility, Options, Symbols, and Collision Detection.

The Export menu allows to export the 3D model of the whole PCB with components to such formats as STEP, ACIS, HMF, HSF, IGES, PDF 2D, PDF 3D, and some others. The exported STEP 3D model can contain the dielectric layers and copper topology with the proper thickness of every layer, according to the PCB stack-up.

The Output menu sets the output file and path for messages (the log file). The Close menu will close the 3D canvas.

The Visibility tab is similar to the Visibility tab in PCB Editor. It allows to turn on or off the different layers and objects in 3D view.

The Options tab allows to enter the options for the active pop-up command, which can be activated by the right mouse click on the canvas. The available commands are: Move, Bend, and Measure path.

With the Symbols tab user can control individually the visibility of any component on TOP and BOTTOM side of PCB, as well as the device case and mechanical parts. To open the list of components you should click on the ">" sign on the left of "All," "TOP," or "BOTTOM" folder.

The Collision Detection tab is a very powerful tool to check any touching of surfaces of different components, or of the component to the rigid or flexible part of the PCB or some mechanical detail. To run the check, enter the Min-spacing value, for example, 0.1 mm, and push the Calculate button. After few seconds, you will see the list of detected collisions as the pairs of component names. A left-click on the name will highlight the component (Fig. 8.54), you can also right click on the name and choose Locate to make the component blink in the canvas.

The Messages window displays all messages from 3D canvas. You can stop displaying of new messages by checking the Silent mode checkbox in the Messages category of the Allegro 3D Canvas Preferences dialog.

If you wish to review the messages later, use File → Output → Messages menu option to save them to a text file.

The messages in 3D canvas follow a special color scheme to easily detect their type:

Messages displayed in red are errors; Messages displayed in pale yellow are warnings; Messages displayed in black are information.

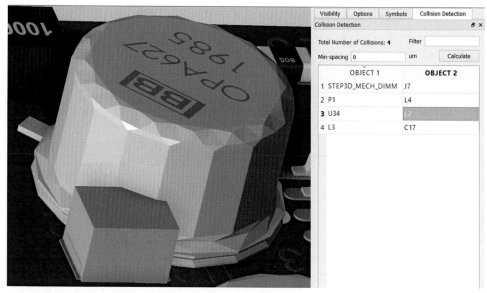

Figure 8.54 Highlight the component after running the Collision detection.

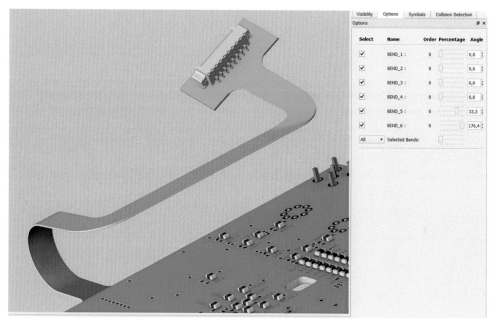

Figure 8.55 Bending the flexible parts.

Bending the rigid-flex printed circuit board in 3D canvas

A right click in the canvas will show a pop-up menu with three options: Move, Bend, and Measure Path. If Bend is selected, you will see the options for the command in the Options tab. Use the sliders to bend the selected or all bending areas (see Fig. 8.55). To enable bending of the flexible sections, first setup several stack-up zones in PCB Editor, then create the bend areas using menu Setup → Bend → Create, and finally set the anchor for 3D bending with the menu option Setup → Anchor 3D View (otherwise the Bend menu will be grayed out). It is possible to combine the bending and collision detection tools to detect collisions when flexible sections of the PCB have bends applied.

References

Coombs, C. F., Jr. (Ed.), (2001). *Coombs' printed circuit handbook* (5th ed.). New York: McGraw-Hill.
IPC-2222A. (2010). *Sectional design standard for rigid organic printed boards*. Northbrook, IL: IPC/Association Connecting Electronic Industries.
IPC-7351B. (2010). *Generic requirements for surface mount design and land pattern standard*. Northbrook, IL: IPC/Association Connecting Electronic Industries.

CHAPTER 9

Printed circuit board design examples

Contents

Complete PCB Design Using OrCAD® Capture and PCB Editor
DOI: https://doi.org/10.1016/B978-0-12-817684-9.00009-6

Introduction

In Chapter 2, Introduction to the printed circuit board design flow by example, the basic PCB design flow was demonstrated, starting from schematic entry with Capture to board routing with PCB Editor, but no attention was given to board design issues, such as component placement and spacing rules, layer stack-up, or trace routing and spacing rules.

Design examples are used here to illustrate the various PCB design considerations described in Chapter 5, Introduction to design for manufacturing, and Chapter 6, Printed circuit board design for signal integrity, and how to use the PCB Editor tools to accomplish the design goals. The part count in each example is kept to a minimum (all designs have fewer than 10 parts and 14 pins/part) so as not to be cumbersome and not to distract the reader from understanding how to route and set up layers. In-depth discussion on PCB design for manufacturability is covered in Chapter 5, Introduction to design for manufacturing, and signal integrity and routing are covered in Chapter 6, Printed circuit board design for signal integrity.

The following examples use parts that may not be included with your version of OrCAD. You can make the parts yourself using the procedures discussed in the previous chapters, or you can copy the parts and footprints provided in the example libraries located on the website for the book.

The *first example* is a simple analog design using a single op amp. The design shows how to set up multiple plane layers for positive and negative power supplies and ground. The design also demonstrates several key concepts in Capture, such as how to connect global nets, how to assign footprints, how to perform design rule checks (DRCs), how to use the Capture part libraries, how to generate a bill of materials (BOM), and how to use the BOM as an aid in the design process in Capture and PCB Editor. The design also shows how to perform important tasks in PCB Editor, such as finding and selecting specific parts and modifying padstacks. Intertool communication (ITC) (such as annotation and back annotation) between Capture and PCB Editor is also demonstrated.

The *second example* is a mixed, digital/analog circuit. In addition to the tasks demonstrated in the first example, the design also demonstrates how to set up split planes to isolate analog and digital power supplies and grounds. Other tasks include using copper pours on routing layers to make partial Ground planes, using copper pours on Plane layers to make nested Power and Ground planes, and defining anticopper areas on plane and routing layers.

The *third example* uses the same mixed, digital/analog circuit from the second example but demonstrates how to use multiple-page schematics and off-page connectors to organize and simplify large circuit designs and to incorporate PSpice simulations into a PCB layer design. It also demonstrates how to construct multiple, separated Power and Ground planes and a Shield plane to completely isolate analog from digital circuitry. The use of guard rings and guard traces is also demonstrated.

The *fourth example* is a high-speed digital design that demonstrates how to design transmission lines, stitch multilayer Ground planes, perform pin/gate swapping, place moated ground areas for clock circuitry, and design a heat spreader.

Overview of the design flow

Regardless of which type of board is made, certain steps must be executed in the design flow process. The following is an outline of the process.
1. Initial design concept and preparation:
   a. Generate initial drawings.
   b. Collect data sheets.
   c. Take inventory of packaging and footprint needs.
   d. Search through the Capture libraries to find the parts. For any parts that are unavailable, construct the parts using the Capture Part Editor or the PSpice Model Editor.
2. Set up the design project in Capture:
   a. Draw the schematic (placing and connecting parts).
   b. Perform an annotation to clean up numbering.
   c. Make sure multipart packages are properly utilized.
   d. Make sure global power nets are properly connected.
   e. Assign related components to groups (rooms) to aid in part placement in PCB Editor.
   f. Perform a Capture DRC to verify that the circuit schematic has no issues. Correct any errors and repeat DRCs as needed.
   g. Generate a BOM to identify PCB assigned and missing footprints.
   h. Search through the PCB Editor libraries to find and assign footprints. For any footprints that are unavailable, obtain the data sheets for recommended

land patterns and design the footprints using PCB Editor and the Padstack Editor.

    i. Generate a PCB Editor netlist (a set of files with description of components and their connections) and open the board design.

3. Define the board requirements:

    a. Board dimensions and mounting hole locations.

    b. Part placement considerations (height restrictions, assembly method).

    c. Noise and shielding requirements.

    d. Component mounting technology [surface-mount technology (SMT), through-hole technology].

    e. Trace width and trace spacing requirements.

    f. Required vias and fan-outs (size and tenting, etc.).

    g. Number of Power/Ground planes and Routing layers.

4. Basic board setup:

    a. Physical:

        i. Create the board outline.

        ii. Place mounting holes.

        iii. Define part and routing restriction areas.

        iv. Add dimension documentation (optional).

    b. Preliminary parts placement:

        i. Use Search and Place tools to place selected parts and groups.

        ii. Perform a board DRC to check for footprint and placement problems.

    c. Layer setup:

        i. Set up Power and Ground planes.

        ii. Set up Routing layers.

        iii. Assign ground and power nets to Plane layers.

        iv. Define thermal relief parameters.

        v. Set which vias to use for fan-outs, routing vias, jumpers, and the like.

        vi. Perform a DRC to check for layer problems.

    d. Final parts placement:

        i. Make sure spacing rules are not violated.

        ii. If using split or moated Plane layers, make sure parts are placed accordingly.

        iii. Check orientation of polarized components (caps, diodes, etc.).

5. Preroute specific nets using manual and restricted autorouting:

    a. Perform power and ground fan-outs.

    b. Preroute critical nets manually.

    c. Perform DRC to make sure that no errors have occurred. Fix errors.

6. Autorouting:

    a. Set up the autorouter.

    b. Run the autorouter.

    c. Perform DRC to make sure that no errors have occurred. Fix errors.

7. Finalizing the design:
  a. Postrouting inspection:
      i. Sharp (acute) angles.
      ii. Long parallel traces (cross-talk issues).
      iii. Via locations.
      iv. Silk-screen markings.
  b. Board clean up:
      i. Unroute and then reroute problem traces.
      ii. Perform a final DRC.
  c. Synchronization with Capture (back annotation).

A detailed example of how to create the artwork and numeric control (NC) drill files for the design is included in Chapter 10, Artwork development and board fabrication.

Example 1. Dual power supply, analog design

The first example demonstrates the basics of schematic entry in Capture, including finding and placing parts, connecting parts and using global power nets, and generating a BOM to assist in the design process. The example continues with PCB design in PCB Editor and demonstrates how to specify board requirements, such as trace width and spacing, and setting up the overall board design; how to define layers; and how to perform manual and automatic routing.

To follow the example exactly as it is presented here, you need to copy the footprints for this example from the website for this book. Copy the contents of `DesignFiles\` `Chap9_DesignExamples\Design_Ex1\Symbols` folder to the user or system footprint *symbols* library folder defined in psmpath and padpath in PCB Editor user preferences. The "system" *symbols* library is located in `\share\pcb\pcb_lib\symbols` (Fig. 9.1).

Although all the schematic and design files for this example are also included on the website, no need to copy them, because all the Capture parts in the example are already included with the OrCAD software. However, if you choose to copy the design files, simply copy the entire `Design_Ex1` folder into your project folder, thus the *symbols* subfolder in it will serve as the local user library for this project. If you have not set up a project folder, now would be a good time to set one up.

Initial design concept and preparation

Before you start a PCB design process, you will likely have some sort of preliminary design concept jotted down. Perhaps PSpice simulations of sections of the design have even been performed. The design concept for this example is shown in Fig. 9.2.

The circuit is very simple, but it contains enough parts that it encompasses the same steps required for larger, more complicated designs. The circuit is a basic amplifier that consists of an active component (the op amp) and several passive components (resistors and capacitors). The circuit also contains an off-board connector that supplies

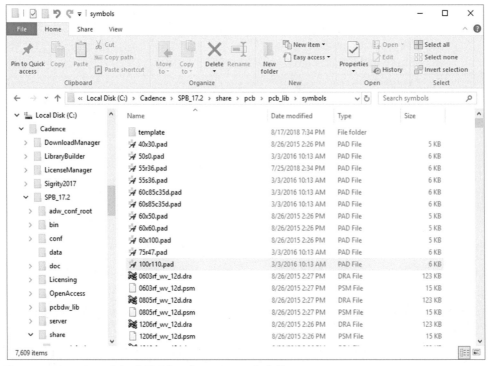

Figure 9.1 Location of the PCB Editor footprint *symbols* library.

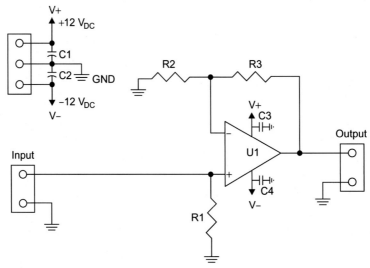

Figure 9.2 Analog circuit for design Example 1.

Table 9.1 Basic parts list and mounting requirements.

Reference	Value	Mounting package
J1	6 pins	Through-hole, 0.200″ pin spacing
C1, C2	10 μF	Through-hole, axial lead
C3, C4	0.1 μF	SMD, size 2010
U1	LM741	SOIC8
R1–R3	1k, 2k	SMD, size 0805

SMD, surface-mounted device.

dual rail power to the board and connections for input and output signals. Through-hole components include the connector and power supply filter capacitors, and surface-mounted devices (SMDs) include the op amp, its bypass caps, and the signal conditioning and gain resistors.

Once you have the basic design down, you need to make a list of all the parts needed to build the circuit, including the off-board connector. Search through your favorite parts catalog to find the parts and download the data sheets from the manufacturers. It is helpful to make a spreadsheet that details the parts and footprints to keep track of the design details throughout the design flow, so that once you receive the board from the manufacturer, the parts fit properly. An example spreadsheet is given in Table 9.1. During the entire design process, you can continually add information to the spreadsheet to document and organize all aspects of the design process. If a problem does occur, the documentation can help identify where the fault occurred.

Setting up the project in Capture

To begin a new design project, start Capture and, from the session window's File menu, choose New → Project. At the New Project dialog box (Fig. 9.3), enter a name for the project. Select the PC Board Wizard radio button. You can also make PCB designs if you select the Analog or Mixed A/D button, which sets up PSpice simulation templates for the project. Since we are not performing PSpice simulations in this example, PC Board Wizard is used. Select the desired location for your project, then click OK.

At the next Project Wizard dialog box, do not enable project simulation, click Next. At the next dialog box, add the Connector, Discrete, and OPAmp part libraries as shown in Fig. 9.4. To add a library, select it (by clicking on it) in the left box and press the Add>> button. After you have added the libraries you want, click the Finish button.

If you are using the demo edition, you will be shown an information box that says, The Demo Edition does not support saving to a library with more than 15 parts.... Click OK each time it asks (you will need to click once for each library that you have added to the project).

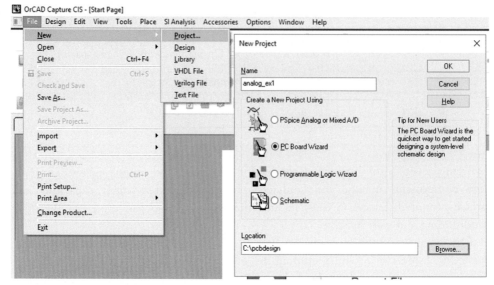

Figure 9.3 Setting up a new project with the PCB Project Wizard.

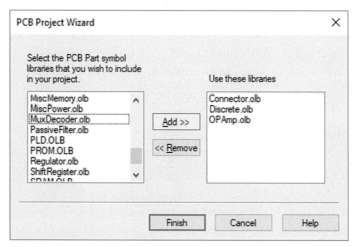

Figure 9.4 Adding parts libraries to the project.

Drawing the schematic with Capture

Fig. 9.5 shows the design goal. You can use it as a reference through the design example or modify it to your liking. If you are using the demo version, remember that you cannot save a design in PCB Editor that has more than 50 parts or parts with more than 100 pins. This design example meets these requirements so that you can save your work.

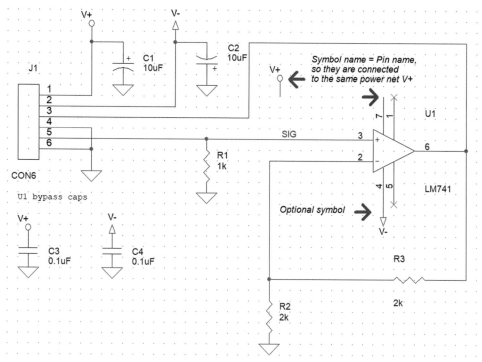

Figure 9.5 Analog circuit schematic design in Capture.

If a schematic page does not open automatically, expand the `Design` icon and the `SCHEMATIC1` folder in the `Project Manager` window, then double click the `PAGE1` icon.

Placing parts

`Place Part tool button`

Before placing any parts onto the schematic page, make sure that the place grid is enabled, which is indicated by the "snap-to-grid" icon: . If parts are placed off grid, you will not be able to connect wires to the part's pins. Table 9.2 lists the parts used in this project and the libraries in which they are located.

To place parts click the `Place Part` tool button, , select `Part` from the `Place` menu, or hit the `P` key on your keyboard. In the `Place Part` dialog box (see Fig. 9.6), select the `OPAMP` library or the `EVAL` library (if using the demo version), and scroll down until you find the LM741 or the uA741 op amp, respectively. Either one will work; it is just a matter of which one you prefer. You can place one of each, compare them, and then delete the one you do not like. If a library is not shown in the

Table 9.2 Capture library parts list.

Reference	Capture part	Capture library
J1	CON6	C:\…\CAPTURE\LIBRARY\CONNECTOR.OLB
C1, C2	CAP POL	C:\…\CAPTURE\LIBRARY\DISCRETE.OLB
C3, C4	CAP NP	C:\…\CAPTURE\LIBRARY\DISCRETE.OLB
U1	LM741	C:\…\CAPTURE\LIBRARY\OPAMP.OLB
R1−R3	R	C:\…\CAPTURE\LIBRARY\DISCRETE.OLB

Libraries: box, you can add it to the list, as explained next. Once you find and select the part, double click on its name with the left mouse button, move to the required location on the canvas, and left click to place the part.

If a library is not displayed in the Libraries: window, you need to add it to the list. *To add a library* to the Libraries: list in the Place Part dialog box, click the Add Library... button, , on the Place Part dialog box (Fig. 9.6). Use the Browse File dialog box to locate the desired OLB library. For building schematics and PCB layouts, you can use parts from either the main tools/capture/library folder or the PSpice subfolder. If you are performing PSpice simulations, select parts only from the PSpice subfolder.

From the DISCRETE library, place two polarized capacitors (CAP POL), two nonpolarized capacitors (CAP NP), and three resistors (R). From the CONNECTOR library, place a six-pin connector (CON6), and from the OPAMP library, place LM741.

Connect parts with wires (signal nets)

Place Wire tool

To wire the circuit, use the Place Wire tool, , select Wire from the Place menu, or hit W on your keyboard to activate the Wire tool. To attach wires between pins, click once to start a wire, move the pointer to the next pin and click once to attach the wire, and continue routing or double click to end the wire. Connect the circuit as shown in Fig. 9.5.

Making power and ground connections

There are three ways of adding power connections to active parts, depending on the part's type of power supply pins. A part's power supply pin can be a power-type pin and nonvisible, a power-type pin and visible, or a nonpower-type pin (such as a passive or input pin), which is always visible. The term *visible* specifically refers to whether the pin is visible to the Wire tool. However, in the general case, a nonvisible pin is also invisible from the user's perspective. Digital parts typically have nonvisible power

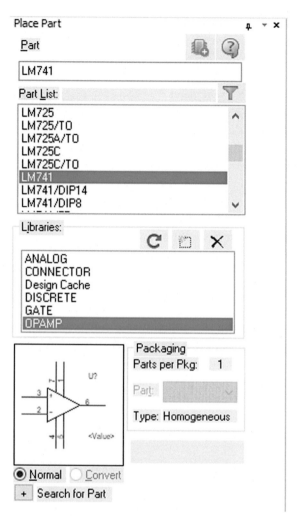

Figure 9.6 Choosing parts from the Place Part dialog box.

pins, while analog parts, particularly op amps, commonly use either visible power pins or one of the nonpower-type pins (which are always visible).

If a part's power supply pin is a power pin and is not visible, you cannot connect a wire (a net) to it directly. A nonvisible power pin is a net and global. You connect a part's power pin to a power symbol by giving the pin and the power symbol the same name. To make the connection, you need to place a power symbol, which is also global, somewhere on the schematic. Power symbols are always visible and are wired to either an off-board connector or a PSpice power supply. To make the names the same, you can change the name of the power symbol, the power pin, or both. An example of how to do this is given next.

If a power supply pin is a power pin and it is visible, you can either take advantage of the power pin's global properties using power symbols or make direct connections to it with wires. If you use the pin's global nature, the pin name and the power symbol name must be the same, as described already. If you make a direct connection to the power pin with a wire, you need not consider the naming convention. If you have a multipart package (e.g., a quad op amp with shared power pins), all the parts within the package, which are placed on a schematic, must have their power supply pins connected in the same way. So either all must be global or must have wires connected to them.

If a part's power supply pin is a not a power pin (see LT1028 in the LIN_TECH library, for example), you must use a wire to connect the pin to some other object, such as a power symbol or an off-board connector. If you place more than one part from a multipart package that has nonpower-type power pins, connections need to be made to only one of the part's power supply pins (although you can make connections to all of them if you want). See Chapter 7, Making and editing capture parts, for more information on pin types.

The LM741 op amp used in this example has visible power supply pins, so we could use their global properties by using power supply symbols to make connections to the off-board connector, but we will add the power supply symbols to the amplifier to make it evident from the schematic what power the amplifier uses. Whether you used the LM741 from the OPAMP library or the uA741 from the EVAL library, both power supply pins are visible power pins. The difference in their appearances is that the uA741 has zero-length power pins and the LM741 has line-length power pins. A zero-length pin is still "visible" to the Wire tool but not to the user.

To place power symbols, click the Place Power tool button and select one of the power supply symbols (VCC in this example) from the Place Power dialog box, as shown in Fig. 9.7. Click OK and place the symbol onto the schematic.

The names of the power pins on the op amp and the names of the power symbols must be the same. You can change the name of the symbol or the pin or both. Some parts have no visible labels for their power pins. *To check a pin's name and type*, select the pin (if it is visible), right click, and select Edit properties at the pop-up menu to get a Property Editor dialog box for the pin (see Fig. 9.8).

To change the name or type of a nonvisible power pin, select the part on the schematic, right click, and select Edit Part from the pop-up menu. You will be given a Capture Part Editor window (see Fig. 9.9). Double click the pin whose name you want to change to bring up the Pin Properties pane in the Property Sheet as shown in Fig. 9.9. Enter the new name in the Name text box and click Apply Pin Changes, if such a button exists in your version of software. Repeat the process for the other power supply pin.

Close the Part Editing window. Click Update Current when Capture asks *Would you like to update only the part instance being currently edited, or all part instances in the design?*

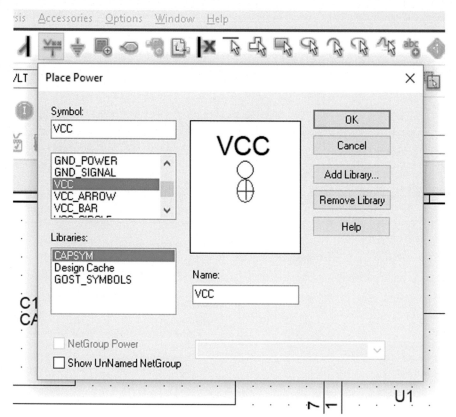

Figure 9.7 Placing global power symbols.

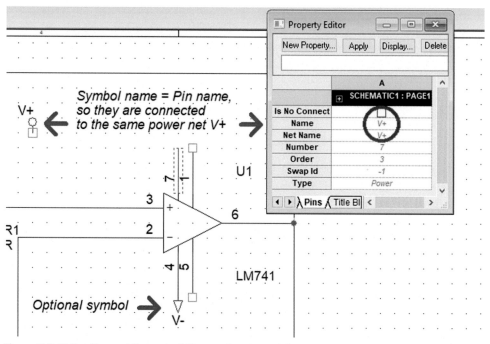

Figure 9.8 Using the part Property Editor to change a pin's name or type.

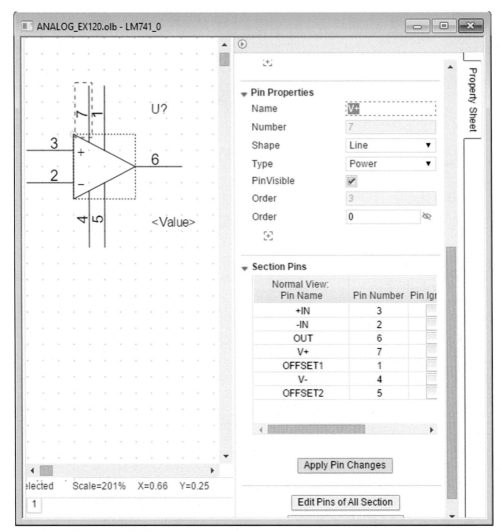

Figure 9.9 Determining a power pin's type and name.

In this case, since there is only one LM741, it does not matter if you click Update Current or Update All. But if you had several LM741s and you did not want all of them to have their power pin names changed, you would click Update Current.

Note: When you change a part on the schematic, the link between the design cache and the part library is broken. See the section at the end of the chapter for details on managing the design cache.

Instead of changing the name of the power pin on the part, you can change the name of the power supply symbol. *To change the name of the power supply symbol*, double

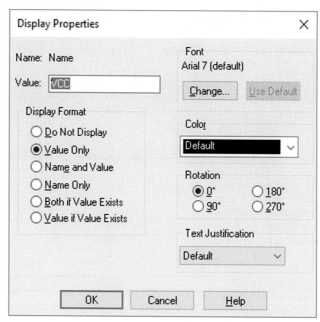

Figure 9.10 Use the Display Properties dialog box to change a power symbol's name.

click the symbol's name on the schematic. The Display Properties dialog box will pop-up, as shown in Fig. 9.10. The op amp's positive power supply pin name is V + , so enter V + in the Value: text box then click OK. Place another power supply symbol and change its name to V − using the same procedure. Copy, place, and connect the V + and V − power symbols as shown in Fig. 9.5.

Next, add a ground symbol. *To place a ground symbol*, click the Place Ground tool button, and select one of the ground symbols from the Place Ground dialog box. Click OK and place the symbol onto the schematic. Place and connect ground symbols as shown in Fig. 9.5.

Note: For designs that will be simulated with PSpice, you must use a ground symbol named 0 for the circuit to simulate correctly. You can use any of the symbols as long as they have the name 0. For PCB designs that will not be simulated, you can use any of the ground symbols and name them whatever you want, e.g. GND in this example. A separate net will be instantiated for each distinct ground symbol name. See Examples 2 and 3 for techniques on how to establish multiple ground systems.

Preparing the design for PCB Editor

Once all of the connections have been made, the next step is to prepare the design for making a netlist.

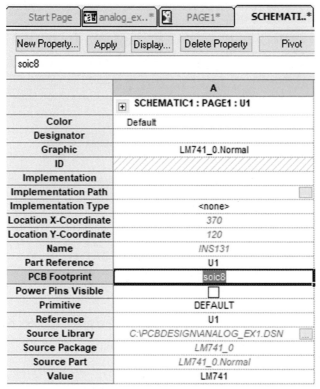

Figure 9.11 Part properties spreadsheet.

Things to do:
- Make sure all of the footprints are assigned.
- Assign parts to groups (called rooms).
- Perform an annotation.
- Clean up the design cache.
- Perform a DRC in Capture.

There are several ways to find out which (if any) parts have footprints assigned to them and if they are the right ones. You can check each part one at a time by double clicking a part to bring up the properties spreadsheet and look at the PCB Footprint cell, as shown in Fig. 9.11. Or you can check all the parts at the same time by selecting all the parts in the schematic (drag a box across the schematic using the left mouse button), then right click, and select Edit Properties... from the pop-up menu. The spreadsheet will show the properties of all the objects that were selected. Select the Parts tab in the bottom. To toggle the spreadsheet view between the column and row organization, right click the upper left corner of the spreadsheet and select Pivot from the pop-up menu.

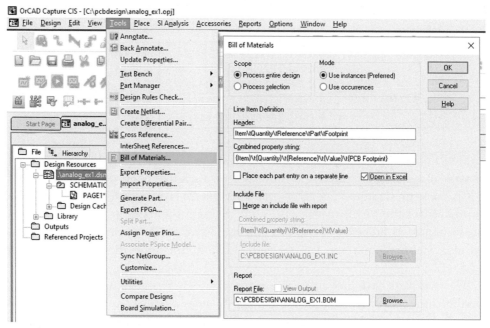

Figure 9.12 Generating a bill of materials to include PCB footprints.

The spreadsheet method is fast and easy for small circuit designs, but it may be impractical for large circuits. An alternative is to generate a customized BOM that lists all the footprints in an Excel spreadsheet. The spreadsheet can be printed and used as a reference while you are searching through the footprint symbol library for the right footprints.

To generate the custom BOM, go to the Project Manager, select the Design icon, and then select Bill of Materials... from the Tools menu (see Fig. 9.12). At the Bill of Materials dialog box, add the text \tFootprint in the text box labeled Header: and add the text \t{PCB Footprint} in the text box labeled Combined property string:. Put a check mark in the Open in Excel box. You can specify the location and name of the BOM using the Browse... button at the bottom. Click OK when you have finished the setup.

The next step is to assign footprints to the components. To find footprints in PCB Editor's footprint libraries, navigate to the \share\pcb\pcb_lib\symbols folder (see Fig. 9.1). Footprints have the *.dra and *.psm file extensions. When you find the footprint you are looking for (see Table 9.3 with a list of appropriate footprints for this project from the Cadence library), either memorize it or copy the name (without the extension) to the Windows clipboard. If you have a license of OrCAD Productivity Toolbox, you can simplify the search of footprints with PCB Library

Table 9.3 Bill of materials with available PCB Editor footprints.

Reference	Capture part	PCB Editor footprint
C1, C2	DISCRETE/CAP POL	ck17–10pf
C3, C4	DISCRETE/CAP NP	smdcap
J1	CONNECTOR/CON6	conn6
R1, R2, R3	DISCRETE/R	smdres
U1	OPAMP/LM741	soic8

Plot function — it documents all the PCB footprints in a library and creates the PDF file with all available footprints drawings based on a selected template.

Go back to Capture. To assign the footprint to a part, display the properties spreadsheet for the part to which you want to assign the footprint (double click the part to display its properties spreadsheet). Type or paste the footprint name (without the file extension) into the PCB Footprint cell in the properties spreadsheet (see Fig. 9.11).

If several parts have the same footprint (the capacitors and resistors, for example), you can change all of them at the same time. *To assign a footprint to multiple parts simultaneously*, hold down the Ctrl key on your keyboard while you select each component, right click, and select Edit Properties... from the pop-up menu. Select the PCB Footprint column (or row if the spreadsheet is pivoted) to select all of the PCB footprint cells then right click and select Edit... from the pop-up menu to display the mini spreadsheet of Edit Property Values shown in Fig. 9.13. Paste the footprint name that you copied from PCB Editor's libraries into the PCB Footprint cell then click OK. Once you have all of the footprints assigned, you can generate another BOM, so that you can see all of the footprints at the same time to make sure you have not missed any. Fig. 9.14 shows the final BOM listing with all the footprints assigned.

You can also use the Find tool to automatically select a part or parts of a certain type. To use the Find tool, select Find from the Edit menu and the cursor will jump to the Find text box as shown at Fig. 9.15. Enter R1 instead of "*," mark Parts in the dropdown list located to the right of the Search icon, and then left click the Search icon. R1 will be highlighted on the schematic and centered on the screen. If you enter R*, then all resistors will be found and could be sequentially selected one by one, or all together if you mark the Highlight choice in the dropdown list. You can use the Find tool to look for other things, such as text and nets, as indicated in Fig. 9.15.

Grouping related components (rooms)
Grouping related components in the schematic can make placing the parts easier in PCB Editor. Parts are assigned a room number in Capture, and the grouping

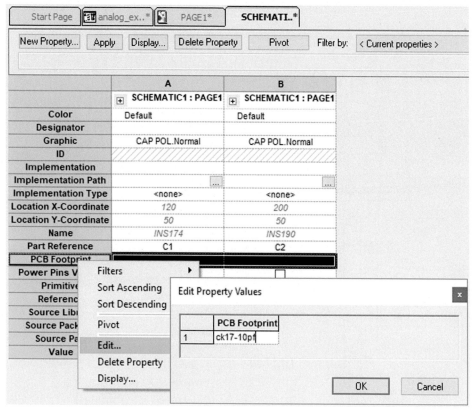

Figure 9.13 Assigning footprints to multiple parts with the Property Editor.

	A	B	C	D	E
1	Revised: Thursday, September 06, 2018				
2	Revision:				
3					
4					
5					
6					
7					
8					
9					
10	Bill Of Mate Page1				
11					
12	Item	Quantity	Reference	Part	Footprint
13					
14					
15	1	2	C1,C2	10uF	ck17-10pf
16	2	2	C3,C4	CAP NP	smdcap
17	3	1	J1	CON6	conn6
18	4	3	R1,R2,R3	R	smdres
19	5	1	U1	LM741	soic8

Figure 9.14 Final BOM listing with all footprints assigned. *BOM*: bill of materials.

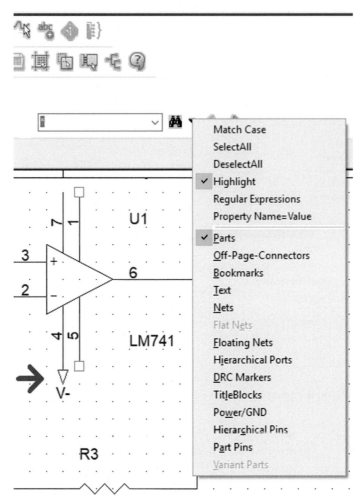

Figure 9.15 Using the Find tool to locate parts.

information is exported to PCB Editor during netlist generation [and engineering change order (ECO) processes]. *To add parts to a group*, select the related parts (e.g., J1 and the two filter caps) on the schematic. Right click and select Edit Properties from the pop-up menu to display the Property Editor spreadsheet (see Fig. 9.16). In the Filter by: dropdown list, select OrCAD PCB Designer Standard to see all available properties. Select the ROOM cell to select that property for all of the components then right click and select Edit from the pop-up menu. In the Edit Property Values dialog box, assign an integer number or any alphanumeric characters for the group (this will be the room name PCB Editor uses). Click OK to close the dialog box. The alphanumeric value will be assigned to all the parts you selected. Close the spreadsheet.

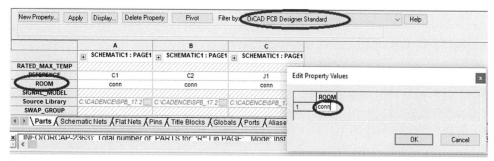

Figure 9.16 Using the Property Editor to assign components to groups (Rooms).

You can also add the room property to the BOM listing by adding \tRoom to the Header: list and \t{ROOM} to the Combined Property String list.

Before making the netlist, it is a good idea to tidy up the design by performing an annotation, cleaning up the design cache, and performing a DRC as described next.

Annotation
Performing an annotation chronologically renumbers the part references in your schematic design from top to bottom. *To perform an annotation*, go to the Project Manager and select Annotate from the Tools menu. The Annotate dialog box shown in Fig. 9.17 is displayed. Select the items you want updated then click OK. Be sure to check multipart components (hex inverters, for example) to make sure that they are annotated correctly. The Advanced Annotation dialog box allows the user to set some additional options, such as starting number of the RefDes on each page, or annotation depending upon the properties attached to the components (it is especially useful for multipart connectors).

Cleanup Cache
Cleanup Cache removes unused parts from the design cache. Unused parts pile up in the design cache when you place parts in the design then later delete them. To clean up the design cache, select the Design Cache folder in the Project Manager, right click, and select Cleanup Cache from the pop-up menu or select Cleanup Cache from the Design menu.

Performing a schematic design rule check in Capture
It is a good idea to run a DRC before generating a netlist. The DRC checks your design for design rule violations and places error markers on the schematic page. *To perform a DRC* in a Capture project, make the Project Manager window active and select the design icon. From the Tools menu, select Design Rules Check.... The Design Rules Check dialog box is displayed as shown in Fig. 9.18.

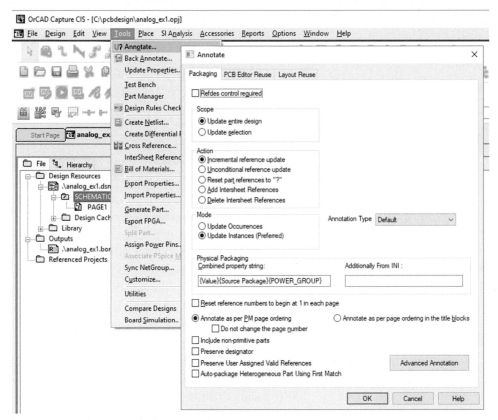

Figure 9.17 Performing a design annotation.

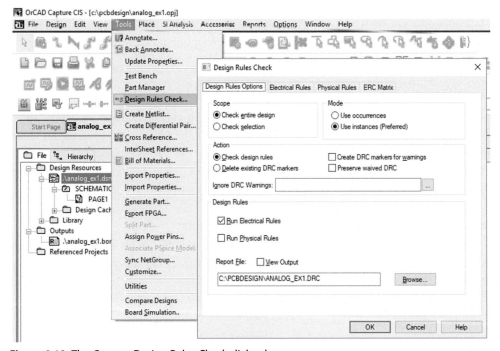

Figure 9.18 The Capture Design Rules Check dialog box.

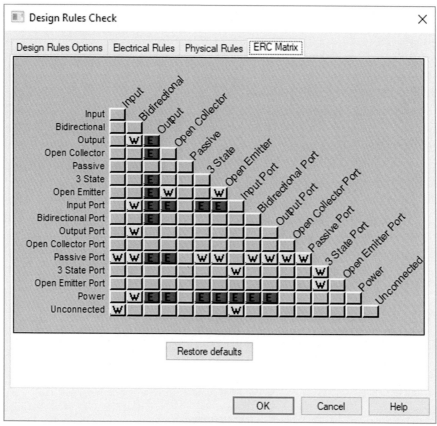

Figure 9.19 The DRC error type selection matrix.

You can choose which physical or electrical rules should be checked in the appropriate tabs of the `Design Rules Check` dialog box. Also you can remove the previously created DRC markers. It is even possible to create and execute your own custom DRC checks using Tool Command Language (TCL) scripting language. For that purpose, there are the sections called `Configure Custom DRC` in both `Electrical` and `Physical` rules tab. There are some ready to use samples available, and to use them, you should mark the required custom checks and set the `Run Custom DRC` check box.

You can refine which Pin Types should not be connected together and what action the DRC takes when it finds such shorts (as errors, `E`, or warnings, `W`) using the `ERC Matrix` tab in the `Design Rules Check` dialog box. As shown in Fig. 9.19, the matrix lists pin types on the left side and along the slanted side. The yellow `W` and red `E` boxes indicate the result of having pins of certain types connected together. For example, an output pin connected to another output pin will produce an error during a DRC, but a passive pin connected to an output pin will not cause an error or a warning.

You can change the boxes by clicking on them with the left mouse button. By clicking on a button repeatedly, you can toggle through the three options: warning, error, or neither.

Generating the netlist for PCB Editor

Once you have all of your connections finished and footprints and rooms assigned, you can create the netlist for PCB Editor. *To create a netlist for PCB Editor*, select the design icon in the Project Manager and select `Create Netlist...` from the `Tools` menu. In the `Create Netlist` dialog box (see Fig. 9.20), select the `PCB` tab and specify a netlist files directory name or accept the default name. Select the `Create PCB Editor Netlist` and the `Create or Update PCB Editor Board (Netrev)` boxes. You can specify an `Input Board File` if you have one (such as a board template or preexisting board design), or you can leave it blank. A default `Output Board File` will be set up for you, or you can specify your own. Select the `Open Board in OrCAD PCB Editor` radio button and then click `OK`. `PCB Editor` will automatically start.

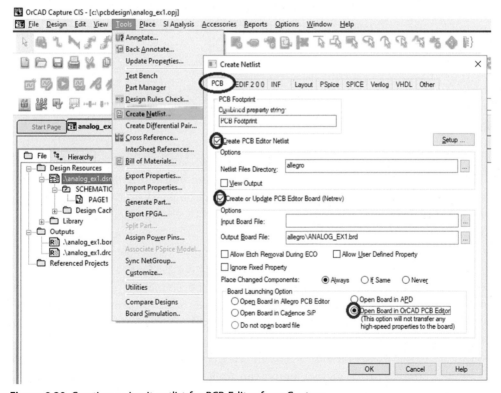

Figure 9.20 Creating a circuit netlist for PCB Editor from Capture.

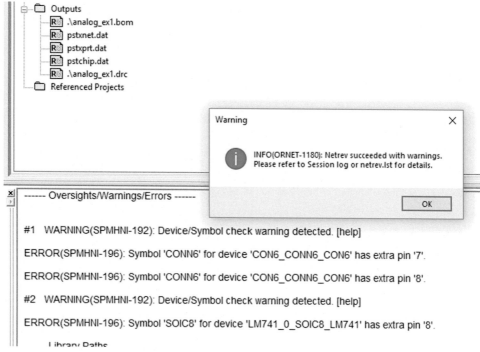

Figure 9.21 Using the netrev.lst file to find netlisting errors.

Problems during netlist creation

During the netlist creation, you may experience success, success with warnings, or failure. If there are problems, OrCAD will list the warnings or errors in the Session Log window as shown in Fig. 9.21. You can scroll up and look for the error or warning description.

In this example an error occurred because some Capture parts have fewer pins than the PCB Editor footprint symbols. Since this is not a fatal error, the netlist process was completed, and PCB Editor will be launched. If a fatal error occurs, PCB Editor will still be launched, but the netlist process will not be completed properly, and no information will be transferred to PCB Editor. In this case, though, the error will prevent placing the U1 and J1, so we need to go back and modify the schematic. We use this error as an opportunity to show how to perform an ECO later in the example.

Setting up the board

The first view you have of your new board will be PCB Editor with a black background in the work area. The next steps are to set up the board outline and define some of the basic design parameters.

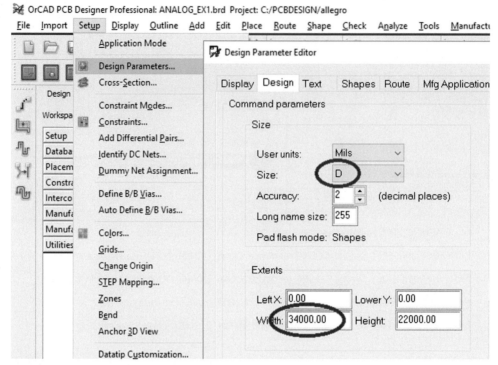

Figure 9.22 Setting up the board extents.

Setting up the extents (work area boundary)

The default size of the work area is 34 in. × 22 in. (size D). If the drawing area is much larger than it needs to be, that can make navigating around and working on your project harder than if the drawing area size is just right. *To change the size of the work area*, go to Setup → Design Parameters, and choose drawing size or change the X and Y Extents on the Design tab of the Design Parameters dialog box, as shown in Fig. 9.22. For our example set the Left X and Lower Y to −1000 to provide the additional working space around your PCB, and set the Width to 10,000 and Height to 5000. Make sure you click Apply to make the changes take effect, or just click OK.

Controlling the Grid

The next step is to turn on the grid. *To turn the grid on*, click the Grid button, ▦ . *To change the grid spacing*, select Grids... from the Setup menu, then the Define Grid dialog box will be displayed (see Fig. 9.23). There are two grid systems: one for copper elements (Etch) and the other for everything else (Non-Etch). You can change the etch grids on each layer individually or change them all simultaneously by using the All Etch entries. Click OK to dismiss the dialog box when you are finished.

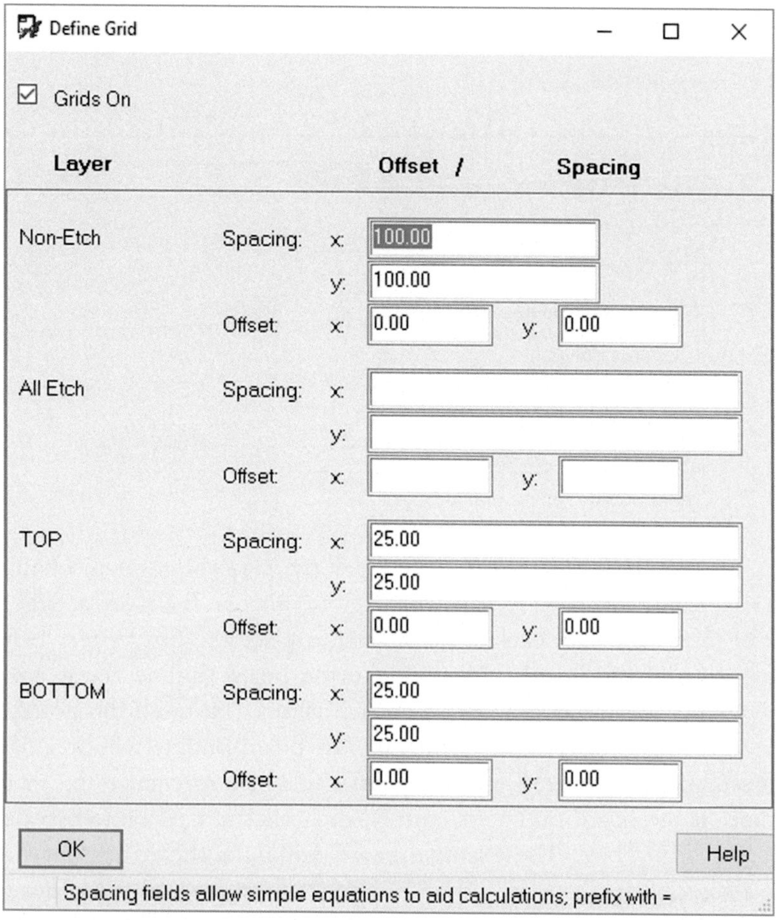

Figure 9.23 Setting the design grids.

Making a board outline

The next step in laying out the board is to make the board outline and add mounting holes so that the boundaries of the board are known.

First, we draw the board outline. Zoom in an appropriate amount so that your final board size will take up about 75% of your display's viewing area (large enough for good resolution but small enough that you can see the whole thing). To change zoom, select one of the zoom options from the Display menu, or select one of the Zoom buttons on the toolbar. You can use the mouse wheel for zooming.

To place a board outline, select Outline → Design... from the menu bar. The Design Outline dialog box will be displayed, as shown in Fig. 9.24. In the basic example in Chapter 2, Introduction to the printed circuit board design flow by example, we just drew a box. This time we draw an irregular shape using the Polygon tool.

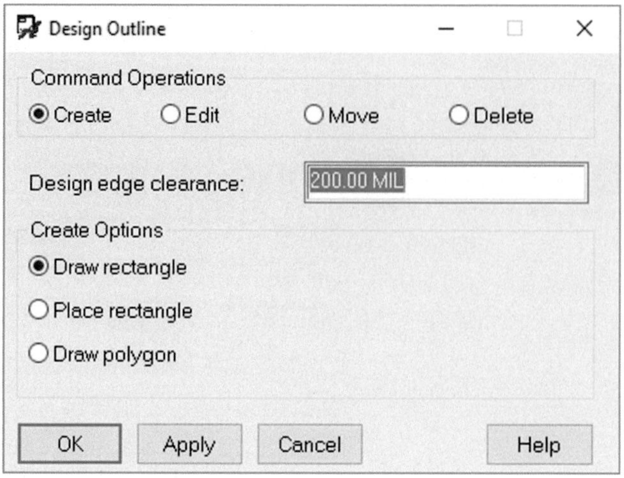

Figure 9.24 The Create Board Outline dialog box.

Set the `Design edge clearance` as you need. Click the `Draw Polygon` radio button in the `Design Outline` dialog box then left click and release once in the work area at the starting vertex of the board. Left click at point 0, 0 to place the first vertex. Left click to place each vertex of the board outline. Make the board outline about 5.0×3.0 in. (5000×3000 mil). Left click and release at each vertex. To finish the board, click on the starting point again to close the polygon. The board outline will be a dashed line with squares (handles). You can use the handles to resize or change the shape of the board outline. If the board outline is correct, then click the `OK` button to dismiss the `Design Outline` dialog box. The example board outline is shown in Fig. 9.25. Now let's create a copy of the shape in another subclass, which we will need later to generate the internal planes. Use menu choice `Shape → Z-Copy`, select `Board Geometry/ Outline` in the `Options` pane and set the `Offset` value to 0. Select the recently created `Design_Outline` shape with left mouse click on its boundary, and it will be copied to a new layer.

Adding dimension measurements

You can add dimension lines and measurements to your board for documentation purposes. You can also add temporary construction lines as guides to mark locations for mounting holes and connectors and the like. Dimensions are placed on the `Board Geometry/Dimension` layer (class/subclass), so once you are satisfied with the board layout, you can turn off that layer to hide the dimension objects or delete temporary construction lines.

For this example, we add dimension lines to mark a mounting hole location. First, set the grid spacing for Non-Etch to 10 mil (see Fig. 9.23). Next you may want to

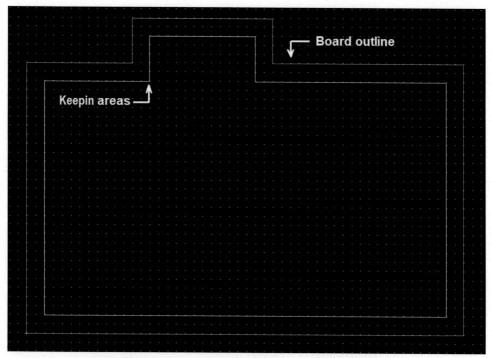

Figure 9.25 The finished board outline and keep-in areas.

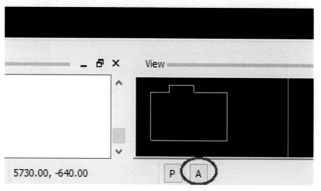

Figure 9.26 Toggle between absolute and relative XY modes.

change the coordinate indication (XYMode) to absolute or relative depending on the dimension you are using. To change the XYMode, click the R (relative) or A (absolute) button at the bottom of the design window (see Fig. 9.26).

To place a dimension line, select Dimension Environment from the Manufacture menu. Right click on the PCB canvas and choose Linear Dimension from the context menu (see Fig. 9.27). Left click to place the first dimension point (point 1 in

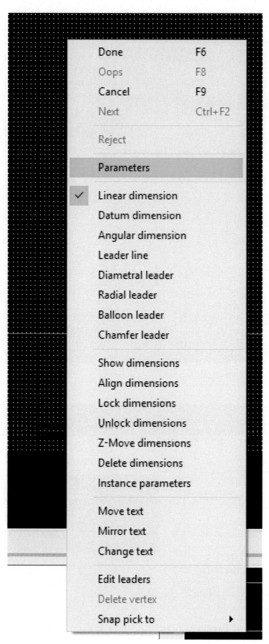

Figure 9.27 Dimensioning tools.

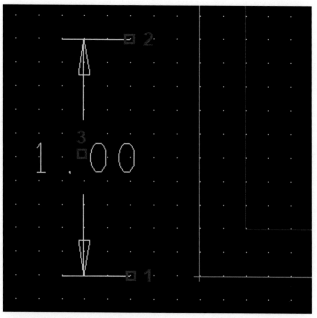

Figure 9.28 Linear dimension line and pick points.

Fig. 9.28), left click to place the end mark (point 2 in Fig. 9.28), and finally left click to place the location of the text (point 3 in the figure). Use the *XY* coordinates to precisely place the dimensions. You can also just pick a line object, and the dimension tool will automatically measure it and place the end points for you. It is very useful to use the "Snap Pick To" function, located in the bottom of the right click menu, to exactly point to the required object. Once you place a dimension and end the command, the dimension object is dynamic. This means that you can move it, and it will be updated with new dimension information.

You can change dimension parameters such as text size and arrow terminations. To change the appearance of the dimension lines, select Parameters from the right click menu to display the Dimensioning Parameters dialog box, from which you can change various aspects of the dimension elements through the tabs shown in Fig. 9.29.

You can also use the Add Line, ＼ , and Add Rectangle, ■ , tools to add construction lines, etc. to your design on the Board Geometry/Dimension layer.

Adding mounting holes

Next we place mounting holes on the board. This should be done before moving any parts into the board outline. *To place mounting holes*, select Place → Mechanical Symbols... from the menu bar, and you will see the Placement list as shown in Fig. 9.30. If no symbols are listed, check to make sure you have the Library box

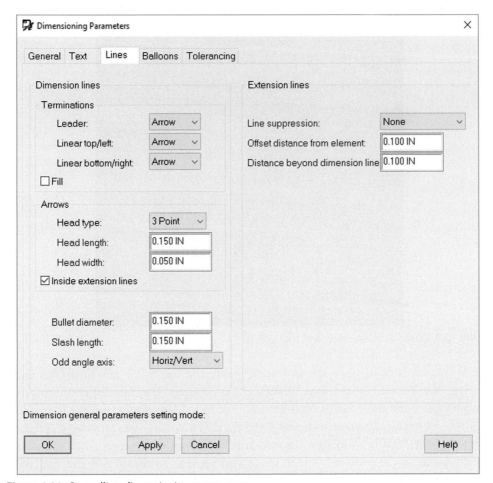

Figure 9.29 Controlling dimensioning parameters.

checked in the Advanced Settings tab, and left click on the plus sign on the left of Mechanical symbols icon to open the list contents.

Select the desired mounting hole symbol (it must be in the library ahead of time). Left click at the desired location in the work space to place the mounting hole. Only one will be placed per click. To add additional holes, recheck the desired symbol and left click again.

To add multiple mounting holes, you can copy and paste instead of using the placement method. To copy a mounting hole, click the Copy button (or type copy in the command line and hit the Enter key). The command window will say Select the elements to copy:. Use the left mouse button to select an existing mounting hole. Left click to place one or more copies of the mounting hole. When you are finished placing the mounting holes, right click and select Done from the pop-up menu.

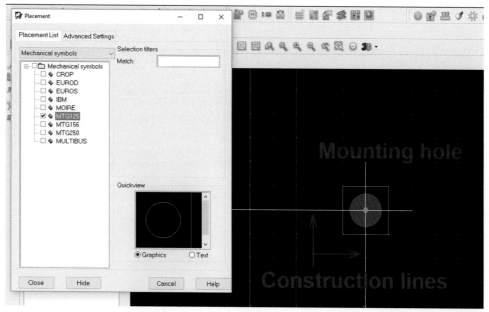

Figure 9.30 Placing mounting holes on the board.

To delete a mounting hole, click the `Delete` button (or type `delete` in the command line and hit the `Enter` key). The command window will say nothing, but you will see the current command name, `delete`, in the lower left corner of the window. Use the left mouse button to drag a box across an existing mounting hole (to select all elements of it). Right click and select `Done` from the pop-up menu. Alternatively instead of using the `Delete` command you can use the left mouse button to drag a box across an existing mounting hole (to select all elements of it). Right click and select `Unplace component`. Note that the Symbols checkbox should be checked in the `Find` pane during the operation with symbols.

The mounting holes supplied with PCB Editor are nonplated. Nonplated holes are drilled separately as one of the very last steps of the board fabrication, and no plating processes occur after the final drilling. See Chapter 8, Making and editing footprints, on designing mounting holes and Appendix D for standard machine screw and drill hole sizes. To make mounting holes plated, select `Tools → Padstack → Modify Design Padstack` from the menu bar, left click the mounting hole pad to select it, right click, and select `Edit` from the `Options` pane; the `Padstack Editor` dialog box will be displayed as shown in Fig. 9.31. In the `Drill` tab, in `Hole plating` section, you can select whether the hole is plated or not. In the `Design Layers` tab you can increase the TOP, INTERNAL and BOTTOM pads to meet the annular ring requirement for the plated holes. Choose `File → Update to Design and Exit` to save the changes to design.

You can leave a mounting hole isolated from the copper on all layers, or if it is a plated mounting hole, you can connect it to a net (e.g., the Ground plane).

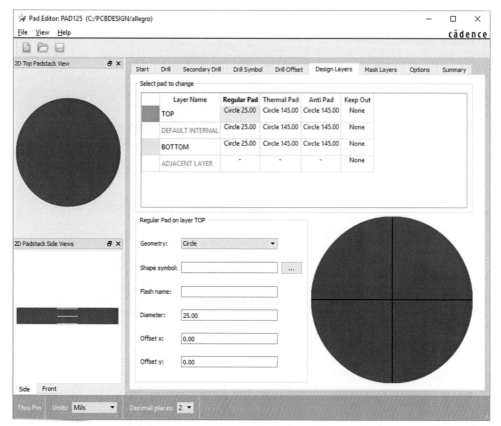

Figure 9.31 Using Padstack Editor to determine hole plating.

To connect a plated mounting hole to a net, you need to add a single pin part to the schematic design in Capture then connect the part's pin to the desired net. The mounting hole will become part of the netlist. Then in PCB Editor the mounting hole can be placed in the same manner as all the other parts, as described later, and it will be connected to the proper net. This is demonstrated in the manufacturing design example in Chapter 10, Artwork development and board fabrication. Along with making a part in Capture, you also need to make a footprint with a connect pin (rather than a mechanical symbol with a mechanical pin) for the mounting hole in the PCB Editor footprint library (see Chapter 8: Making and editing footprints, for details).

Placing parts

Placing parts is part art and part science. How you ultimately place the parts on the board depends on both mechanical and electrical factors. Mechanical factors include designing for manufacturability (assembly and soldering processes) and physical board constraints (size, shape, etc.). Electrical factors include functional signal flow, thermal management, signal integrity, and electromagnetic compliance. Usually all these considerations are important, and in some cases, they can conflict with one another. These issues are discussed in greater detail in Chapters 4—6 and in the IPC standards. In this example the parts are simply placed so that the board layout is similar to the flow of the schematic.

Before you begin placing parts, you may want to adjust the placement grid resolution. Go to Setup → Grids and change the Non-Etch Grid to 50 mil (or 25 mil for greater resolution, as suggested by IPC).

The three basic ways to place parts are manually, quick place, and through ITC (Intertool communication). The easiest method for small projects is to use the manual method, as was shown in Chapter 2, Introduction to the printed circuit board design flow by example, (i.e., select Place → Components Manually from the menu bar as shown in Fig. 9.32B, select the part or parts you want to place, and left click to place them). In this example, we see how to use the quick-place method and the ITC tool to select parts for placement.

Note: Regardless of how you place the parts, you will likely have a problem placing U1 and J1 in this example if you used the default op amp and connector parts in Capture and default footprints in PCB Editor. If this occurs, we will see later how to fix the problem using the ECO function.

To use the quick-place method, select Place → Quickplace... from the menu bar; the Quickplace dialog box will be displayed (see Fig. 9.32A). There are many ways of selecting and filtering parts for placement. When using the Placement Filter options, a property has to be assigned to a part or parts for that filter to work (e.g., a room has to be assigned to use the Place by room filter). Note that there are similarities between the Quickplace dialog box and the Placement dialog box. When using the Place by REFDES filter, the Placement dialog box works better (it has more details options and allows you to see what parts match the filters), but the Quickplace dialog box gives more control with the Placement Position section options.

In this example, we use the Quickplace dialog box to place parts by rooms. Recall that rooms were assigned to parts in Capture; this is where we use that property. Before we do, though, we need to define areas on the board where the parts with common room numbers will be placed. *To define a room on the board*, select Outline → Room... from the menu bar. The Room Outline dialog box will be displayed (see Fig. 9.33). Also, if you look at the Options tab, you will notice that the default class/subclass is Board Geometry/Both_Rooms (also indicated in the Side of Board section, the Both radio button is checked). In the Room Outline dialog box, you can specify the room (group) that was assigned in Capture (i.e., the Room Name). You can also select the room type. The options are Hard, Soft, Inclusive, Hard straddle, and Inclusive straddle, which determine how the DRCs are issued (see *Using the ROOM and ROOM_TYPE Properties* in Allegro Help). With the Room Outline dialog box displayed, draw the outline for each room name from the list. When finished, click the OK button to dismiss the Room Outline dialog box.

Once the room outlines are drawn, you can then use the Place by room option to place parts. In the Quickplace dialog box (or the Placement dialog box), select the Place

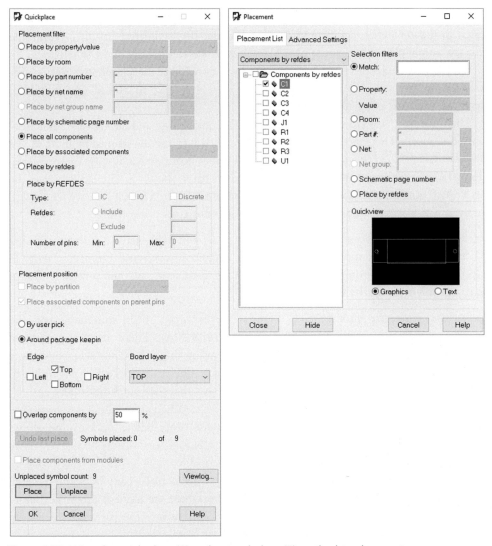

Figure 9.32 Using the quick-place (A) and manual place (B) method to place parts.

by room option and select the desired room name, then click the Place button. The parts that were assigned to that room name will be placed in the room outline (see Fig. 9.34). You can then use the Move tool, ⊕, to move the parts around within the room.

Note: Remember that U1 and J1 will probably not place properly because of the problem mentioned previously. This will be addressed later.

If you select a part and do a Show Element, you will see the ROOM property included as shown in Fig. 9.35.

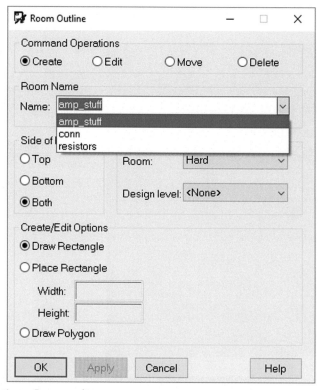

Figure 9.33 Creating a Room outline.

Intertool communication

ITC is a communication link between Capture and PCB Editor. This link allows you to perform two functions: cross-probing (also called *cross-selection*) and cross-highlighting. Cross-probing is used as an aid in placing components in PCB Editor (communication is from Capture to PCB Editor only). Cross-highlighting is a tool that allows you to visually highlight components, pins, or nets in one application and simultaneously highlight it in the other application (communication is bidirectional). To enable intertool communications, in Capture in the Options/Preferences dialog box, select the Miscellaneous tab and check Enable Intertool Communication. Click OK (see Fig. 9.47). Cross-selection is possible if PCB Editor is running a command, like Highlight.

Using cross-probing to place parts

Depending on the complexity of your design, it may be necessary to layout specific parts of the board in a particular order or location on the board. In this case, it might be easier to pick the specific parts from the schematic and have PCB Editor place those parts, and this is exactly what cross-probing does. For cross-probing to be

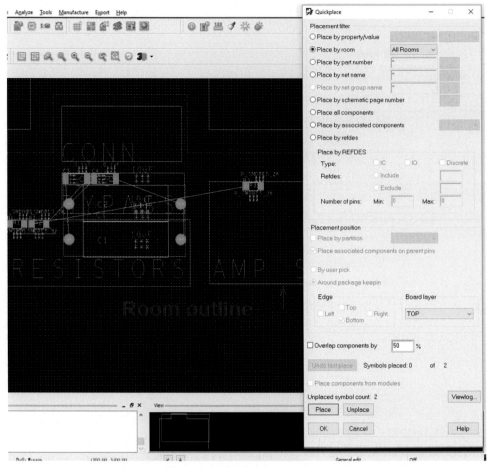

Figure 9.34 Parts placed in a room using quick place.

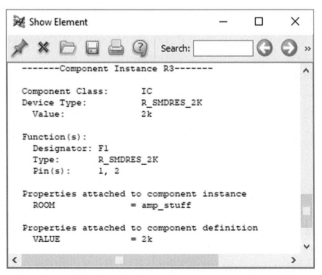

Figure 9.35 Using Show Element tool to display a component's room assignment.

enabled, you have to have your schematic open in Capture, your board layout open in PCB Editor, and be in manual placement mode in PCB Editor.

Prepare for cross-probing in PCB Editor by selecting `Place → Components Manually...` from the toolbar to display the `Placement` dialog box. From the dropdown menu, select `Components by refdes` and, in the `Selection filters` section, make sure the `Match:` radio button is selected (see Fig. 9.37 later), but don't select any component.

To begin cross-probing from Capture to PCB Editor, go to your schematic in Capture. Left click the part you want to place to highlight it (C1 in Fig. 9.36).

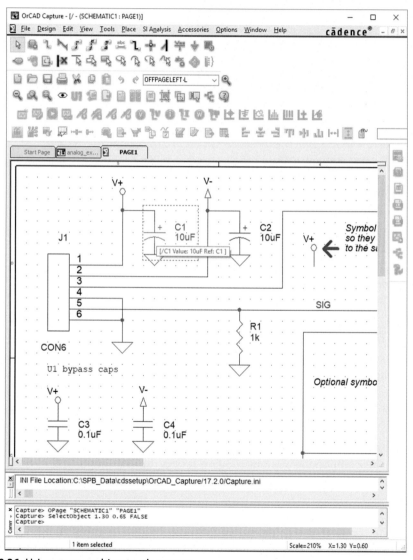

Figure 9.36 Using cross-probing to place parts.

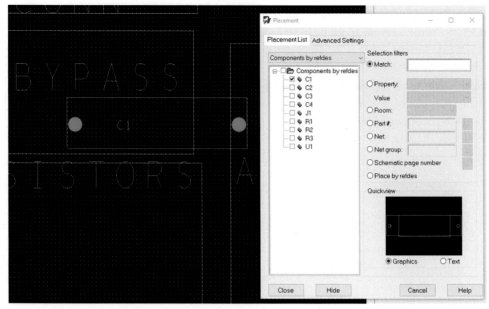

Figure 9.37 The part selected in Capture is also selected in PCB Editor.

Now go back to PCB Editor. That part will be checked in the dialog box, and the part will be attached to your cursor, as shown in Fig. 9.37. Left click to place the part. You can either click `Close` (or right click and click `Done`) to quit or go back to the schematic and select another part or several parts as just described.

Rotating parts

To rotate a placed part, select the `Move` button, left click to select the part, then right click; select `Rotate` from the pop-up menu. The part will pivot around its origin, which may be pin 1 or the body center. Move the mouse around the pivot point until it is in the correct orientation. Use the `Options` pane to select a different rotation criterion. Left click to place it, then right click and select `Done` from the pop-up menu. If you choose to rotate the part around the particular `Symbol Pin #`, you can select any pin number among existing pins of the component. But be careful next time when you select a component with other pin numbers—it will be impossible to rotate it until you change the `Symbol Pin number` to the one which exists in that part.

To spin the component without moving it you can use menu `Edit → Rotate` then left click on the part. You can choose the rotation center point in the `Options` pane.

Mirroring parts

If you want to move a part to the backside of your board, you mirror it. To mirror a part select the `General Edit` button, check `Symbols` in the `Find` pane, left click to select

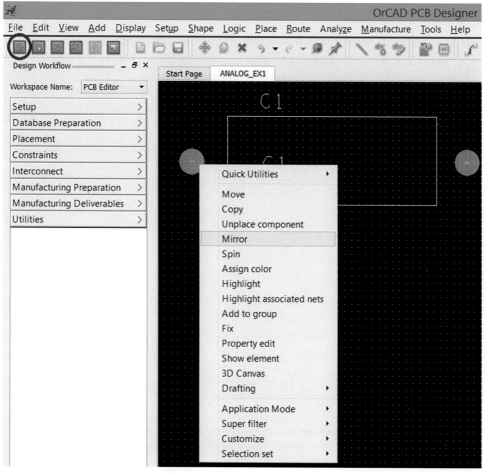

Figure 9.38 Mirroring a placed component.

the part, right click and select Mirror from the pop-up (Fig. 9.38). While dragging you also can click right button and select Mirror in the pop-up menu. Release left mouse button to place the mirrored part to correct position. You can also use menu Edit → Mirror and select the object or group of objects. If you select several objects using rectangle window, Select by ..., or Temp Group, you will need to select a pick point.

As mentioned earlier U1 and J1 may not place properly because of the netlist warning that occurred during the netlist creation. The problem is that the part U1 in Capture has only seven pins and the footprint in PCB Editor has eight pins. Every pin in a PCB Editor footprint must have a corresponding pin in the Capture part that is using it. So we either need to add a pin to the Capture part in the schematic or delete a pad in the PCB Editor footprint on the board. While it is possible to do the latter by deleting the connection type pin and installing a mechanical type pin, this would make that footprint useful for only that part or parts like it. You would need a different footprint for other op amps. This is true for CONN6 package for connector as well, which actually has eight pins in the footprint. So we should change the parts in Capture or create new footprints in PCB library then perform an ECO to update the board design.

Changing a part in capture

Place pin button

Go back to the schematic in Capture. Left click the op amp to select it. Right click and select Edit Part from the pop-up menu.

In the part editor (see Fig. 9.39), click the Place pin button, . In the Pin Properties dialog box, enter NC for pin name and the number 8 under pin number. Select the Short and Passive options. Click OK and place the pin somewhere on the part boundary, as shown. You can also add a line to make the pin touch the body if you like or leave it floating, since it is a NC pin. And, in that case, it might be a good idea to add text indicating that.

Note: If a part has more than one NC pin, you have to give them different names (e.g., NC1, NC2, etc.) because a unique name is required for each pin; otherwise you

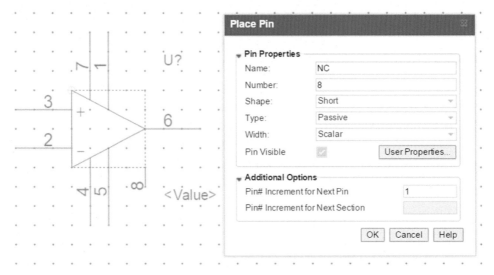

Figure 9.39 Adding a no-connect pin to a part in Capture.

will get a DRC error. Once you finish modifying the part, close it by clicking the X in the upper right corner of the part editor window. Click Update Current (or Update All if there are more than one of this type of part) when Capture asks Would you like to update only the part instance being currently edited, or all part instances in the design?

Make sure that you place a NC symbol (Place → No Connect) on all unused pins on the schematic.

If you have a number of not-connected Pins and would rather not display them on the Part, you can add a property named NC (using right-click → Edit Properties on the selected part, and then the New Property button). The NC property takes a comma separated list of the not-connected Pin Numbers (If required, the NC property can be added to the Part in a Library). Do it for the J1 connector, create the NC property containing pins 7,8.

Performing an engineering change order

Once the part has been changed on the schematic, the changes need to be updated in the board design. In PCB Editor, an ECO is used to update the board design with changes that were made to the schematic in Capture, whereas an update from PCB Editor to Capture is called *back annotation*. Performing an ECO is a two-step process: (1) recreate the netlist in Capture and (2) import the changes into the board in PCB Editor.

To perform an ECO, begin in Capture by making the Project Manager window active. Highlight the Design icon and then select Tools → Create Netlist from the menu bar to display the Create Netlist dialog box (Fig. 9.40). So far this is the same process used to create the original netlist, but the next step is slightly different. In the Options section below Create or Update PCB Editor Board (netrev), use the Input Board File: Browse... button, ▢, to locate the current board you are working on (the one you want updated). Then use the Output Board File: Browse... button to choose the location and name of the new updated board. You can update a board into itself, so the input board and the output board are the same, as shown in the figure. Note that the PCB Editor board project should be closed before this operation. In the Board Launching Option section, select Open board in PCB Editor to open your board file. Click OK.

If you relaunched PCB Editor from Capture, the board will be automatically updated with the new information. If you selected Do not open board file (see Fig. 9.40), then you need to import the changes to update the board. *To update the design in PCB Editor*, select Import → Netlist from the menu bar. The Import Logic dialog box will be displayed. Select Design entry CIS (Capture) in the Import logic type section. Select the desired option in the Place changed component section. Use the Browse... button to locate the directory listed in the Import directory: (see Fig. 9.41) if it is not already displayed. Click the Import button to finish the ECO process. Your board should now have the updated information.

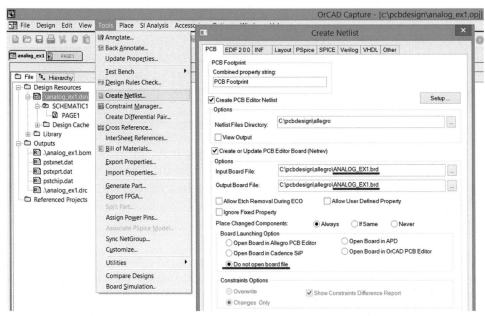

Figure 9.40 Setting up for an ECO. *ECO*, engineering change order.

Figure 9.41 Import ECO information with the Import Logic dialog box. *ECO*, engineering change order.

Note: Depending on which version of the software you have, you may experience difficulties running the autorouter if you performed the ECO using the import logic method. If that occurs, save your design and close PCB Editor. Go back to Capture and perform another ECO, but launch PCB Editor from the `Create Netlist` dialog box. You will not need to import logic again, and the autorouter should work properly.

If you were not able to place the op amp, U1 (and may be J1), you should be able to place it now. The quickest way to place the part is to select `Place → Components manu-ally` from the menu (or click the `Place Manual` button on the toolbar). U1 should be on the list, check it and place the part on the board, and click Close when you are finished.

Once the parts have been placed on the board, you need to position them. If all of the layers and nets are visible, it can be difficult to identify which parts are which. You can do a couple of things: turn off all or most of the nets and make only the necessary layers visible (e.g., silk-screen layers and pins).

Controlling layer visibility

There are two methods for controlling layers (class/subclass) visibility. The first method is the `Options` tab (Fig. 9.42), and the second is the `Color` dialog box (Fig. 9.43). *To change the visibility of a layer using the* `Options` *tab*, move your mouse pointer over the `Options` tab (if the pane is not displayed) and select the class and subclass of interest. Toggle the colored button to the left of the subclass to show or hide that subclass. Or right click on it to assign the new color.

To change the visibility of a layer using the `Color` *dialog box*, click the `Color` button, ▦ , on the toolbar. Most of the layers are common to both the `Options` tab and the

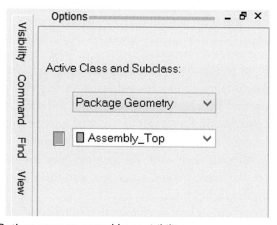

Figure 9.42 Use the Options pane to control layer visibility.

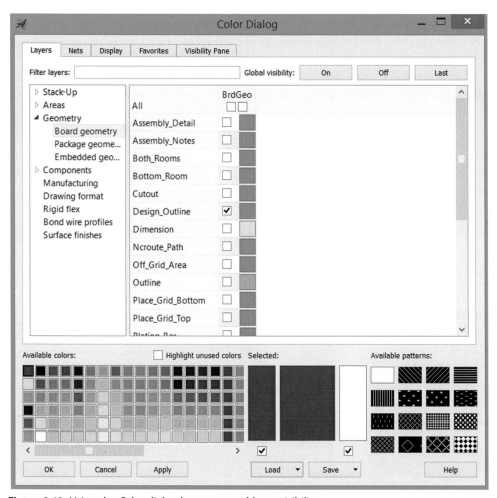

Figure 9.43 Using the Color dialog box to control layer visibility.

`Color` dialog box, but the `Color` dialog box gives you access to more objects. Besides letting you change the visibility of objects and layers, it lets you change their colors. *To hide (or show) a layer*, find the class (the list of folders) and subclass you are interested in. Toggle the check box beside the object or layer of interest (blank means it is off). *To change the color of an object or layer*, find and click on the color of choice in the color palette located in the bottom, then click the colored box for the object of choice. The box will change to the desired color. Click `Apply` to make the changes take effect, then click `OK` to dismiss the dialog box.

You can use the `Color` dialog box to change the visibility of one subclass or object at a time, an entire class, or the entire project simultaneously. *To change an entire*

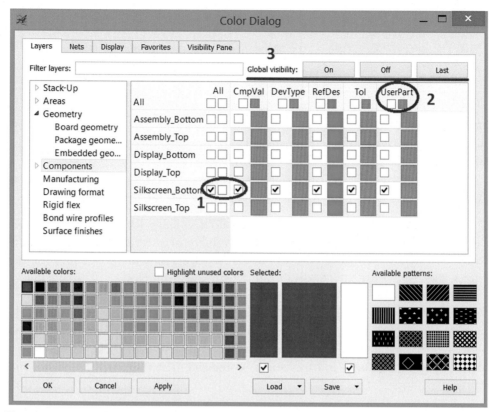

Figure 9.44 Controlling the visibility of multiple subclasses simultaneously.

subclass, toggle the check box in the All row or column as indicated by points 1 and 2 in Fig. 9.44. *To change the visibility of all classes and subclasses simultaneously*, click the Global visibility buttons (point 3 in Fig. 9.44). Make sure you click the Apply and then OK buttons to make the changes take effect.

When placing parts and routing traces, it is sometimes easier to turn off all but a few essential subclasses. For example, when you first start to move parts around, you may want only the board outline and the component reference designators and outlines visible. In that case, it is easier to turn off all classes, using the Global Off button, then turn on only the ones you want to be able to see. Some layers and classes/subclasses that you will likely need are

```
Board Geometry/Design_Outline
Areas/Package Keepin
Areas/Route Keepin
Package Geometry/Silkscreen_Top (for component outline)
Components/RefDes/Silkscreen_Top (for component reference designator, R1, U1 etc.)
```

Optional ones are

```
Board Geometry/Dimension
Board Geometry/x_Room (where x is Top, Bottom, or Both)
Areas/Package Keepout
Areas/Route Keepout
```

Controlling net visibility

There are three ways to turn nets on and off. The first method controls global rats visibility and the others control individual rat visibility.

To simultaneously make all rats invisible, click the Unrats All button, ⊞ , and *to make them all visible* again, click the Rats All button, ⊞ .

There are two methods to control individual rat visibility (or the visibility of specific groups of rats). One way to control the visibility of individual nets is to use the Constraint Manager. To launch the Constraint Manager, click the Cmgr button on the toolbar, ⊞ .

Note: The Constraint Manager takes a second to start, but if it doesn't seem to do anything for a long while, check the command window at the bottom. If it says Cannot launch Constraint Manager while a command is active, right click and select Done from the pop-up menu to end the current command. Then relaunch the Constraint Manager.

The Constraint Manager is shown in Fig. 9.45. Select the Properties tab, select the General Properties icon under the Net folder. Locate the net or nets of interest and click inside the cell under the No Rat column. Select the On choice to apply the no rat property and make the net invisible. You can either close the Constraint Manager or leave it open. Changes made in the Constraint Manager are effective immediately.

The last method for controlling the visibility of specific nets is through the use of the Find filter pane. First, click the Unrats All button to turn off all the rats. Next select Display → Show Rats → Nets from the menu. Display the Find pane and check only the Nets box. In the Find By Name list, select Net and click the More... button. In the Find by Name or Property dialog box (see Fig. 9.46), left click the desired nets in the left box to move them to the right box. When you finish selecting the nets, click the Apply button to make the changes take effect. Click the OK button to dismiss the dialog box. Only the selected nets will be visible.

Cross-highlighting between Capture and PCB Editor

When working with large projects, it can be difficult to find specific nets or parts in your board design, or you might want to know where a particular trace belongs in the

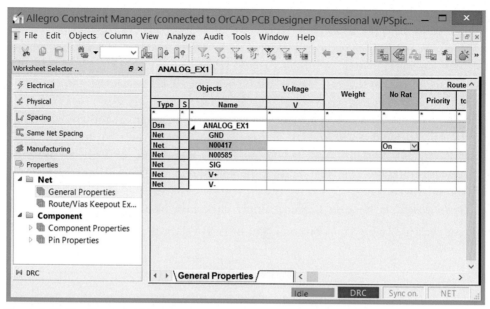

Figure 9.45 The Constraint Manager.

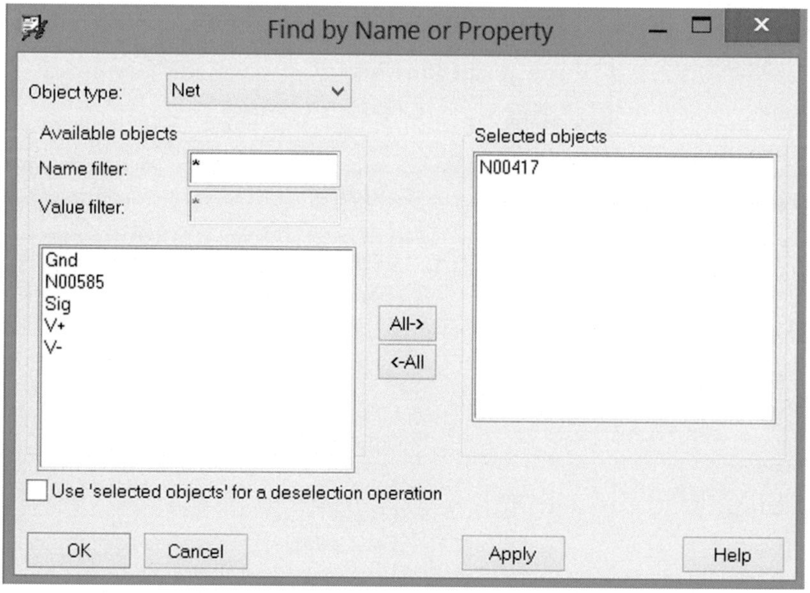

Figure 9.46 Using the Find filter to select specific nets.

circuit. Cross-highlighting allows you to select a part in one application and see it highlighted in the other application.

To enable cross-highlighting, you need `Intertool Communication` enabled in Capture. *To enable* `Intertool Communication` *in Capture*, select `Options` → `Preferences` from the Capture menu bar. In the `Preferences` dialog box (Fig. 9.47), select the `Miscellaneous` tab and check `Enable Intertool Communication`. Click `OK`.

Now if you select a net, a part, or a pin on a part in Capture, the corresponding objects will be highlighted in PCB Editor, so you will be able to apply the PCB Editor commands to them. And, if you highlight a net, a part, or a pin on a part in PCB Editor, using the `Highlight` button, the corresponding object will be highlighted in Capture as well.

To dehighlight an object in Capture, simply click in a blank area in the schematic page, or click the `Dehighlight` button, 🔅, in PCB Editor then click that object.

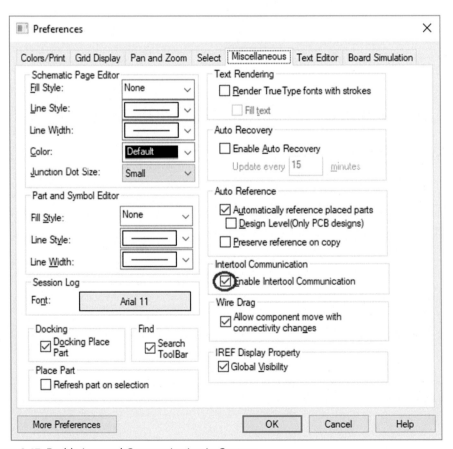

Figure 9.47 Enable Intertool Communication in Capture.

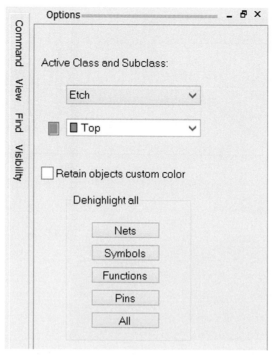

Figure 9.48 Use the Options pane to select types of objects to dehighlight.

To dehighlight an object in PCB Editor, you have to click the Dehighlight button and select that object. Deselecting an object in Capture does not dehighlight the object in PCB Editor. Also, selecting the Dehighlight button does not dehighlight all components; you have to click on each one individually or use the Options filter tab to select certain objects (see Fig. 9.48). Once you select the Dehighlight button, cross-highlighting ceases. To restart cross-highlighting, you have to finish the current command using right click and Done.

Finding parts using the Find filter pane

Once the parts are in their relative positions, you can begin to carefully and precisely place the components. It is often a shuffle game at first, until you get a feel for the parts and the space you have to work with. Besides the cross-selection and cross-highlighting tools, another method for finding parts or nets is to use the Find filter tab. To use the Find filter to highlight a part, display the Find pane (see Fig. 9.49), select the object(s) you want to find (point 2), click the More... button (point 3) to display the Find by name or Property dialog box. In the list of components (or whatever you are looking for), select the item of interest (point 4) to move it to the Selected

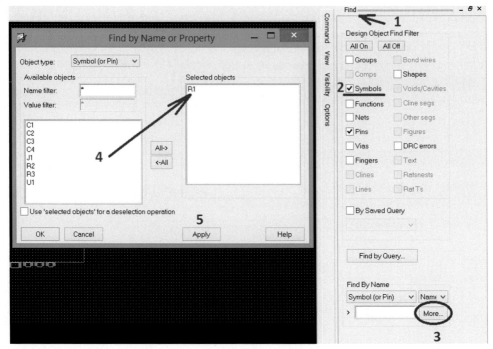

Figure 9.49 Using the Find filter to highlight a part.

objects window and click Apply (point 5). The object will be centered in the screen and highlighted.

Once all the parts have been placed, your design might look something like that in Fig. 9.50.

Some of the reference designators on the Silkscreen layer might be upside down or in an undesirable location. *To rotate the reference designators*, select the Move button, check only the Text box in the Find filter tab. Move your mouse to the Options tab and select the desired angle from the list. Move your mouse back over to the board area and select the reference designator you want to rotate, right click and select Rotate from the pop-up menu. Left click to finish the rotation, then left click again to place the text, right click, and select Done from the pop-up menu. Note: See also the Autosilk tool described in Chapter 10, Artwork development and board fabrication.

Design rule check and status

The next steps in the design process are to define the layer stack-up and begin routing. Before we do that, though, it is a good idea to perform a DRC to check for any problems with the board and how the components are placed. By default the design rule checker is always on, so you do not usually have to specifically perform one unless it

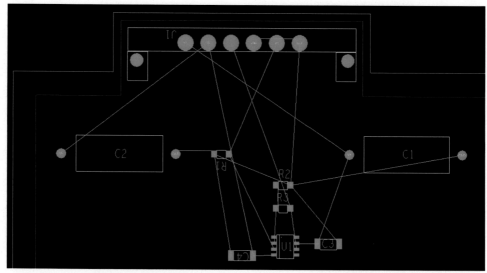

Figure 9.50 The placed components.

is out of date. May be you will want to automatically setup the Manufacturing constraints using Setup → DesignTrue DFM Wizard menu. There are two ways to update the DRC. The first way is to click the Update DRC button, ⬚ , on the toolbar (if that toolbar is displayed). The second way is to use the Status dialog box.

Using the Status dialog box

To check for DRC errors through the Status tool, select Check → Design Status from the menu bar. The Status dialog box (see Fig. 9.51) shows information about how many nets are routed, status of shapes, and if there are any DRC errors. If the DRC Errors box is green, then there are no errors. If the DRC Errors box is red, then there are errors in the design, but the box does not indicate what they are. If the DRC Errors box is yellow, then either the DRC is out of date (click the Update DRC to update it) or there are DRC *warnings*. Warnings may be flagged rather than errors, depending on what the problem is and how the DRC parameters are set in the Constraint Manager (described shortly). To actually view what DRC errors are, you need to generate a DRC report.

Generating design rule check reports

A DRC report is also used to find out the location of DRC errors and warnings as flagged by the Status tool. *To generate a DRC report,* select Export → Quick reports → Design Rules Check Report from the menu bar. As shown in Fig. 9.52, a DRC report tells you how many errors there are, where they are located, what type of errors they are, and who the offending parties are.

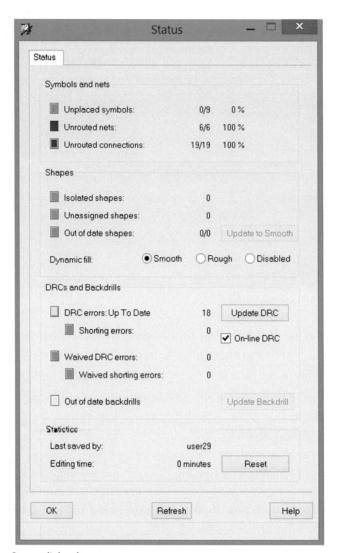

Figure 9.51 The Status dialog box.

You can also generate a DRC report by selecting Export → Reports from the menu bar. A Reports dialog box will be displayed as shown in Fig. 9.53. From the Reports dialog box, you can select the type of report you want to see. It allows you to select additional options, such as generating multiple reports simultaneously and writing reports to files. To select a report, double click on the desired report in the Available Reports section and then click the Generate Reports button. A DRC report is the same with either method.

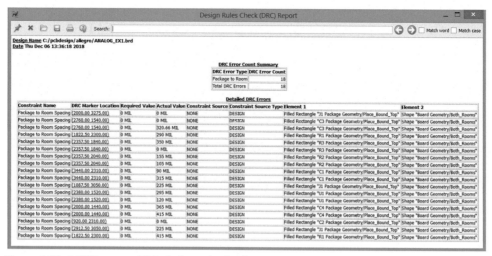

Figure 9.52 A DRC report. *DRC*, design rule check.

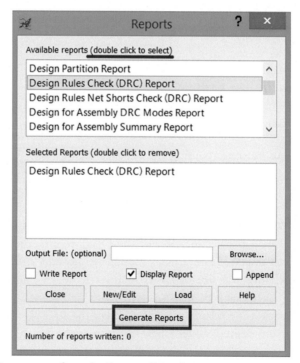

Figure 9.53 Selecting a report from the Reports dialog box.

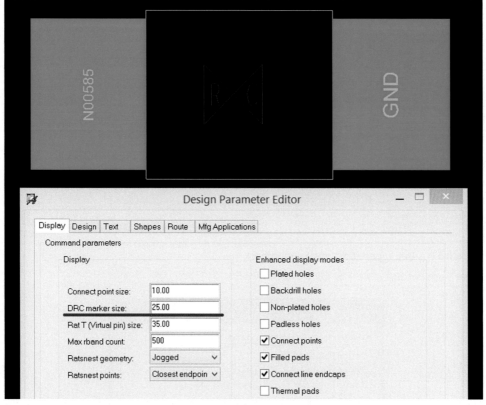

Figure 9.54 A DRC marker (A) and its properties (B). *DRC*, design rule check.

You can create custom reports using the Reports dialog box. An example of how to create a pick and place file is given in Chapter 10, Artwork development and board fabrication.

Design rule check error markers

If you click on the DRC marker location in the DRC report, the error marker will be centered in your display. The DRC marker is bowtie shaped, as shown in Fig. 9.54. You can change the size of DRC markers to make them easier to see. *To change the size of the DRC markers*, select Setup → Design Parameters from the menu bar. In the Design Parameter Editor dialog box, select the Display tab and change the DRC marker size to the desired value. Click Apply to apply the change and OK to dismiss the dialog box.

You can find out additional information about a particular error with the Show Element window. *To display additional information on a DRC marker*, select the Show Element button, ⓘ , on the toolbar, select DRC Errors in the Find filter tab, and then

Figure 9.55 Reviewing a DRC error with the Show Element tool. *DRC*, design rule check.

left click the DRC marker. A text box will be displayed that provides in-depth information about the error, as shown in Fig. 9.55.

Once all DRC errors related to placement are corrected, the layer stack-up can be defined if it has not already been done.

Defining the layer stack-up

The op amp used in this example requires a dual rail supply and at least one Ground plane. Two Routing layers also are needed. Therefore a minimum of five layers is required. Since most multilayer boards are made up of double-sided cores (and therefore, usually have an even number of layers), we use the extra layer as an additional ground plane. The layers could be stacked up in several ways (see Chapter 6: Printed

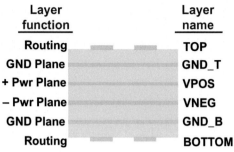

Layer function		Layer name
Routing		TOP
GND Plane		GND_T
+ Pwr Plane		VPOS
– Pwr Plane		VNEG
GND Plane		GND_B
Routing		BOTTOM

Figure 9.56 Layer stack-up for Example 1.

circuit board design for signal integrity, for other examples). The stack-up shown in Fig. 9.56 is used in this example because it provides the opportunity to demonstrate two methods for defining Plane layers.

Setting up the Layout Cross Section

Once you have the parts in place, the next step is to set up the layers. In this design we said that we needed six layers: two power planes, two ground planes, and two routing layers.

First, let's take an inventory of what we have. *To review or modify the layer stack-up*, open the stack-up editor (Cross-section Editor dialog box) by selecting the Xsection button, , on the toolbar. The Layout Cross Section dialog box is shown in Fig. 9.57.

The Layout Cross Section dialog box lists the layers in the stack-up, their type, material properties, electrical properties, and image polarity.

By default the stack-up is defined by the outer surfaces (air), two Routing layers (top and bottom copper), and a dielectric layer (FR-4) between the copper layers. We now add four Plane layers and four Dielectric layers.

To add new pair of layers, that is, dielectric and conductor, right click on the name of any one of the existing layers and select Add Layer Pair Above (or Add Layer Pair Below as appropriate) from the pop-up menu (see Fig. 9.58). Do this four times.

When you have finished, you will have four new dielectric and four new conductive layers. Rename the layers as you want. In this example we use the names GND_T, VPOS, VNEG and GND_B to indicate the purpose of each layer. Convert every other conductive layer to a Negative Plane layer by selecting Plane from the Types section Layer list, and checking the Negative Artwork checkbox in the Physical section. If you don't see the Negative Artwork column, double click on the section header. The finished stack-up is shown in Fig. 9.59. Click Apply and then OK.

Now if you display the Visibility pane, the new layers will be listed too. They all are given a default color, so you can customize your stack-up by changing the colors.

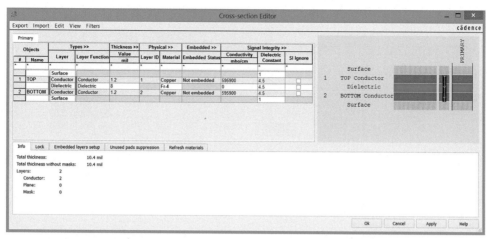

Figure 9.57 The layer stack-up as shown in the Layout Cross Section dialog box.

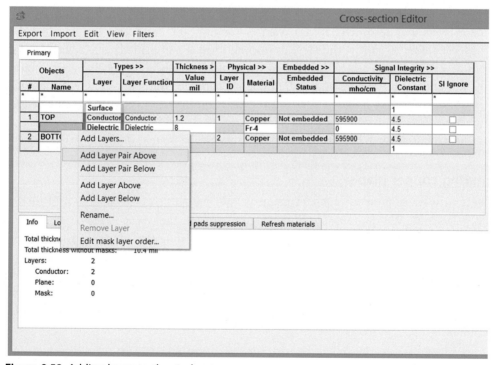

Figure 9.58 Adding layers to the stack-up.

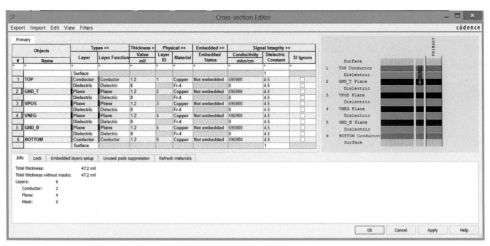

Figure 9.59 The finished layer stack-up.

To change the color of a layer, select the Colors button on the toolbar to display the Colors dialog box. Select the Stack-Up folder; the colors of all conductor and nonconductor layers will be shown. If you want to see only the conductor layers, select the Conductor folder to narrow the view to just the routing and plane layers. Select a color from the palette then select the box in the matrix of subclasses to change its color. Fig. 9.60 shows an example of a custom color scheme for the stack-up. Generally, it is a good idea to select light, neutral, shades for Plane layers, so that they are not too distracting, since they typically cover a large area.

Once the stack-up is defined, the copper planes can be poured onto the proper layers.

Pouring copper planes

The two ways to pour a copper plane are to use the Setup Plane Outline tool or the Add Shape tool. We demonstrate both, beginning with the Setup Plane Outline tool. Afterward how padstacks are connected to the planes using the Add Shape tool will be demonstrated.

Before pouring the copper, set the etch grid to the proper resolution (100 mil is suggested for this exercise). As you are placing pick points, you can pan the view around by using the arrow keys on the keyboard or by pressing the mouse wheel and dragging.

Pouring a copper plane using the setup → Outlines Option

To pour the copper, select the plane layer from the Options pane (Etch/GND_T, for example). Then select Outline → Plane... from the menu bar, as shown in Fig. 9.61. The Plane Outline dialog box will be displayed (see Fig. 9.62).

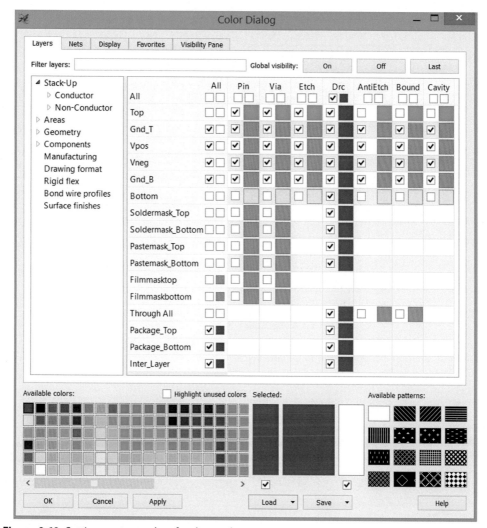

Figure 9.60 Setting custom colors for the stack-up.

Using the Plane Outline dialog box, connect the proper net to the layer by select-ing the Browse... button. In the Select a Net dialog box, select the desired net(GND or GND_SIGNAL in this case) and click OK. If you are making a simple rectangle or box, you can select the Draw or Place Rectangle radio buttons in the Create Options section. Otherwise select the Draw Polygon radio button (do not click OK or Apply yet).

Move your mouse pointer to the design area, left click, and release to place the first vertex of the copper pour plane. If using the Polygon tool, place vertices (corners) to define the outline of the plane. Left click again on the beginning point to complete

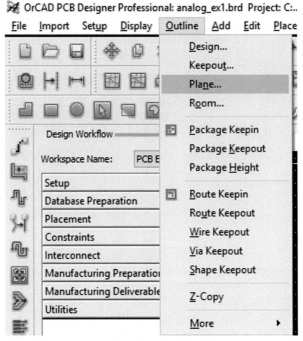

Figure 9.61 Starting a copper pour on a Plane layer.

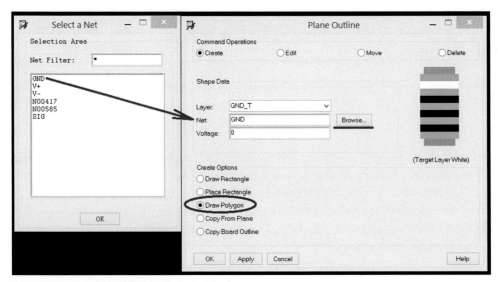

Figure 9.62 Drawing a plane area and assigning a net.

the plane. If you are just drawing a rectangle, then left click at the corner opposite the starting point. Click OK button to close the dialog box.

Before pouring copper areas on the next plane layers, check to see what type of shape was placed on the plane. *To determine what type of shape the copper is,* hover your pointer over the plane border; it should become highlighted, and a data tips box should pop up with a description of the shape. If it does not, make sure that the Shapes option is checked in the Find filter pane. The poured copper will be one of three types and displayed as (1) Filled Rectangle "Class/Subclass, Netname", (2) Shape, "Class/Subclass, Netname", or (3) Shape(auto-generated) "Class/Subclass, Netname". For example, if a copper pour area is dynamic copper on the Vpos layer and is connected to V + net, the data tips pop-up box should say Shape(auto-generated) "V + , Etch/Vpos".

If the copper pour object is the wrong type of shape, you can usually change it without having to delete and redraw it. To change a shape's type, select Shape → Change Shape Type from the menu. Check the Options menu and make sure the To dynamic copper option is selected in the Type: list. Left click the shape to select it then right click and select Done from the pop-up menu. PCB Editor will display the warning Conversion will result in the loss of voids within this shape. — Continue? Click Yes. The object should be converted to dynamic copper and all flashes, route keep-in, and route keep-out areas should be automatically updated. If problems persist, delete the object and redraw the copper pour area using the Add Shape tool, as described next.

Pouring a copper plane using the Shape Add Tool

To pour a copper plane using the Shape tool, select the target plane layer from the Options pane (Etch/Vpos, for example). Then click the Shape Add (to place a polygon) or the Shape Add Rect button (to place a rectangle). Go to the Options pane again and make sure the shape fill type is Dynamic copper. In the Assign net name: area, click the Browse... button to display the Select a net dialog box, as shown in Fig. 9.63.

Left click the net to select it (e.g., V +) then click the OK button to dismiss the dialog box. Left click in the work area to place each vertex of the plane outline. When you reach the last vertex, right click and select Complete from the pop-up menu. The outline will be drawn but still highlighted with the editing handles (boxes) visible. You can modify the shape if necessary or right click and select Done from the pop-up menu. Display the data tips box again to make sure the correct shape was drawn.

Once the first plane outline (copper pour area) is drawn, you can copy and paste it onto the other layers rather than drawing each one individually. *To copy a plane outline,* start a new outline (Outline → Plane... from the menu). As described previously, select the target layer (e.g., GND_B) and assign a net to the plane using the Browse... button. In the Create Options area, select the Copy from Plane radio button, as shown in

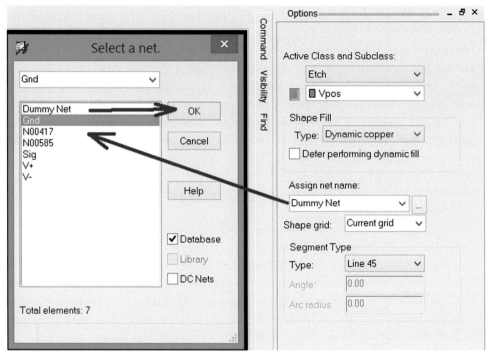

Figure 9.63 Drawing a copper pour shape and assigning a net.

Fig. 9.64. Select the existing plane layer from the list. At the command prompt it will say select a plane to copy. Select the plane on its edge. This is done since there could be more than one plane on a layer and PCB Editor needs to know specifically which plane to copy. Click OK when you have selected the plane. You should now have two Routing layers (top and bottom) and four Plane layers (two Ground and two Power planes). Again, make sure that the new copper pour areas are dynamic copper shapes, using the data tips box, and that connectivity is made from each plane to the correct pins as described next. You can also use the Copy Board Outline radio button which will allow you to create the plane shape as a copy of Board Geometry/Outline shape with a desired offset of its boundary.

Verifying connectivity between Pins and Planes

A padstack that connects to a plane usually does so through a thermal relief, while padstacks that are not connected to a plane are isolated from the plane with a clearance area around the padstack. Now that the planes are in place from the previous step, we need to make sure that connections to the plane and isolations from it are done properly. If you are using a version of PCB Editor that has had its libraries modified, then some of the following discussion may not apply. The original library that comes with

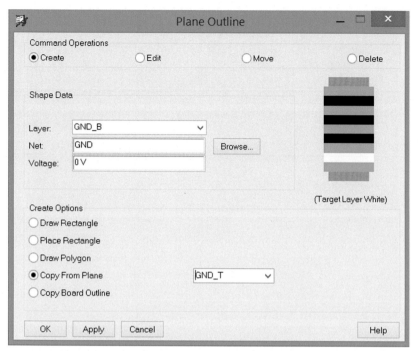

Figure 9.64 Drawing a new plane outline by copying an existing one.

PCB Editor contains some parts that lack the proper thermal reliefs and clearances assigned to the padstacks. If you zoom into J1 and turn off all the layers except for the top Ground plane, you should see one of the four views shown in Fig. 9.65. If you see the bottom one, then the next section does not apply to you, so you can skip it and go to the section after that.

If you see the view shown at the top of Fig. 9.65, you need to change the display options before you can see the thermal reliefs. To show thermal reliefs, select Setup → Design Parameters from the menu. At the Design Parameter Editor, select the Display tab. In the Enhanced Display Modes area, check the Thermal pads option (see Fig. 9.66). Click Apply and then OK.

When viewing thermal pads is enabled in the Design Parameter Editor, you should see either the third or fourth image in Fig. 9.65. If you see the first or the second image, then the plane outline is the wrong type, or you didn't check the Negative Artwork checkbox in the layer stackup for this layer. See the previous section to determine the type of outline and how to change it if it is wrong.

If you see the third image, then no thermal flashes have been assigned to the padstack. The next section explains how to modify your design to use the thermal reliefs and clearances.

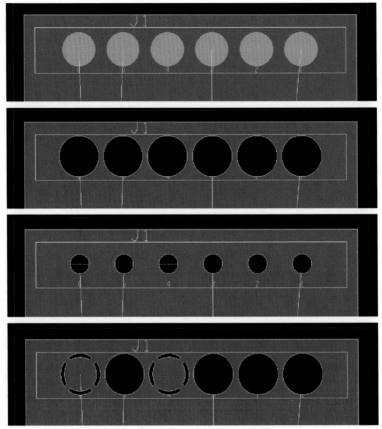

Figure 9.65 Four possible views after placing a new plane outline. (A) layer artwork type should be changed to negative, (B) shape type should be changed to dynamic copper, (C) no thermal flashes were assigned, (D) correct negative layer is set with thermal flashes assigned.

If you see the bottom view, then your pads have been set up with thermal flashes and clearances, so you can skip down to the "Defining trace width and spacing rules" section.

Adding thermal reliefs and clearances to padstacks

To add the thermal reliefs to or change clearances on a padstack, you need to modify the padstack design using the Padstack Editor. You can modify just the padstack in the design or you can modify the padstack definition located in the *symbols* library. We will modify just the design padstack in this example. *To launch the Padstack Editor from PCB Editor*, select Tools → Padstack → Modify Design Padstack... from the menu bar. Move your pointer to connector J1, right click on pin 1, and select Edit from the pop-up menu. The Padstack Editor dialog box will be displayed. Select the Design Layers tab as shown in Fig. 9.67.

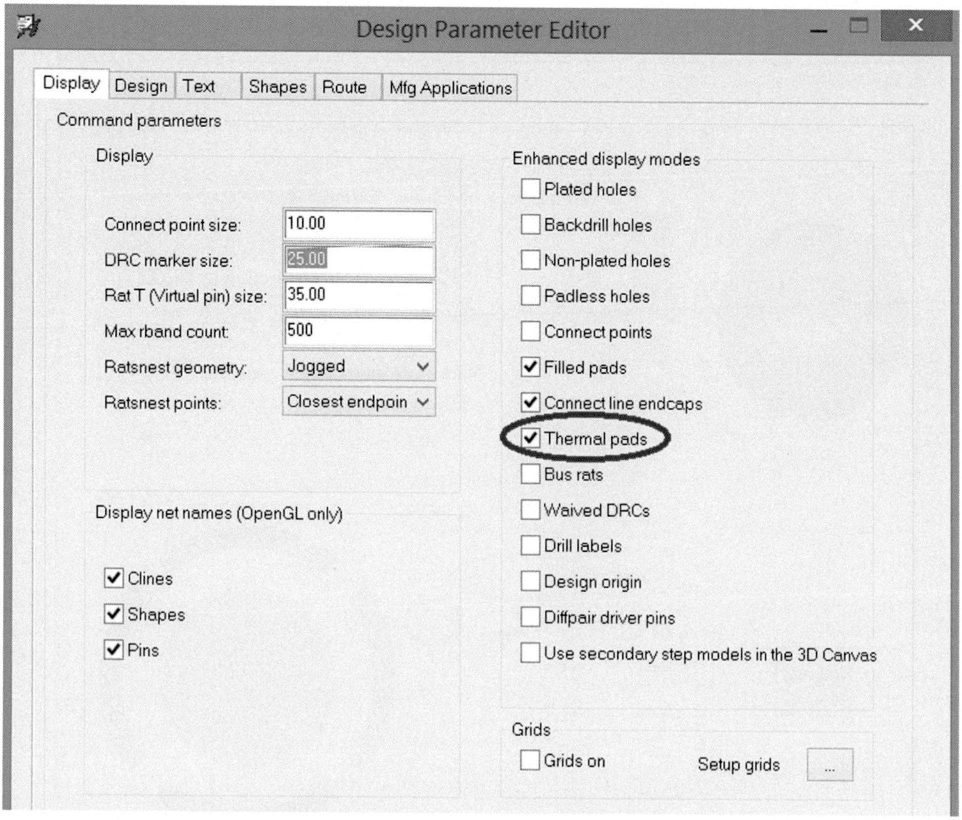

Figure 9.66 Use the Design Parameter Editor to show thermal reliefs.

On each of the layers the `Thermal Pad` and clearance `Anti Pad` needs to be changed. Select a layer row and `Thermal Pad` cell by clicking on it. Select `Flash` from the list of available geometries and either type `F2_3X2_7` in the flash symbol name box or use the `Browse (...)` button beside it to use the `Library Shape Symbol Browser` dialog box to choose from a list as shown in Fig. 9.68.

Note: To do this, you must have the `F2_3x2_7` flash symbol copied to the PCB Editor *symbols* folder with the other footprints or in the working allegro folder. You can copy the one from the OrCAD examples located in the `\share\pcb\toolbox\getting_started\panelization\pcb_context\symbols` folder in the OrCAD file path or from the website for this book. Or you can click `Create New Flash Symbol` button, assign the symbol name (e.g. `F2_3x2_7`), and use `Add → Flash` in the flash symbol editor window to create the new thermal flash symbol with desired shape and size (set Inner diameter 90 mil / 2.3 mm, Outer diameter 106 mil/2.7 mm and Spoke width 20 mil/0.5 mm). Save the created flash symbol.

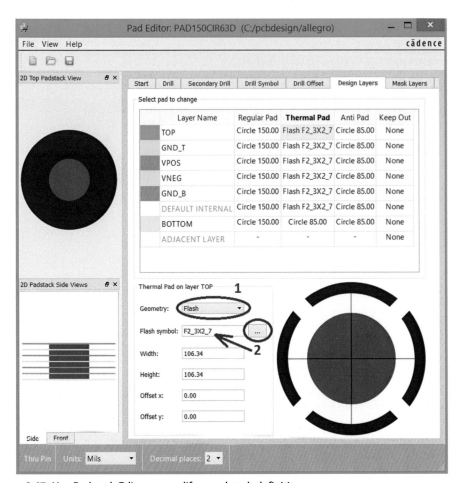

Figure 9.67 Use Padstack Editor to modify a padstack definition.

Assign the flash symbol and change the Anti Pad clearance to 155 on each layer. Once the changes have been made, select File → Update to Design… from the menu bar. Close the Padstack Editor to return to the board design. Right click and select Done from the pop-up menu. You should see the thermal reliefs now (make sure you have the Thermal pads box checked in the Design Parameters dialog box).

Editing one pin actually edits all pins with the same name (even ones on different parts), so as you make each plane layer active, you should see the various clearances and thermals for the pins connected to that layer. If you want to change only one pin when there are many of that type of pin, select Tools → Padstack → Modify Design Padstack… from the menu bar as before. Left click on the pin and in the Options pane select the Instance radio button. PCB Editor will automatically append a number to

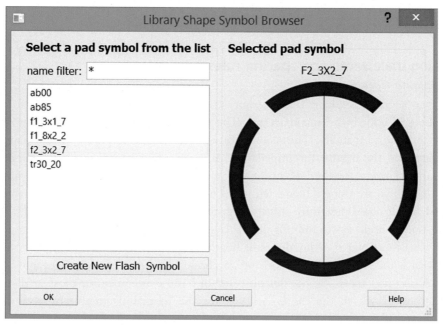

Figure 9.68 Selecting a flash symbol for a pin.

the name of the padstack to differentiate it from the others. Click the Edit button at the bottom of the pane to display the Padstack Editor.

You will probably need to change the padstacks on part symbols C1 and C2, which could use flash symbol F1_8X2_2 from OrCAD examples. The remaining components in this design are surface mount and have no internal layers. And as we will see shortly, the vias are full contact type and do not use thermal reliefs.

Unless you modify the actual library parts, you will have to do this every time you start a new design and use the parts in the library. Instead of selecting the Update design option, you can select the Save as option prior to closing the Padstack Editor. This saves a copy of the padstack (*name*.pad) in the working directory. You can copy the padstack and put it in the *symbols* library, so that it can be used in other projects or you can select Tools → Padstack → Modify Library Padstack... from the menu bar to modify it directly in the *symbols* library. You need to know the name of the padstack (s) that you want to modify to use this option. Over time many of the parts in the library will be modified, and you should rarely have to do these modifications. However, you may want to back up the original library before you make a bunch of changes to it, and it is a good idea to make a backup of custom padstacks, flash symbols, and footprints that you develop as well.

Once the plane layers are properly set up, you can fan out the surface-mount components to their respective Plane layers. To do this however, we need to assign or verify which via will be used for fan-outs and routing and that trace width and spacing requirements have been defined, as described in the next section.

Defining trace width and spacing rules

Determining trace width

The trace width depends on three design considerations. The first consideration is the capabilities of your board manufacturer. The traces need to be wider than their minimum fabrication capability. The second consideration is the required current handling capability, and the third is the impedance. According to the data sheet, the short circuit output current for the 741 op amp is 64 mA. If we use a safety margin of, say, 100 mA and we are using a board with 1 oz copper, then per Eq. (6.17) in Chapter 6, Printed circuit board design for signal integrity (or the graph from Fig. 6.38), the minimum trace width is 1.3 mil for inner layers and 0.5 mil for outer layers. From Table 6.8 in Chapter 6, Printed circuit board design for signal integrity, you can see that, with 6-mil traces, you can run up to about 300 mA on inner traces and about 600 mA on outer traces. So the default value of 5 mil is adequate for most small-signal applications. However, the chosen width must also be manufacturable by your board house. Five mil is fairly narrow for many board houses, so we will change the widths.

For this example, we specify that all signals be of very low frequency (~ 20 kHz) so that trace impedance is not a concern. An example of designing controlled-impedance transmission lines is given later in the digital design example (Example 4).

Trace width and spacing rules are set using the Constraint Manager. *To set trace width rules*, open the Constraint Manager by clicking the Cmgr button, ![CM], on the toolbar. Select the Physical worksheet and select the All Layers icon in the Net folder, as shown in Fig. 9.69. The default minimum width for all nets is 5.00 mil and

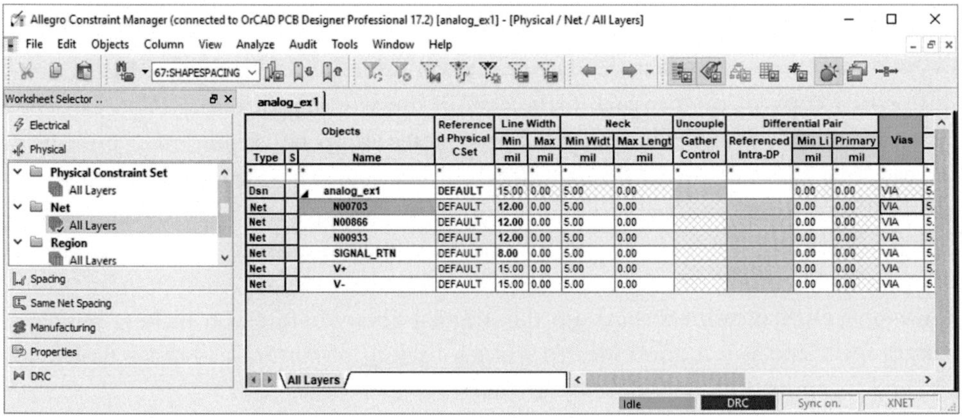

Figure 9.69 Use the Constraint Manager to set trace width rules.

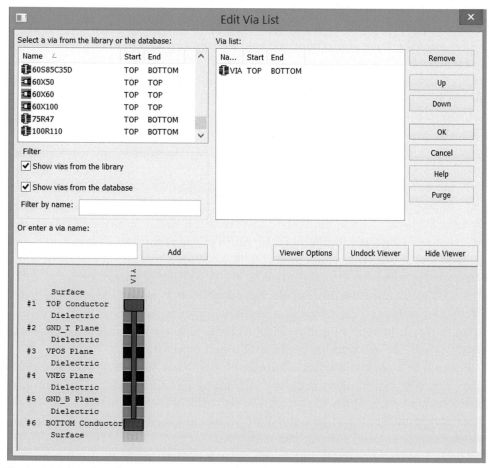

Figure 9.70 Use the Edit Via List dialog box to establish the allowed vias.

maximum is set at 0.00 (which means there is no maximum limit). Set the line widths to different values and note the differences when the board is routed later. Changes made in the Constraint Manager are effective immediately.

Assigning vias

Notice also in Fig. 9.69 that you can *specify particular vias for each net* on the Physical worksheet of the Constraint Manager. To select a different via, left click inside the cell for the via you want to change. In the Edit Via List dialog box shown in Fig. 9.70, select the via that you want to use on that net and remove unwanted via(s). Click OK when your selection is complete. If you need a via that is not listed, you can make a custom via using the Padstack Editor, which is discussed in Chapter 8, Making and editing footprints.

Complete PCB Design Using OrCAD® Capture and PCB Editor

Determining trace spacing requirements

The op amp used in this example requires a dual rail supply, which can be up to ± 15 V (and the input and output voltages must be between the rails). The greatest voltage difference possible between any two traces then is 30 V peak to peak, so the spacing between the traces must be able to support this. If we decide to play it safe and use the 31−50 V range and we plan on using a soldermask, then, using Table 6.8 in Chapter 6, Printed circuit board design for signal integrity, the route spacing should be greater than 4 mil on internal layers and greater than 5 mil on external layers. Again the selected spacing rules must fall within your board manufacturer's capabilities also.

To set the spacing rules, select the Spacing worksheet in the Constraint Manager. You can set spacing rules by layers, by nets, or by region. *To set the rules by layers*, select the All Layers icon in the Spacing Constraint Set folder. In the Objects column, you can double click the + sign near the name of constraint set to expand and view the entire list of objects included to DEFAULT Constraint Set (layers in this case). *To set the rules by net*, select the All Layers icon in the Net folder, as shown in Fig. 9.71. To set specific rules (e.g., line-to-line spacing), double click on the Line to >> in the table header to expand or contract the available object types. In Fig. 9.71, Line To is circled in red at the top of the spreadsheet. The spreadsheet is a matrix that gives you control over all the spacing rules between all object types.

It would be a good idea to add all similar Nets to Net Classes (e.g., you may create power nets class, digital-signal nets class, and analog signal nets class). Then you may create the appropriate CSets (constraint sets that will contain the set of rules specific for each type of Nets), and then apply the CSets to those Net Classes, rather than applying values, or CSets even, to individual Nets.

Once all of the trace width and spacing rules are set we can begin routing fan-outs and traces.

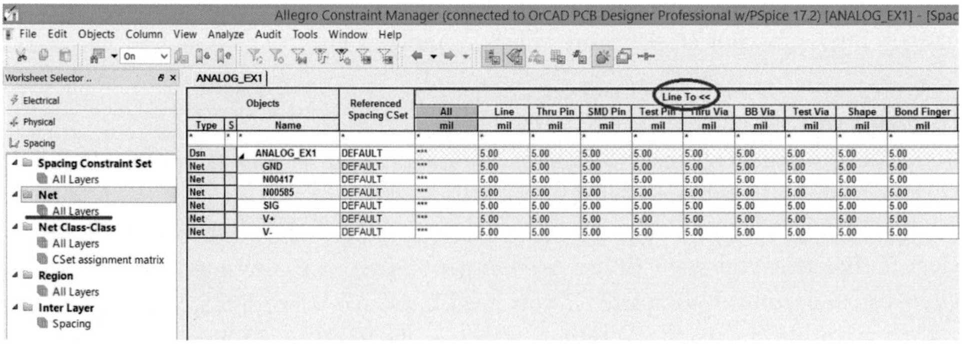

Figure 9.71 Setting trace spacing rules in the Constraint Manager.

Prerouting the board

Routing fan-outs for power and ground

We want to fan out only power and ground, so we need to disable all the other nets. In particular, we start with the GND net on the ground layers. First, we make the general nets invisible by turning on the `No Rat` property. To make nets invisible, open the `Constraint Manager` and select the `Properties` tab and select the `General Properties` icon under the `Net` folder. *To make a net invisible*, select the `On` option in the `No Rat` column as shown in Fig. 9.72. *To disable a net from being routed*, select the `On` option in the `No Route` column in the `Route Restrictions` section. You can set a property on many nets simultaneously by dragging a box across the nets then setting the property. For this example, make all nets invisible except for the GND net. For demonstration purposes, we also want to prevent the power nets from being fanned out, so the `No Route` restriction is applied. Note that the signal nets will not be fanned out because we can control that by configuring the autorouter as shown next.

Select `Route` → `PCB Router` → `Route Automatic` from the menu bar. Select the `Routing Passes` tab in the `Automatic Router` dialog box (see Fig. 9.73). Select the `Fanout` box and deselect the other boxes.

Click the `Params...` button to display the `SPECCTRA Automatic Router Parameters` dialog box (see Fig. 9.74). Here we can control what type of nets get fanned out, in what direction the fan-outs are routed, and how pins and vias are shared. Note that not all options are functional in the OrCAD version of PCB Editor. Under `Pin Types` check the `Specified` radio button and the `Power Nets` box. This allows only nets assigned to Plane layers to be fanned out and, because the V + and V − nets have a `No-Route` restriction (in Constraint Manager), only the GND net will be fanned out to the ground planes. Click `OK` to dismiss the `Router Parameters` dialog box. Click the `Route` button on the `Automatic Router` dialog box to begin the fan-out. Only the GND net should be fanned out.

To complete the fan-outs for the power nets, go back to the Constraint Manager and clear the no-route restriction; repeat the fan-out procedure for the V + and V − nets.

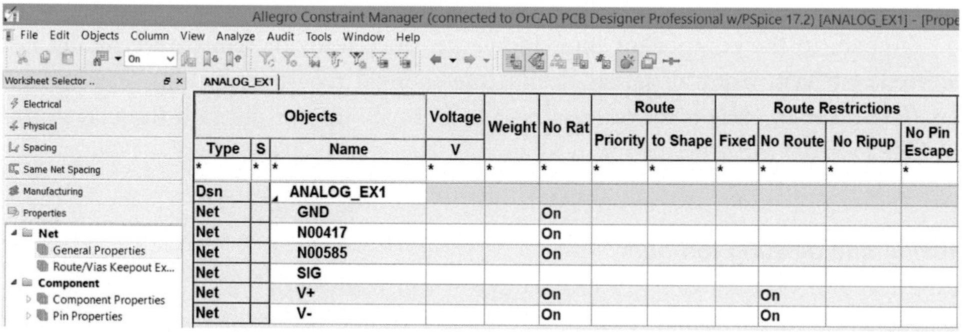

Figure 9.72 Using No Rat and No Route to control net routability.

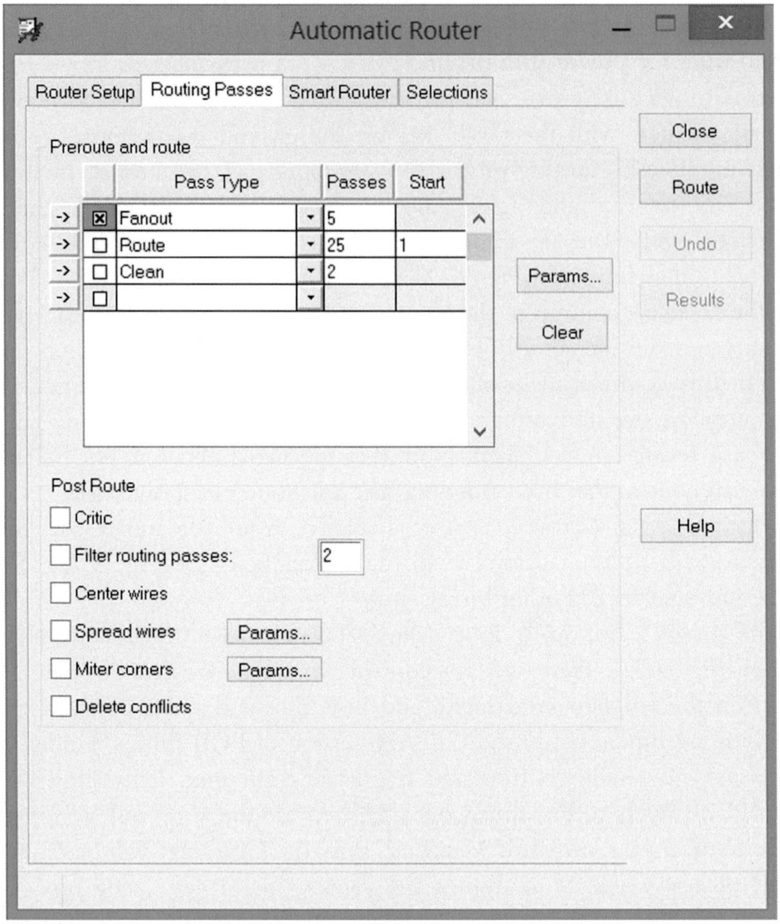

Figure 9.73 Using the autorouter to perform fan-outs.

If you look closely at the vias used for the fan-outs (and routing by default), you will notice that they have no thermal reliefs to the Plane layers. If you want thermal reliefs for the vias, you can add them using same procedure as described previously to add thermals to the component pins. However, since leads or wires are typically not soldered into vias, many designers do not use thermal reliefs on routing and fan-out vias. There are arguments for both sides, but thermals are not used on the vias in this example.

Moving and deleting fan-outs

If the location of a fan-out is a problem, you can easily move it. *To move a fan-out*, select the Slide tool, , left click on a via to select it. Move the pointer to where

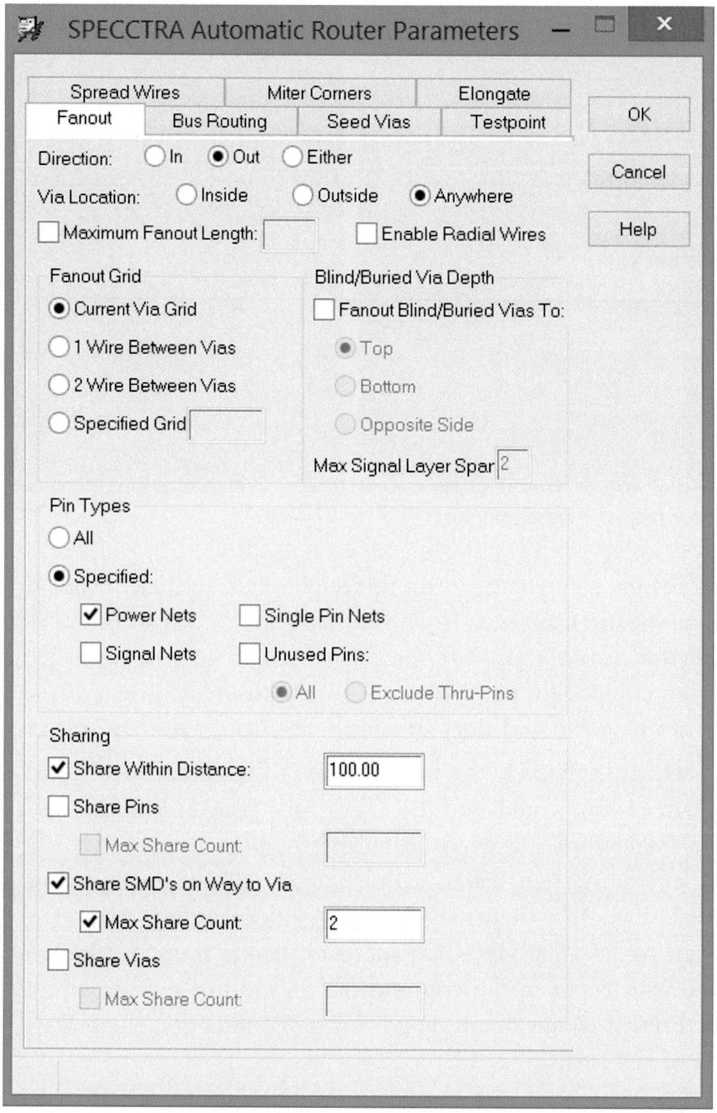

Figure 9.74 Setting up fan-out options.

you want the via and left click to place the via. Set the etch grid to smaller values to allow smooth slide.

If for some reason you want to delete a fan-out and route it manually, you can do that too. *To rip up a fanout,* select the `Delete` tool, , from the toolbar, select only the `Net` in the `Find` filter, and select only `Symbol Etch`, `Clines` and `Vias` in the `Options`

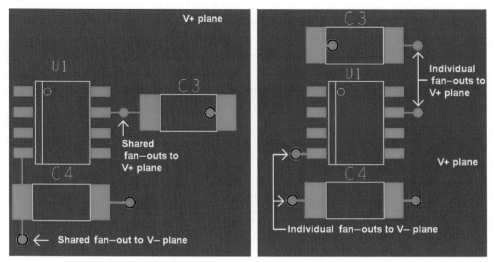

Figure 9.75 Three fan-out methodologies.

tab. Click the net trace you want to rip up. Right click and select Done from the pop-up menu when you are finished.

To manually redo a fanout, use the Add Connect tool, , to route a trace and fan-out via from the component pin (on the top or bottom layer) to a Plane layer. Make sure that the net is visible and does not have a No-Route restriction on it. Select the Add Connect tool, and left click the rat's nest near the component's pin. Left click at the location where you would like the via. Right click and select Add Via from the pop-up menu. Right click and select Done from the pop-up menu.

Fig. 9.75 (left) shows two shared via techniques, where the variation is a result of the via location. Fig. 9.75 (right) shows using individual vias instead of shared vias, where each part has its own vias. This approach makes routing a little easier, since it leaves an open path between the components on the top routing layer. As described in Chapter 6, Printed circuit board design for signal integrity, there are arguments for all three types of fan-outs.

At this point, it is a good idea to update the DRC. After all errors are fixed (there should be none), we can route the rest of the board.

Manually routing traces

Traces (clines) can be routed manually or by the autorouter. We begin by looking at manual routing.

To begin manually routing traces

Add Connect tool

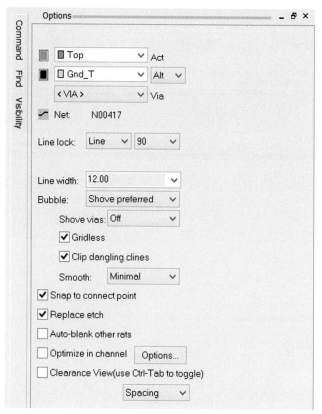

Figure 9.76 Using the Options tab to control manual routing properties.

To manually route a trace, select the Add Connect tool, 🖱 , left click on the desired trace, and left click again to place a vertex and route the trace. The default trace characteristics will be determined by the Constraint Manager. You can make changes to the trace as you are routing by using the Options tab (see Fig. 9.76), where you can change routing layers, line width, and other properties as shown in the figure.

Locking (fixing) traces

Once the fan-outs and specially routed traces are completed, you can lock (fix) them to prevent them from being inadvertently unrouted or shoved by the autorouter. *To fix traces*, select the Fix button, 🖱 , from the toolbar and left click the desired trace.

To unlock (unfix) a trace, select the Unfix button, 🖈 , and left click a trace to unfix it.

To find out if a trace is fixed or not, use the Show Element button, ⓘ , and left click on a trace to find out the status. You can also use the Constraint Manager to fix traces

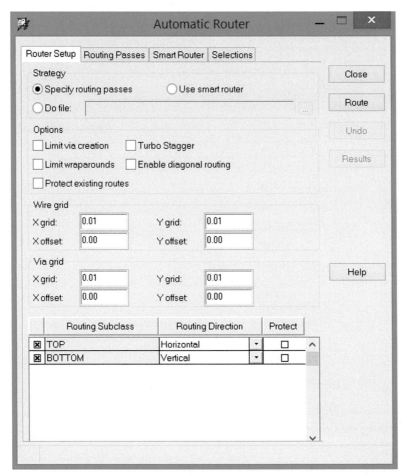

Figure 9.77 Setting up the autorouter strategy and options.

(select the `Properties` tab and set the `Fixed` properties in the `Route Restrictions` section).

Running the autorouter

Once the fan-outs and manual routing tasks are completed, you can quickly route the rest of the board using the autorouter. *To start the autorouter*, select `Route → PCB Router → Route Automatic` from the menu bar. As shown in Fig. 9.77 (`Route Setup` tab, `Strategy` options area), there are three ways to use the autorouter: (1) specify routing passes, (2) use the smart route, and (3) set up a `SPECCTRA` (Allegro) `DO` file. You use the `Route Setup` tab on the `Automatic Router` dialog box to select which method to use. Methods 1 and 2 are both easy to use, method 3 is not discussed here. The smart

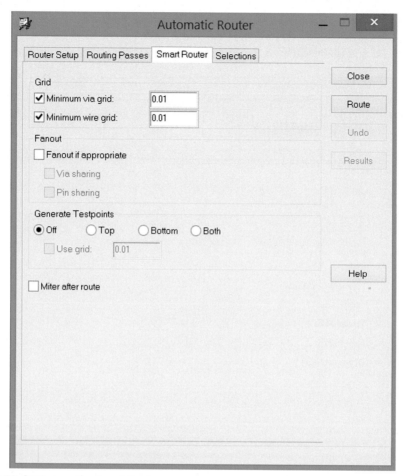

Figure 9.78 Setting up the Smart Router.

router has fewer options to set than the routing passes method, but the routing passes method gives you more control over the autorouter. Methods 1 and 2 are described next. Click the radio button for the method you want to use then click the corresponding tab to change the route settings for that method.

Use the Smart Router tab to set up the autorouter for that method. As shown in Fig. 9.78, you need to specify a minimum set of options, and they are pretty straightforward.

With the routing passes method, you have greater control over the routing process. In addition to the settings on the Router Setup tab and the Routing Passes tab, you also have access to the Params... dialog box, as was shown during the fan-out description. Some of the features available in the routing passes method are used in later examples. For this example, just uncheck the Fanout box (since we have already done

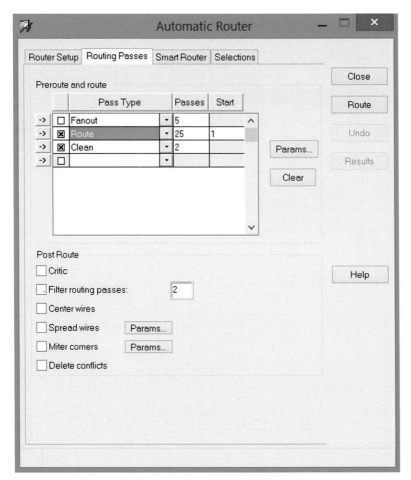

Figure 9.79 Setting up the Routing Passes options.

that) and check the Route and Clean boxes. Click the Route button to finish the board (Fig. 9.79).

If your design is large and complicated, you can limit routing to specific sections of the board using the Selections tab filter (see Fig. 9.80). As shown, you can select (or restrict) specific nets for (from) routing. Using the Selections tab in addition to the Constraint Manager No-Route restrictions gives you extensive control over the autorouter.

Finalizing the design

Postrouting inspection

Depending on how you placed your parts, your board might look something like Fig. 9.81 (not all layers shown). After the autorouter has finished routing the board, it should be checked for problem areas, such as acute angles, long parallel traces, and bad

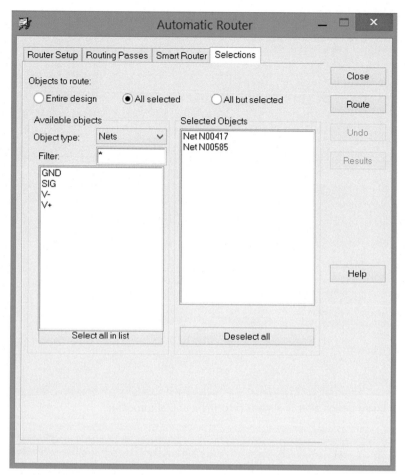

Figure 9.80 Selecting which traces to route.

via locations. Chapter 6, Printed circuit board design for signal integrity, discusses several of these and other routing issues. It is also necessary to check for new DRC errors that may have been caused by the routing process and check for any nets that the autorouter could not route.

Checking routing statistics

Once you have all the traces routed, you can check for new DRC errors and unrouted nets using the Status dialog box (Fig. 9.82). *To view routing statistics*, select Check → Design Status... from the menu bar. The Status dialog box provides information about the number of unrouted nets as well as unplaced components.

If the routing statistics are less than 100% but you cannot easily locate unrouted nets, use the Visibility tab to turn off all Etch layers and make sure that the Rats All

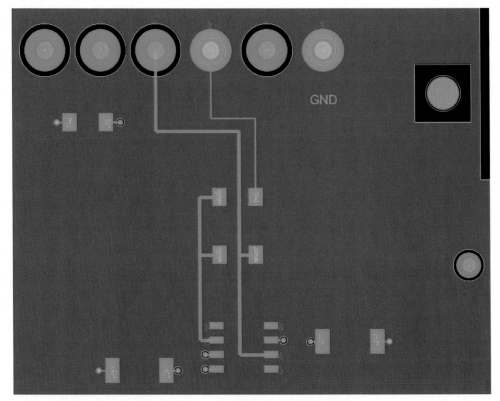

Figure 9.81 Board design after final parts placement and autorouting.

button is active, . Use the Find filter (look for nets only) and the Show Element tool and drag a box across the board area. Any unrouted nets should be displayed in the Show Element text box with information about the net's name, to which pins the net belongs, and its location. An example is shown in Fig. 9.83.

Cleaning up a design by glossing

After the autorouter has finished routing the board (or after performing any last-minute manual routing tasks), you can have PCB Editor go over the design and perform a cleanup using the Gloss command. *To clean up (gloss) a board design*, select Route → Gloss → Parameters... from the menu. At the Line Smoothing dialog box (see Fig. 9.84), check the options you would like PCB Editor look for and fix. When you have made your selections, click the Gloss button.

You can attach either the NO_GLOSS property or the FIXED property to the nets so that gloss does not modify the routing of these nets. Use the Constraint Manager to apply the NO_GLOSS property to a net. Select the Properties tab and the

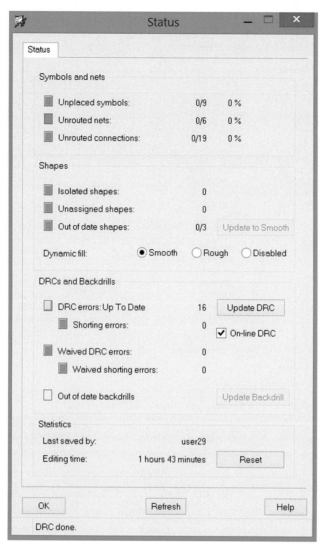

Figure 9.82 Use the Status dialog box to check for unrouted nets and DRC errors. *DRC*, design rule check.

Net/General Properties icon then set the No Gloss property to On. To apply the FIXED property, you can also use the Constraint Manager or use the Fix button on the toolbar.

Synchronizing the design with Capture (back annotation)
Once the board design is finished, you can export information from your PCB design to your schematic by performing a back annotation. Back annotation in PCB Editor is

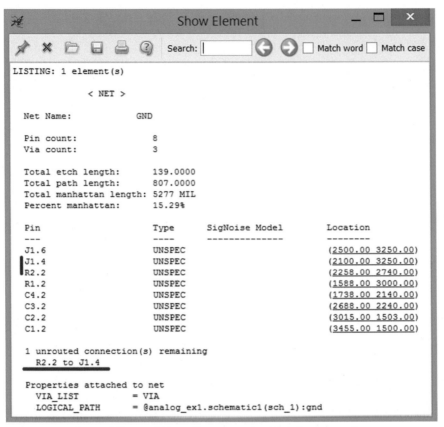

Figure 9.83 Using the Find filter and Show Element tool to identify unrouted nets.

called feedback (search for it in *PCB Editor Command Reference*). Also see *Backannotating to Allegro Design Entry HDL or System Connectivity Manager* in *Transferring Logic Design Data*, and the *Capture User's Guide*. See *Back annotation from PCB Editor*. To find the needed information, select Help → Documentation to run Cadence Help, press Show Documentation Browser icon, and choose the folder Allegro Design Entry CIS → User Guides → OrCAD CIS User Guide.

Information that is back annotated has to be common to PCB Editor and Capture. To see some examples of these properties, go to the schematic page and double click a part to display the properties spreadsheet. Select the Cadence-Allegro filter. Some examples are HARD_LOCATION, NO_SWAP_GATE, NO_SWAP_PIN. The configuration file allegro.cfg, which is located in the C:\Cadence\SPB_17.2\tools\capture by default, lists all of the properties included during back annotation.

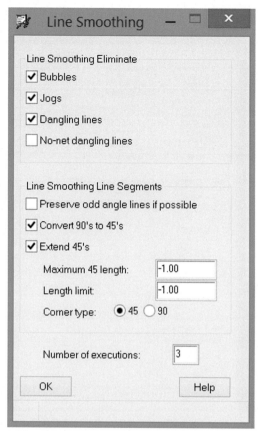

Figure 9.84 Use the Gloss command to add finishing touches to the board design.

To perform a back annotation from PCB Editor to Capture
- save your PCB design and exit from PCB Editor.
- open you schematic project in Capture and select the design icon.
- select Tools→Back Annotate (Fig. 9.87).
- check the path to your board file in the PCB Editor Board File window.
- check the path to your Netlist file location. The Netlist file is used to store the exchange data between Capture and PCB Editor.
- press Setup and enter the path to the configuration file allegro.cfg (Fig. 9.88).
- check on the box Update Schematic in the bottom of the dialog.
- press OK.

You can also create the back annotation netlist directly from PCB Editor. *To create back annotation netlist files directly from PCB Editor*, select Export → Back Annotation Netlist... (Fig. 9.86). You may choose the default schematic editor type in User Preferences (Fig. 9.85). Then you will be able to import the generated back

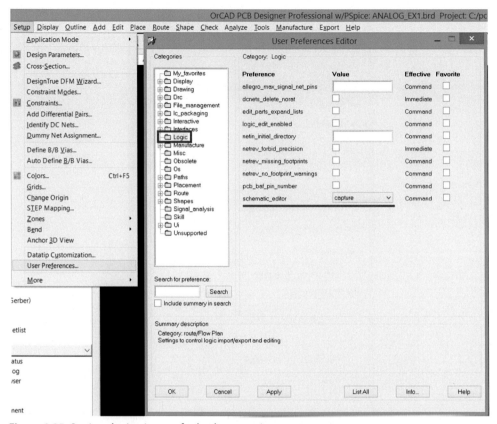

Figure 9.85 Setting the Logic type for back annotation.

annotation netlist files into OrCAD Capture schematic editor at another workstation. To see what changes were made, double click a part in the schematic to view the properties sheet. Select `Cadence-Allegro` from the `Filter by:` list. Fig. 9.89 shows the trace width properties that were set in the Constraint Manager in PCB Editor (A) are imported into the `Edit Properties` spreadsheet in Capture (B).

This completes the first example. At this point, you would produce the manufacturing files for the board and send them off to a board house for manufacturing. We look at this in greater detail in Chapter 10, Artwork development and board fabrication.

Example 2. Mixed analog/digital design using split power, Ground planes

This example shows how to design a circuit with a mixture of analog and digital parts, multiple power planes, and a single Ground plane split into analog and digital sections that have a common reference point.

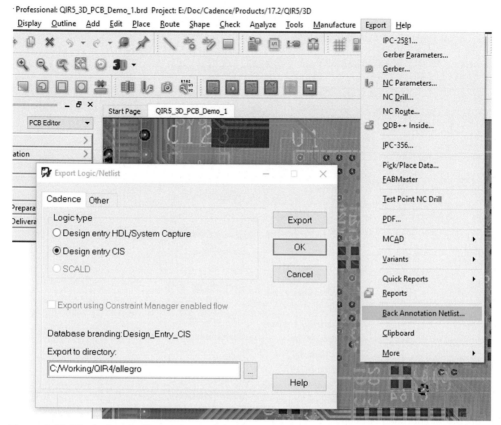

Figure 9.86 Displaying the Export Logic dialog box.

Mixed-signal circuit design in Capture

Fig. 9.90 shows the circuit design example. The circuit consists of a mixture of analog and digital parts, multiple power sources, and separate analog/digital Ground planes. It has an off-board connector, an analog signal conditioning circuit (the op amp U1), an analog-to-digital converter (ADC) (U2), a digital microcontroller (U3), and a digital serial-to-parallel shift register (U4). Note that the bypass capacitors have been omitted to accommodate the part count limitations of the demo version of the software.

The circuit has five global power nets, which include analog power (V + , V −) and analog ground (AGND) for U1 and U2, and 5 V digital power (VCC) and digital ground (GND) for U3 and U4. Analog and digital grounds are often kept separate to help prevent digital switching noise from affecting the analog ground. But, for the circuit to work (especially when using analog-to-digital and digital-to-analog converters), the two grounds must have a common reference point. The two grounds are tied together through a low-impedance, single point connection at the connector.

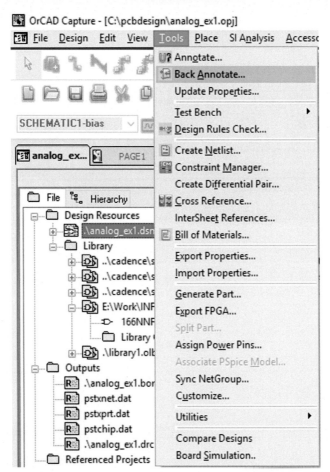

Figure 9.87 Performing back annotation in Capture.

Figure 9.88 Setting up the back annotation path and allegro.cfg file.

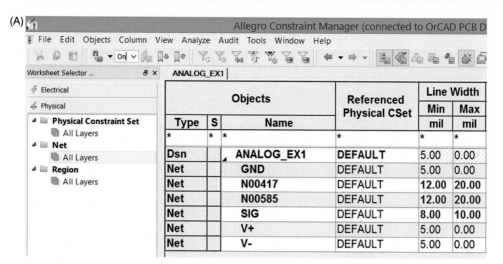

Figure 9.89 Net properties back annotated to (A) PCB Editor (B) Capture from.

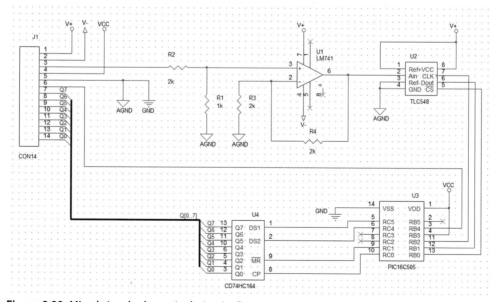

Figure 9.90 Mixed-signal schematic design in Capture.

Table 9.4 Design parts list and schematic and footprint libraries for design Example 2.

Reference	Part	Source library	PCB footprint
J1	CON14	CAPTURE/LIBRARY/CONNECTOR.OLB	conn14
R1	R 1k	CAPTURE/LIBRARY/DISCRETE.OLB	smdres
R2, R3, R4	R 2k	CAPTURE/LIBRARY/DISCRETE.OLB	smdres
U1	LM741	ANALOG_EX1.DSN	soic8
U2	TLC548	CHAP9EXAMPLES.OLB	soic8
U3	PIC16C505	CHAP9EXAMPLES.OLB	dip14_3
U4	CD74HC164	CHAP9EXAMPLES.OLB	soic14

The procedures and tools described in the first example are used to start a new project; place the parts onto the schematic page, connect the parts, and assign footprints. Table 9.4 lists the parts used and the footprint assignments. The conn14 footprint symbol is a modified version of the CONN20 included with the software (it is not included with the demo version); a copy of the modified symbol is included on the website that accompanies this book. You can either get it from the website, or create it yourself from scratch.

There are a couple of significant differences between this and the last example. The differences are as follows:

1. All analog parts and some digital parts have explicit connections to power and ground nets, while the shift register is connected to the digital power and ground globally, as described later.
2. Two different ground symbols (AGND and GND from CAPSYM library) are connected to one net.
3. Some of the connections are made with a bus rather than wires.

Power and ground connections to digital and analog parts

Digital gates from the Capture libraries have nonvisible (invisible) power and ground pins. Their power pins are named *VCC*, and their ground pins are named *GND*. Connecting a wire to an invisible power pin does nothing. Connections from nets to invisible power pins can be made only through global net names (made global by CAPSYM power, ground symbols, or off-page connectors). So the power pin VCC on U4 (CD74HC164) is connected to the schematic's VCC symbol by the name *VCC*, and the ground pin is connected to the digital ground symbol by the name *GND*.

To connect a digital gate's ground pin to a ground net in your schematic, you have to either name the digital ground net *GND* or change the name of the digital gate's ground pin to match the name you assigned to the digital ground net (e.g., *DGND* instead of *GND*)—this is required for all digital gates. It is easiest to use the default *GND* name for the digital ground.

Most other Capture parts use either visible power pins or input or passive pins for power and ground. And instead of using *VCC* and *GND* as names of the power pins, they often use *V+* and *V−* or *VDD* and *VSS*. When parts use visible power and ground pins, you can either take advantage of their global nature (and simply use their names to make connections to global nets) or make direct connections to the pins with wires. When parts use input or passive pins for power and ground, you have to make explicit connections (using wires) between the pins and the power or ground nets. Said another way, you cannot make connections to invisible power pins, and you have to make connections to pins that are not power pins; but visible power pins may have either wired or global connections to them. When you make an explicit connection to a visible power pin with a wire, the pin becomes connected to that net (the name of the wire) regardless of the name of the pin.

Connecting separate analog and digital grounds to a split plane

The CAPSYM power and ground symbols are global in nature and share characteristics of nonvisible power pins (i.e., connections to nets can be made by name or by physical connections). Fig. 9.91 shows how the analog and digital grounds are named and displayed. Two CAPSYM symbols are used with distinct names to differentiate the two ground systems. The analog parts of the circuit are referenced to the analog ground (AGND) symbol, and the digital components are referenced to the digital ground (GND) symbol. At some point the two grounds need to be connected together for the ADC to work correctly (particularly with single-ended inputs). At the connector the

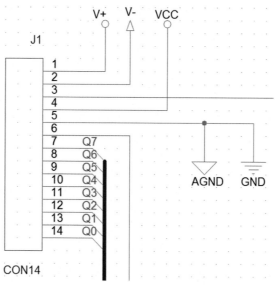

Figure 9.91 Connection of the analog and digital grounds and bus connections.

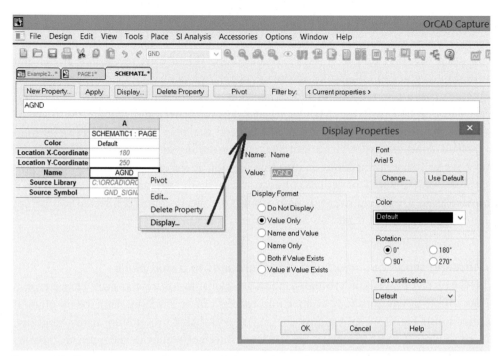

Figure 9.92 Changing a ground/power symbol's name and displaying it.

two grounds may leave the board separately and be connected at the power supply, or they may be connected at a pin (or pins) and leave the board together as one common ground. In this example the two grounds are connected at pin 5.

To place the analog and digital ground symbols, use the Place Ground tool, ⏚, and select the desired ground symbols. To change the symbol's name (e.g., from GND_SIGNAL to AGND), select the symbol on the schematic page to highlight it, right click, and select Edit Properties from the pop-up menu to display the Properties spreadsheet (see Fig. 9.92). Select the Name row, right click, and select Edit from the pop-up menu to display the Edit Property Values dialog box. Enter the new name of the ground symbol and click OK. To display the name of the symbol on the schematic page, select the Name row, click the Display button on the top of the spreadsheet or from the right click pop-up menu, and select the appropriate radio button on the Display Properties dialog box (Fig. 9.92).

Place the second ground symbol and rename it (if applicable). Use the Wire tool to connect the two ground symbols together then to the connector pin. The two ground symbols were distinct global nets until they were connected by the wire (net). Nets typically have names such as N04432 unless they are connected to a global net or symbol, then they take on the global name. A net can have only one name, and since the net just placed is connected to two global symbols with different names, it has to pick one of them (it decides alphabetically).

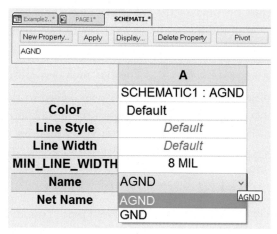

Figure 9.93 Selecting the ground net name.

To select the name of a global net, double click the net (or select the net and right click and select Edit Properties) to display the Schematic Net Properties spreadsheet (Fig. 9.93), select the Name row, right click, and select Edit to display the Edit Property Values dialog box. Use the dropdown selection list to choose which name the net should go by. Click OK and then close the spreadsheet.

It appears on the schematic that the analog and digital grounds are separate ground systems, connected only at J1. As far as Capture is concerned, they are actually the same net—the name of which was decided in the dropdown list in Fig. 9.93. Once we get to PCB Editor, we will separate the ground systems using a split plane.

Using buses for digital nets

Digital circuits often have multiple, related input/output wires. The serial-to-parallel shift register in Fig. 9.90 is an example, in which Q0 through Q7 are data lines that represent the analog voltage sampled by the ADC as an 8-bit word. Rather than draw all eight lines throughout your schematic, you can group the eight members of the word into a single bus (the *blue line* in Figs. 9.90 and 9.91).

A bus is a solid line that represents a group of wires or signals. Buses are not used to make connections to pins, but buses contain the signals connected to pins. Signal lines are added to a bus through aliases. Nets that are contained in buses are named, for example, A0, A1, A2, and A3, and the bus that contains them is named A[0..3].

Buses are placed on the schematic using the Place Bus tool, ⌐. After you have placed the bus, you define (name) it with an alias using the Place Net Alias tool, abc. Buses are named by a specific convention, busname[n..m], in which busname is the alias (a name you give it), and [n..m] is a member list made up of integers. The integer *n* in busname[n..m] is the first member of the list, and integer *m* is the last member of the bus list. The integers within the list can be separated with two periods, [n..m], a colon, [n:m], or a dash, [n–m]. The nets (wires) that make up the members are placed on the schematic using the Place Wire tool, ⌐. The bus members are placed on the schematic as you would place any other wire, but they are added to the bus by giving the wire a name (a member alias) that associates it with the bus. The members are named as busname *n* through busname *m* with the Place Net Alias tool. So in this example, the bus alias is named Q[0..7], and the member aliases are named Q0, Q1, to Q7. The short, slanted lines that are used to connect the wires to the bus are placed on the schematic using the Place Bus Entry tool, ◢.

The microcontroller (U3) has a couple of unused pins. To indicate that the pin is a "no-connect," place a NC symbol on the pins by selecting the Place no connect button, ☒, from the toolbar and place NC symbols on the unused pins.

Note: Any pin in a schematic used to make a PCB can have a NC symbol, but when PSpice simulations are performed, some pins are not allowed to "float," and the NC symbol has no effect on this as far as PSpice is concerned. In that case, enter RtoGND in the FLOAT property in the Properties spreadsheet, which will virtually connect this pin of PSpice model to ground via the virtual resistor. For more information, see the *Capture User Guide*, and search for RtoGND.

Once the design is completed, it is a good idea to do some housecleaning and design checking before making the netlist. Such tasks (as described in the previous example) are as follows:

1. Perform an annotation to clean up part numbering.
2. Make sure that global power/ground nets are properly named.
3. Make sure all parts have the correct footprints assigned to them (use the BOM to assist).
4. Perform a DRC to verify that the circuit design has no issues.
5. Correct any errors and reperform the DRC as needed.

Once the design has been second-checked and is free of errors, create the PCB Editor netlist as described in the first example (from the Project Manager go to Tools → Create Netlist and select the PCB Editor tab).

Defining the layer stack-up for split planes

As discussed in the first example, it is a good idea to have the board constraints defined prior to starting the board layout. Items to consider are the board technology, board size constraints, noise and shielding requirements, part placement restrictions, the number of Power and Ground planes and Signal layers, and trace width and spacing requirements.

The voltage and current constraints are similar to those of the analog design, except that the analog parts should be segregated from the digital parts on the board. In the

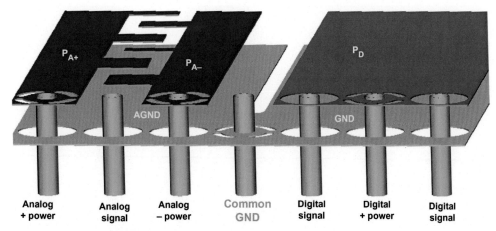

Figure 9.94 A mixed-signal power and ground system.

Table 9.5 Mixed-signal printed circuit board design constraints: component and board requirements.

Component mounting type	Mixed SMDs and THDs
Number of sides with parts	1
Number of plane layers	2
Number of routing layers	2
Total number of layers (min)	4
Smallest leads	SOIC
Pad spacing (min)	50 mil
Pad width (max)	25 mil
Maximum current	0.1 A
Inner trace width (min)	1.3 mil
Outer trace width (min)	0.5 mil
Maximum voltage	10 V
Inner trace spacing	4 mil
Outer trace spacing	5 mil

SMD, surface-mounted device; *THD*, through-hole device.

first analog design example, we used four planes for V + , V − , and ground. A similar approach could be taken here, but in this example, we use one plane for V + , V − , and VCC, and one plane for both the analog and the digital grounds, as shown in Fig. 9.94, to demonstrate the process of working with split planes. The default PWR layer is divided into three completely separate Power plane areas, using three copper shapes to supply analog V + (P$_{A+}$), analog V − (P$_{A−}$), and digital VCC (P$_D$), as indicated in the figure. The Ground plane is partially divided into two Ground planes for analog ground (AGND) and digital ground (GND) but left connected by a small section of copper near the connector's ground pin. This arrangement segregates the

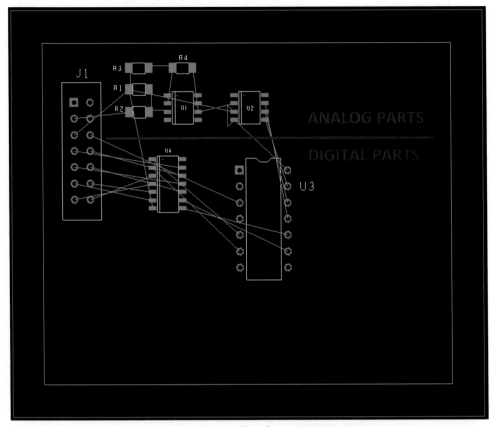

Figure 9.95 Initial parts placement for the mixed analog and digital design.

analog components from the digital components and therefore reduces switching noise in the analog circuitry. To accomplish this the board requires two Plane layers and at least two Routing layers. Table 9.5 lists the design constraints.

To begin the board layout, we follow the same procedure as outlined at the beginning of these examples and discussed in detail in the analog design example.

From Capture, perform the Create Netlist procedure and launch PCB Editor.

As described in the first example, draw the board outline, use one of the place tools to find and place parts into the board outline, and perform a DRC to check for footprint problems prior to doing anything else. Fig. 9.95 shows an example of how the parts might be placed. Notice that the analog parts are segregated from the digital parts.

Defining the layer stack-up

When the board constraints were defined earlier, we said that we wanted two Plane layers and two Routing layers, so we need to add the Plane layers. Follow the

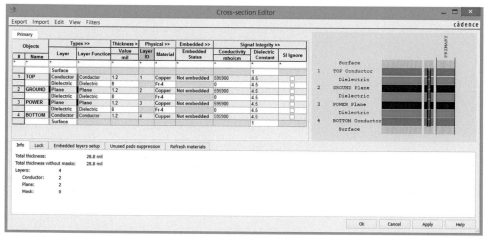

Figure 9.96 Layer stack-up for the mixed-signal board design.

procedure from the first example (right click and add layers) and add the two plane layers, as shown in Fig. 9.96.

Setting up a split GND Plane

The objective in splitting a Plane layer into analog and digital sections is to remove copper from the plane using a shape void. Fig. 9.97 shows the desired split Ground plane. Electrically the AGND and GND nets are the same but physically we want to separate them. To do this, a copper shape, which is attached to the GND net, is added to the Ground layer. This makes them electrically connected. To physically separate the analog from digital circuitry, we create a void strip between the analog and digital areas of the board.

Before placing the plane, you may need to adjust your route grid. Set it to about 5 mil, because you need the resolution to properly place the vertices (corners). *To add the GND shape to the ground plane*, use the procedure used in the first example. In short, select Outline → Plane... from the menu bar (or use the Shape Add tool and select Dynamic copper in the Options pane). Select the Ground plane, the GND net, and the Draw Rectangle option in the Plane Outline dialog box. Draw the Ground plane then click OK to dismiss the dialog box. To view the thermal reliefs, follow the procedure used in the first example (set Thermal pads in the Design Parameters dialog box, and assign flashes to the pad-stacks as necessary). Next we split the plane into analog and digital areas.

Note: Ground planes can be drawn with filled rectangles, static shapes, or dynamic copper shapes. For best results, use dynamic copper shapes. To identify the shape type of existing shapes, hover your mouse over the shape edge (make sure the Shapes option is selected in the Find filter pane). To change a shape to dynamic copper, select Shape

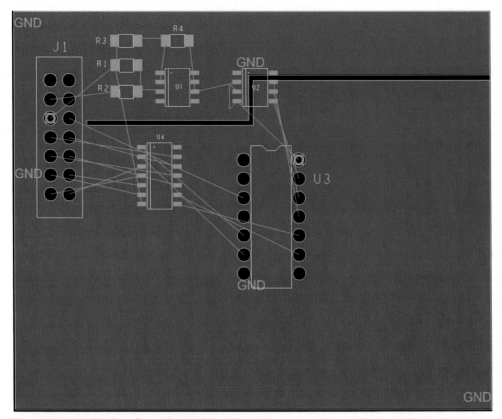

Figure 9.97 A split place for analog and digital circuits.

→ Change Shape Type from the menu, check Options, then left click the shape you want to change, right click and choose Done.

To split a plane, select the plane layer from the Options tab by selecting the Etch class and Ground subclass. Select the Shape Void Polygon button, ⟳ , on the toolbar, or from the Shape → Manual Isolation/Cavity menu. At the command prompt, PCB Editor asks Pick shape or void to edit. Left click the GND plane. PCB Editor then asks Pick void coordinates. Left click at the vertices of the void area you want to define, as shown in Fig. 9.97. Draw the void so that the analog and digital areas are separate but leave a connected area by the connector, J1, so that the two Ground planes remain electrically connected. The void between the ground areas should be about 50 mil wide.

U2 is the ADC and has both analog and digital properties. Some ADC manufacturers specify that the analog and digital grounds should be common near the integrated circuit (IC). Right now the analog and digital grounds are separate, but we will connect them directly under the IC by placing a small patch of copper, which will be

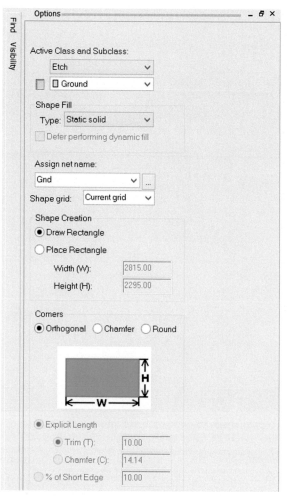

Figure 9.98 Setting the properties of the copper bridge.

connected to the GND net, to bridge the void separating the analog and digital planes, then merge the patch with the main Ground plane.

To create a copper bridge, select the `Shape Add Rectangle` button, ![icon], and use the `Options` tab to define the shape on the Etch Ground plane and connect it to the GND net, as shown in Fig. 9.98. Note that, if the Ground plane was made with a dynamic copper shape, then the bridge also needs to be dynamic copper rather than static filled.

Place the bridge as shown in Fig. 9.99. Left click and release at the upper left and lower right corners to draw the shape. Make sure to overlap the shape and the Ground plane to allow the merge process to work properly.

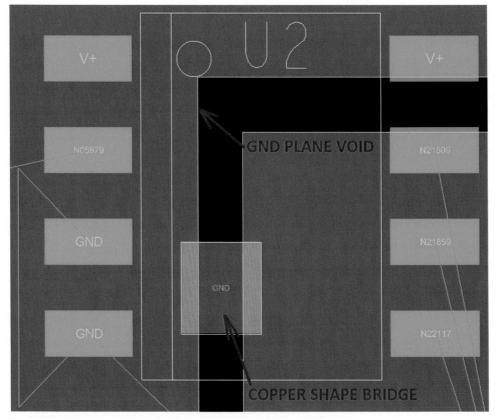

Figure 9.99 Placing a copper shape to bridge the two ground planes.

To merge copper shapes, select Shape → Merge Shapes from the menu bar. At the command line, PCB Editor will ask you to Pick primary shape to merge to. Select the Ground plane. Then PCB Editor will ask Pick shapes to be merged. Select the bridge, right click, and select Done from the pop-up menu. The bridge is now part of the Ground plane.

Notice that at the J1 connector we have the "common ground area" in Fig. 9.97, but that the analog and digital ground areas are separated throughout the rest of the board. The area under the ADC does not defeat the "split" plane. Recall from Chapter 6, Printed circuit board design for signal integrity, that return currents follow the path of least impedance, so return currents related to U3 and U4 do not go out of their way to pass through the area under the ADC.

Setting up separate planes on a single plane layer

Unlike the ground system, in which there is really only one ground net and one Ground plane, the power system has three distinct power nets (V + , V − , and VCC)

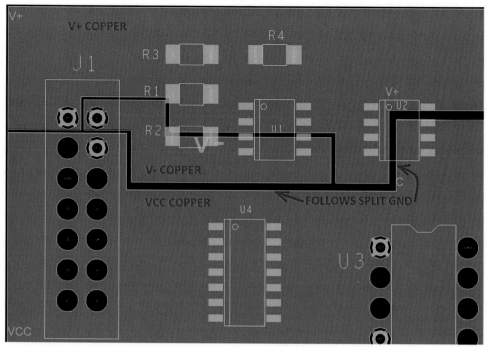

Figure 9.100 Setting up multiple power areas on a single plane.

that need to share the Power plane. To accomplish this, distinct copper areas are drawn for each net on the one Power plane. The goal is shown in Fig. 9.100.

We begin by placing the VCC Power plane in the digital section of the board. As described in previous sections, we start by selecting Outline → Plane from the menu bar. At the Plane Outline dialog box, select the Power plane and select the VCC net. Select Draw Polygon and left click to place the eight vertices that make up the VCC plane area. Click OK to complete the process and close the dialog box. Keep the VCC plane above the digital Ground plane and away from the void area. Once you have the VCC plane area drawn, there should be a thermal relief around the connector pin attached to VCC (pin 4 in this example).

Repeat this process for the V + and V − plane areas. Note that the area between the digital and analog plane areas should match up with the void that was placed in the GND plane until you get close to the connector. Also note that we need not place voids on the Plane layer to split the plane, since the copper areas are distinct objects and do not need to be split. When drawing the V + copper pour, make sure to include J1 pin 1 and the V + fan-outs for U1 pin 7 and U2 pin 8 and make sure the copper pour stays on the analog side of the split Ground plane. When drawing the V − copper pour, make sure to include J1 pin 2 within the boundary and stay on the analog side of the Ground plane, as shown in Fig. 9.100.

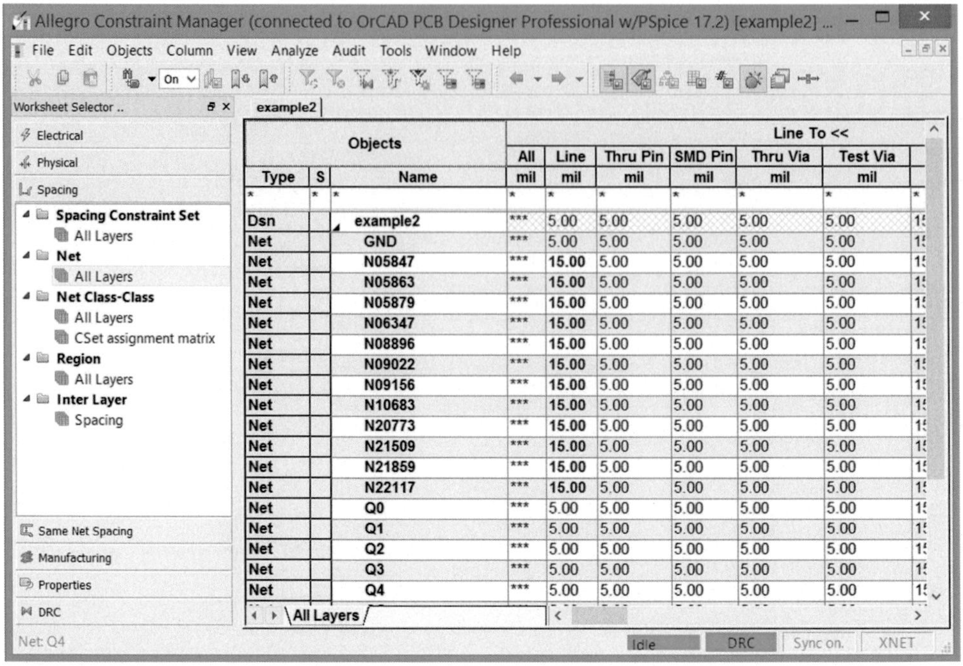

Figure 9.101 Setting the trace spacing rules.

Setting up routing constraints

Before fanning out or routing the board, we need to set up the design constraints using the Constraint Manager. Launch the Constraint Manager by selecting the Cmgr button, ![CM], on the toolbar. *To change the spacing constraint*, select the Spacing tab and the Line to / Line column in Net / All Layers folder. Change some of the nets to whatever you like, just to see what happens. Fig. 9.101 shows the line-to-line spacing of all the nonbus nets set to 15 mil.

To change the line width constraint, select the Physical tab and the All Layers icon under the Net folder. Change a couple of the minimum line widths to see the effect. The GND and a couple of the analog nets were changed, as shown in Fig. 9.102. Note that changing the GND line thickness changes the width of the fan-out traces (the short trace going from an SMD pad to the via to GND) and thermal spokes on positive Plane layers, as we will see.

Next we begin the process of fanning out power and ground. We could disable all the nets except for the power and ground nets to do the fan-out, as we did in the last example, but we will force the autorouter to ignore the signal nets, so that we need not do that extra step. Perform the fan-out as described in the first couple of steps of the first example. Select Route → PCB Router → Route Automatic from the menu. In the

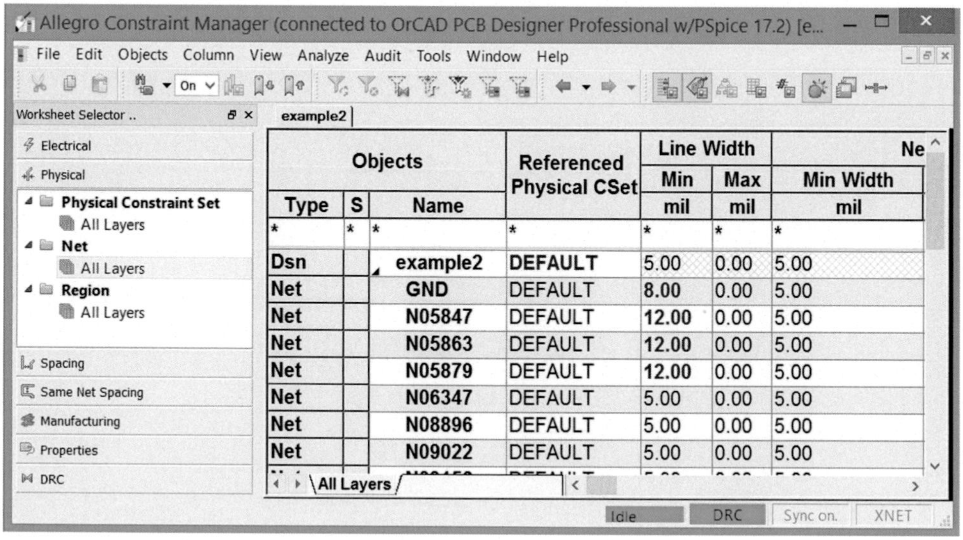

Objects			Referenced Physical CSet	Line Width		Min Width	Ne
				Min	Max		
Type	S	Name		mil	mil	mil	
*	*	*	*	*	*	*	
Dsn		example2	DEFAULT	5.00	0.00	5.00	
Net		GND	DEFAULT	8.00	0.00	5.00	
Net		N05847	DEFAULT	12.00	0.00	5.00	
Net		N05863	DEFAULT	12.00	0.00	5.00	
Net		N05879	DEFAULT	12.00	0.00	5.00	
Net		N06347	DEFAULT	5.00	0.00	5.00	
Net		N08896	DEFAULT	5.00	0.00	5.00	
Net		N09022	DEFAULT	5.00	0.00	5.00	

Figure 9.102 Setting the trace width rules.

Automatic Router dialog box, select the Router Setup tab and select Specify routing passes in the Strategy section. Select the Routing Passes tab, select Fanout option, and deselect the Route and Clean options. Click the Params... button. In the SPECCTRA Automatic Router Parameters dialog box, select only the Power Nets option under the Pin Types section. Click OK then click Route to generate the fan-outs.

Check the fan-outs to make sure the router did its job and also check that there are no DRC errors by checking the Status dialog box (Check → Design Status...) or generating a DRC report (Export → Quick Reports → Design Rules Check Report). If everything looks OK, then we can route the rest of the board.

Note that Display → Assign Colors can be used to override the Layer colors and assign colors to an individual Net, or group of Nets. Also note that it is possible to create fan-outs individually using Route → Create Fanout and setting its parameters in the Options pane.

To make sure that the fan-outs or any manually routed traces remain intact, they should be locked in place (fixed). *To fix traces*, display the Constraint Manager and select the Properties tab. Under the Net folder, select the General Properties icon. Set the On flag in the Fixed column for any trace you want to protect. Alternately you can use the Fix button, 🔘 , and the Unfix button, ✈ , on the toolbar to select traces and components to fix or unfix.

After performing the fan-out and fixing routed traces to protect them, we can *set up to route the signal nets*. Select Route → PCB Router → Route Automatic from the menu. In the Automatic Router dialog box, select the Router Setup tab and select Specify

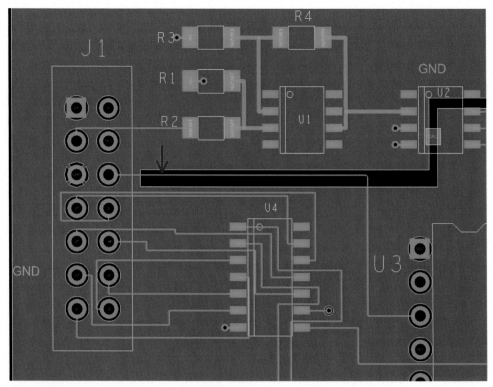

Figure 9.103 The circuit after fan-out and routing.

routing passes in the Strategy section. Select the Routing Passes tab and select the Route and Clean options and deselect the Fanout option; click the Route button to route the rest of the board. If you want to restrict which signal nets get routed, use the Constraint Manager to set routing constraints. Use the Route Restrictions on the Properties tab of the Constraint Manager.

An example of a routed board is shown in Fig. 9.103.

As in the earlier example, a postrouting inspection is performed after autorouting to look for problem areas, and offending traces are unrouted then rerouted. Fig. 9.103 shows an area that could use some work (Fig. 9.104 shows a close-up view of the area). The red arrow points to a trace that was routed over the void in the GND plane.

As described in Chapter 6, Printed circuit board design for signal integrity, we do not want this to occur, as it increases the loop inductance of the trace (an impedance-matching issue), which can lead to ringing problems and EMI. In this case the easiest fix is to simply move the trace down. *To move a trace*, select the Slide tool, 🖳 , left click and release on the trace to select it. Move the mouse to the desired location, left click and release to place the trace. To prevent this from happening on subsequent

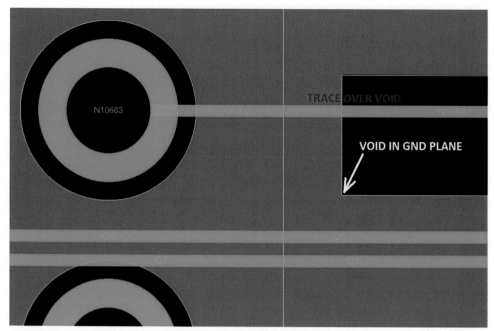

Figure 9.104 Look for problem areas after autorouting.

routing actions, we can also place a Route Keep-out area in the void to keep the auto-router from routing other traces over the void.

There is a new function Display → Segments over Voids, which can help detect such locations by creating a report to navigate and displaying the segments in a different color in the canvas.

Adding Route Keep-Out areas

To place a route keep-out shape area, select the Route Keepout class and Top subclass from the Options tab. Select the Shape Add tool and place a polygon on the top layer that overlaps the void by 10–20 mil, as shown in Fig. 9.105. This will keep traces away from the void when you re-route the board.

Another potential problem area is the traces that run between the microcontroller (U3) and the ADC (U2). The traces run side by side in close proximity for a relatively long distance. The traces should be moved farther apart (use the Slide tool as described previously) or guard traces can be added to provide isolation between the traces (discussed later and in the next example). The traces that run from U4 to J1 could also be a problem. A simple solution there would be to move some of the traces to the bottom layer, as described next.

There is a new tool in Analyze → Workflow Manager, the Coupling Workflow, which can help to find such locations with excessive coupling between the signal traces.

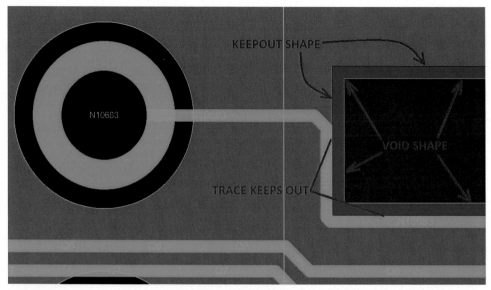

Figure 9.105 Add a route keep-out shape to control routing.

Moving a routed trace to a different layer

As long as a trace has a through hole or via on each of its ends, you can easily change which layer the trace is on without having to rip it up. *To move a routed trace to another layer,* select the `General Edit` button, ⬜ , select the trace you want to move (use the `Find` filter if necessary), right click, and select `Change to layer` from the pop-up menu then select the layer to which you want the trace to move, as shown in Fig. 9.106.

Adding Ground Planes and guard traces to Routing layers

As discussed at the beginning of these examples and in Chapter 6, Printed circuit board design for signal integrity, you can reduce noise levels on signal lines by surrounding traces with copper planes and guard traces. Not everyone agrees with this practice, but it is demonstrated here in the interest of completeness. The following procedures demonstrate how to use the `Add Connect` tool to add ground nets and use shapes to add copper planes to Routing layers.

Routing guard traces and rings

If you have one or two traces that could have cross-talk problems, you can add guard traces between them that can be attached to the Ground plane with vias. You can also add guard rings around component pins. Like the guard traces, the guard rings are attached to the Ground plane with vias. It should be noted though that their usefulness is debated in the literature, because they can cause more problems than they fix if not applied correctly (see Appendix E for references). Fig. 9.107 shows examples of

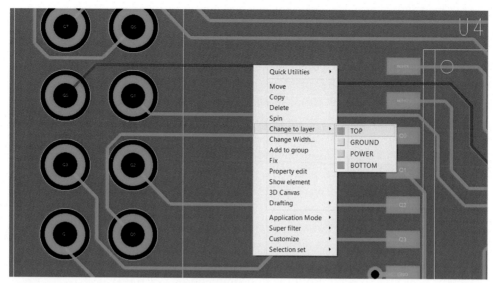

Figure 9.106 Moving a trace to a different layer.

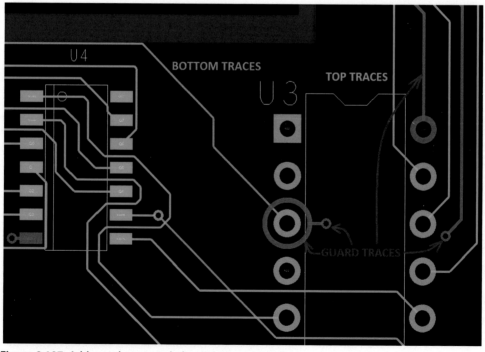

Figure 9.107 Add guard traces to help minimize cross talk.

guard traces between the control and the data lines and guard rings around one of the microcontroller pins.

To place guard traces that are attached to the Ground plane, select the Ground plane on the `Options` tab and select the `Add Connect` tool. Left click on the Ground plane where you want to begin the guard trace. Right click and select `Add Via` from the pop-up menu. The `Add Connect` tool should still be active, and you should now be on the Top layer (but check the `Options` tab to make sure). Draw the trace where you want the guards to be. At the end of the trace, left click to finish the trace then right click and select `Add Via`. At various intervals along the trace, add vias to securely stitch the guard trace to the ground plane. *To add the vias*, select the `Add Connect` tool, left click on the trace where you want the via, right click and select `Add Via`. Repeat this where ever a via is needed.

To add a guard ring around a component pin:

1. Select `Shape → Circular`.
2. Right click over the pin.
3. Choose `Snap pick to → Pin` from the pop-up menu (Fig. 9.108A).
4. Draw the round shape connected to `GND` net, and left click to finish the shape (Fig. 9.108B).
5. To create the void in it, select `Shape → Manual Isolation/Cavity → Circular` (Fig. 9.108C).
6. Left click on the recently created shape to select it. It should become highlighted.
7. To select the pin as the center of the created void right click over the pin and choose `Snap pick to → Pin` from the pop-up menu, and draw the void around the pin. Left click to finish the void.
8. Select `Route → Connect` to create the via that will connect the guard ring to the `GND` plane (Fig. 9.108D).

It's also possible to add the arrays of vias around the object using the tool `Place→Via arrays→Boundary`. If you use this tool, you should set up the parameters in the `Options` pane:

- Via net and Padstack.
- Cline: On single/both sides of Cline.
- Via object offset.
- Maximum via-via gap.

Adding Ground planes to Routing layers

Next we add Ground planes to the top and bottom Routing layers. *To add Ground plane areas to a Routing layer*, make the Top (or Bottom) Etch layer active and select either the `Shape Add` button, ▣ , or the `Shape Add Rect` button, ▣ . Display the `Options` tab and select `Dynamic copper` and assign the GND net to the shape, as shown in Fig. 9.109.

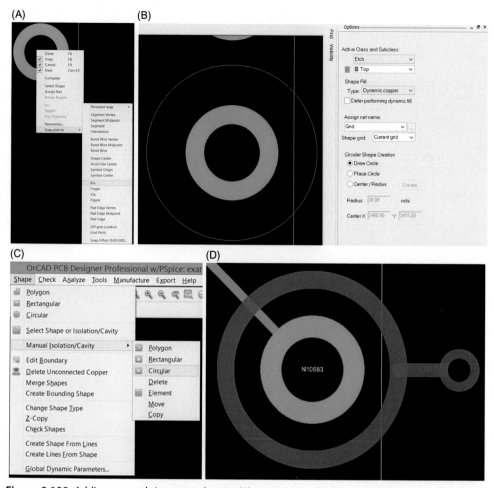

Figure 9.108 Adding a guard ring around a pin. (A) snap to pin, (B) draw round shape, (C) draw cavity.

An example of a copper plane is shown in Fig. 9.110. Notice that thermal reliefs are automatically placed on pins and pads. The spacing of the plane to the pad (pin) and the spoke width are set using the rules in the Constraint Manager.

Thermal reliefs are set to orthogonal (+) by default, but you can change them to diagonal (×) if you prefer. *To change the direction of thermal reliefs on a positive Plane* (Routing) layer, select `Shape → Global Dynamic Parameters` from the menu bar. In the `Global Dynamic Shape Parameters` dialog box (see Fig. 9.111), select the `Thermal relief connects` tab and choose the desired direction.

To change the spacing of the Ground plane to pins on a Routing layer, launch the Constraint Manager and select the `Spacing` tab (Fig. 9.112). Select the `Shape to` section in the `Spacing Constraint Set` folder and select the `TOP` layer row (or whichever layer

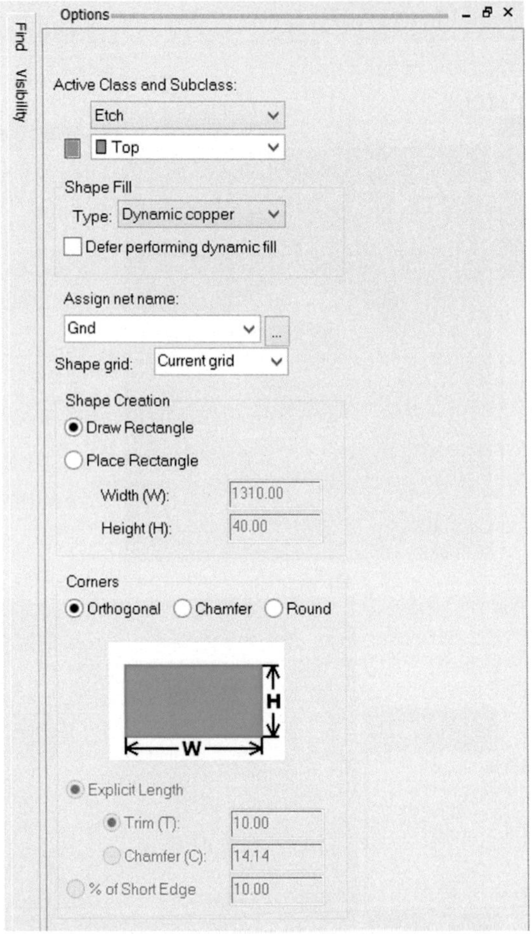

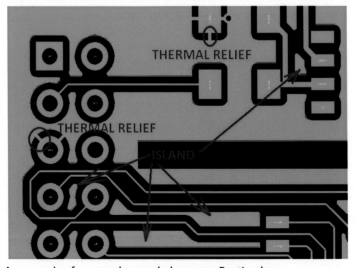

Figure 9.109 Selecting Etch properties for a ground area on a Routing layer.

Figure 9.110 An example of a poured ground plane on a Routing layer.

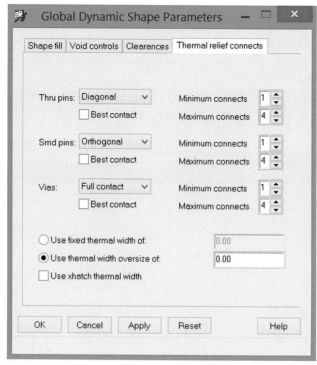

Figure 9.111 Setting thermal relief properties for positive planes.

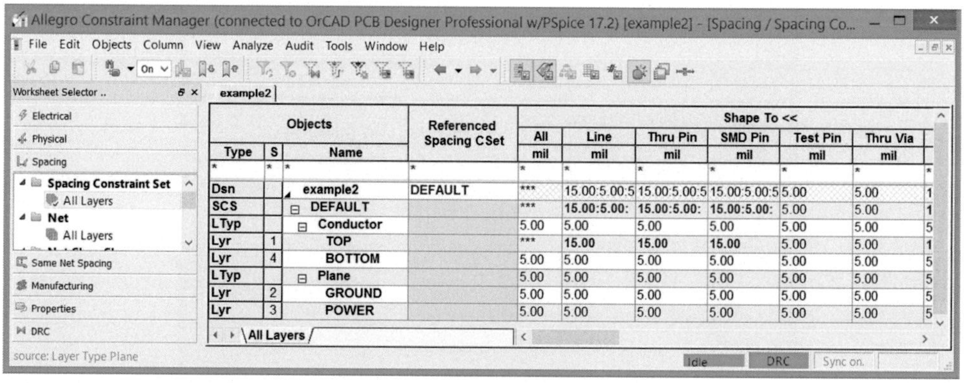

Figure 9.112 Setting trace to copper pour spacing rules.

you want to change) in the DEFAULT spacing CSet. Change the `Shape to Line`, `Shape to Thru Pin`, or even `Shape to All`, and so forth settings to the desired width. Changes are effective immediately; however, if your plane areas were not drawn using dynamic copper, your design may look as if something has gone horribly wrong.

If the ground plane on the Top layer suddenly takes over, the design is actually OK, the shapes just need to be updated. *To update the Ground plane shape*, select `Check` → `Design Status` from the menu bar. In the `Status` dialog box (Fig. 9.113), select the `Update to Smooth` button. The Ground plane should now look correct again and have the proper spacing.

You can enable the immediate update of Shapes to the Smooth state by setting the `Smooth` radio button in `Shape` → `Global dynamic parameters` → `Shape Fill`.

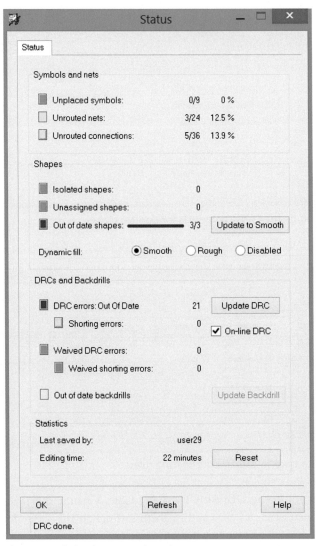

Figure 9.113 Use the Status dialog box to update shapes.

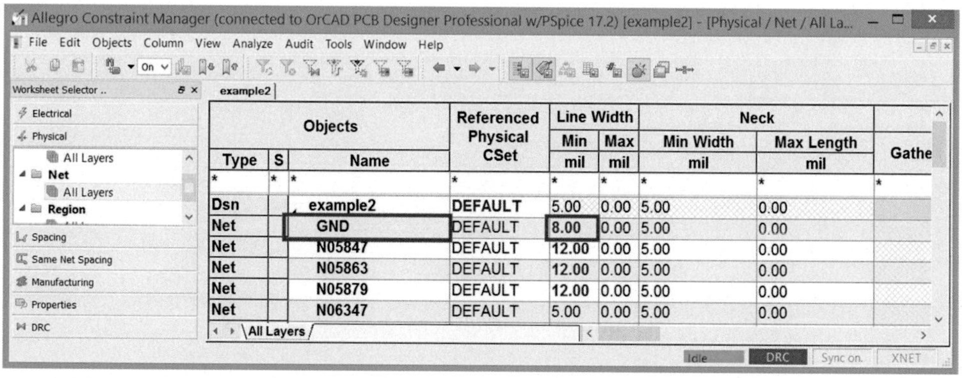

Figure 9.114 Use the Constraint Manager to control thermal relief spoke width on Routing layers by changing the trace line width.

To change the spoke width of thermal reliefs on a Routing layer, select the `Physical` tab in the Constraint Manager (see Fig. 9.114). Select the `All Layers` icon under the `Net` folder. Change the line width of the net that has the thermal reliefs you are interested in. The spokes and traces will now be the same for the nets you modify. Perform the Update Shapes step as described previously to update your board design. If you need to have certain trace widths for specific segments on a net and you do not want to have these changes affect them, you must go back and change those segments back to the way you want them using the `Change Objects` command (in the `Edit` menu) and the `Options` and `Find` filter tabs. Once the changes have been made, you can then `Fix` them so that the changes are not undone.

Once your plane is established, remember to place voids and merge copper areas and the like on the new plane to match the inner Ground plane as appropriate. Update the DRC to make sure the copper pour did not cause any errors.

You can add a Ground plane to the Bottom layer (or any other Routing layer for that matter) using the same procedures used for the Top layer. When you have several Ground planes, it is important that they have many low-impedance connections to help keep the planes at the same potential throughout the board area. This is done by placing vias, which are connected to the net for those planes, at various places. These vias are sometimes called *stitching vias*. *To place stitching vias on a board* to connect multi-layer Ground planes, select the `Add Connect` tool, left click on the plane that needs the via (a trace will be started), right click and select `Add Via` from the pop-up menu. *To place multiple vias*, use the `Copy` button, with checked `Vias` in the `Find` filter pane to copy the via and left click wherever you want a via placed (right click and select `Done` to quit). An example is shown in Fig. 9.115. It is also a good idea to place a couple of vias underneath the ADC (U2). To add the array of stitching vias, you can also use the `Place → Via Arrays → Matrix` tool.

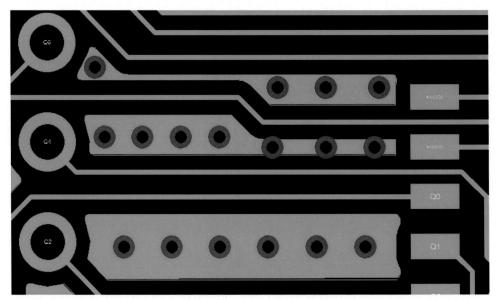

Figure 9.115 Use stitching vias to provide good connections to ground planes.

If you look closely at the ground strips between some of the traces and between some of the pins on the connector, J1, in Fig. 9.110, you will notice that some areas are isolated from the rest of the Ground plane. These isolated strips are called *islands*. Islands can act like antennas, which can pick up high-frequency noise (EMI) and cause problems for the rest of the circuit. To solve this problem, the strips and islands need to be either tacked down to the Ground planes using the stitching vias, trimmed, or removed altogether.

Use vias as just described to tack the strips to the underlying Ground plane. For parts of strips or islands that you want to trim from larger, stitched sections, use one of the Shape Void tools to remove unwanted sections of the strip. *To trim islands and isolated strips*, select one of the Shape Void tools and use the Options tab to select the correct layer and net to void. Select the shape, then draw the void shape over the area you want removed.

To completely remove strips or islands, use the Island_Delete tool, ⬛, or select Shape → Delete Unconnected Copper. By default all islands on a layer associated with a particular net will be deleted. Use the Options tab to select whether to delete all the islands on that layer or just specific ones. *To delete only specific islands*, select the First button on the Options tab then left click on the island you want to delete. Right click and select Done when finished.

This completes Design Example 2.

Example 3. Multipage, multipower, and multiground mixed A/D printed circuit board design with PSpice

Introduction

Multipage schematics can be used to organize and simplify large circuit designs and to incorporate PSpice simulations prior to laying out the board design. The mixed analog/digital circuit from the last example (see the schematic in Fig. 9.90) is reused in this example but is modified to demonstrate how to route a single PCB from a multipage schematic project and add PSpice simulation capabilities. The example also demonstrates two methods used to establish isolated Ground planes using blind vias and a buried chassis shield. The technique allows quiet circuits to be placed on one side of the board and shielded from noisy digital circuits, which are placed on the other side of the board.

Multiplane layer methodologies

In the previous example, a single Plane layer was used to produce a digital ground and an analog ground even though there was really only one ground net. The two ground systems were produced by physically segregating the parts on the board and removing a strip of copper from the one plane (creating a split plane) between the two circuit areas.

In high-density, high-frequency digital designs, multiple Ground planes are often used even when there is only one type of circuit ground (as demonstrated in Example 4). This helps reduce loop inductance when using multiple Routing layers and control trace impedance (see Chapter 6: Printed circuit board design for signal integrity, for details). When two Plane layers are used for a single ground net, connections are made to both planes simultaneously via plated through holes (whether for through-hole leads or fan-outs from SMDs) anytime a connection is made to the net. This is shown in Fig. 9.116.

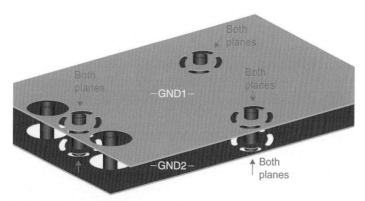

Figure 9.116 Multiple Ground planes and multiple connections for one ground net.

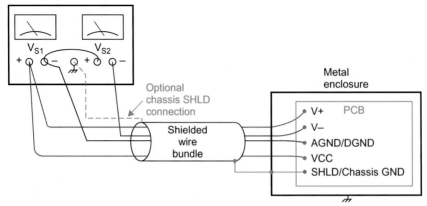

Figure 9.117 A multipower/multiground system with a chassis shield.

In this example, continuous Plane layers are desired for both the analog and the digital parts of the circuit. As described in Chapter 6, Printed circuit board design for signal integrity, significant cross talk can occur between adjacent planes if there is any overlap between the two plane areas. Since both the analog and digital planes extend out to the limits of the board, there is complete overlap. One way to minimize cross talk is to insert a shield between them that carries no signal currents.

Fig. 9.117 shows the system design concept for this example. This is just one of many possible types of PCB power distribution schemes. The system uses a dual power supply for ± analog power for op amps and a single power supply (VCC) for digital circuits. The analog and digital systems each has its own ground system on the PCB; however, a common reference voltage is required for the ADC. To facilitate both requirements the grounds are tied together at a single point on the PCB before returning to the power supply. To keep the two ground systems from experiencing cross talk on the PCB, they are separated by a Shield layer buried inside the PCB, which is connected to the chassis ground and the shielded wire bundle. Extensive coverage of noise reduction and shielding is provided in the literature (see Ott, 1988 especially for detailed coverage of this topic).

The two grounds are connected at a controlled point (or points) in several ways. The simplest is to use a jumper wire at the connector. However, the ADC in this example also requires a common ground area under the IC package. A jumper wire is not practical in this case.

The challenge in setting up a ground system like this is that the two ground systems must be kept separate everywhere except at the tie point. This is not possible in Capture (at least not in a straightforward manner). As soon as the two ground nets are connected on the schematic (even if only at one point), the two nets become one everywhere in the netlist and cannot be separated in PCB Editor, since it is a single

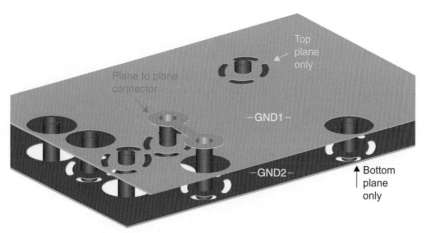

Figure 9.118 A shorting strip or jumper wire can be used to make plane-to-plane connections.

net. Therefore the two distinct ground nets (or three, counting the shield) are kept separate in Capture and only made to appear connected in the schematic (for documentation purposes). The individual nets are then tied together at the common reference point in the board layout.

Several methods can be used in PCB Editor to tie the distinct grounds together at the common reference point. The first method uses a plane-to-plane connector (a shorting strip), as shown in Fig. 9.118. The shorting strip is a footprint with two padstacks that can be shorted together with a wire jumper or a copper strip (a line or shape on the `Top Etch` class) placed in the footprint symbol or on the board design. To use the shorting strip a Capture part must be made that has two pins but no electrical connection between them, then the PCB Editor footprint is assigned to the Capture part. This method is demonstrated in Example 4.

Note: The shorting strip can be replaced with a jumper wire or ferrite bead soldered into the footprint padstacks. The copper strip method is demonstrated in the example, since it lowers cost and simplifies assembly.

The second method uses a specialized padstack that forces the planes to be shorted together by the padstack definition, as shown in Fig. 9.119. Normally, clearances are specified on Plane layers when there is no connection to the layer and if the netlist specifies a connection to the Plane layer, PCB Editor knows to insert a thermal relief (if a flash is assigned) in place of the clearance. However, if you explicitly specify that the clearances on Plane layers are smaller than the drill diameter, then the clearance will be drilled out and shorted to the padstack barrel during the plating process. If you force a connection to a Plane layer, PCB Editor will generate a DRC error, which can be waived. This method also is demonstrated in Example 4.

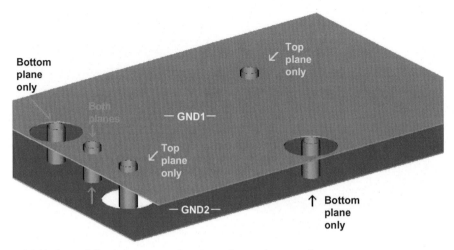

Figure 9.119 A specialized shorting padstack used as a plane-to-plane connector.

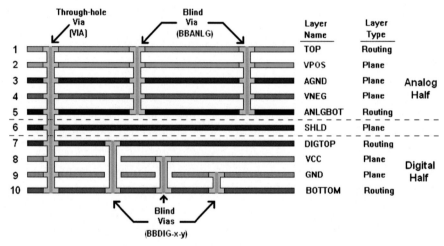

Figure 9.120 Layer stack-up for the shielded dual plane example with blind vias.

One more method of shorting two nets was added in OrCAD 17.2. You can place the small static shape with two properties attached to it: DYN_DO_NOT_VOID, to prevent the voiding of overlapped dynamic shapes, and NET_SHORT, to list the nets shorted by this shape (e.g., GND:AGND). You can add the DYN_DO_NOT_VOID property by right click on the shape boundary and selecting Property edit, and you can add the NET_SHORT property by right click and selecting Net Short from the pop-up menu.

The stack-up in this example demonstrates how to use PCB Editor to implement one method of EMI and cross-talk reduction. A 10-layer board and blind vias are required. The layer stack-up is shown in Fig. 9.120. Since two methods can be used to construct blind/buried vias, both are discussed. The first method is demonstrated on the analog half of the board using a blind via named BBANLG, and the second method (which makes several blind/buried vias simultaneously) is used on the digital side of the board. Since the second method produces multiple vias, the names of the vias have a base name (BBDIG in this example) and a suffix that describes which planes are connected (e.g., BBDIG-x-y). So the middle blind via on the bottom of Fig. 9.120 is BBDIG-VCC-BOTTOM. Both methods are explained in greater detail.

Capture project setup for PSpice simulation and board design

Setting up a Capture project for PCB design and PSpice simulations at the same time has not been widely covered in the literature, so this example addresses that point. Extensive coverage of PSpice simulations in general is not provided in the example, but the process of setting up a project that allows PSpice simulations and the result of a basic simulation are included. To begin a PCB design project that can be simulated with PSpice, start Capture, and from the File menu, select New → Project. When the New Project dialog box is displayed, select the PSpice Analog or Mixed A/D option as shown in Fig. 9.121. Enter a name for the project, use the Browse... button to set up a new folder for the project, and click OK.

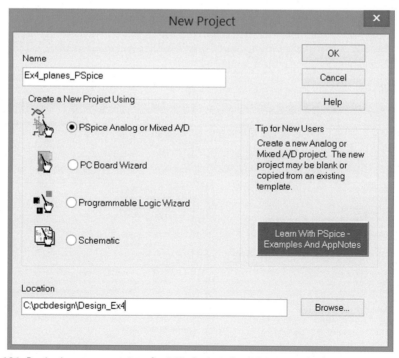

Figure 9.121 Beginning a new project for PCB design plus PSpice simulation.

When the `Create PSpice Project` dialog box is displayed, select the `Create based upon an existing project` radio button and select the `empty.opj` project template as shown in Fig. 9.122. Click `OK`. You will need an OrCAD PSpice license to use the PSpice simulator.

Adding schematic pages to the design

Three schematic pages are needed for this project: one for analog circuitry, one for digital circuitry, and one for PSpice simulation sources. A new project initially contains one schematic page, `PAGE1`. This page will be renamed, and two more pages will be added. *To change the name of an existing schematic page*, select the `Schematic Page` icon, right click, and select `Rename` from the pop-up menu. Enter the name `Analog` in the dialog box and click `OK`. *To add a schematic page to a schematic folder*, select the `SCHEMATIC1` folder, right click, and select `New Page` from the pop-up menu, see Fig. 9.123A. Enter a name for the

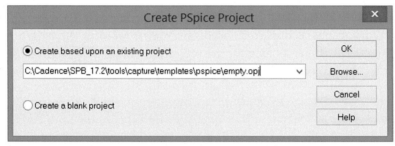

Figure 9.122 Selecting a PSpice project template.

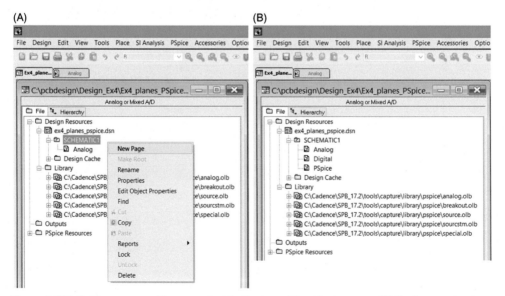

Figure 9.123 Setting up a multipage project in Capture. (A) add new page, (B) final page structure.

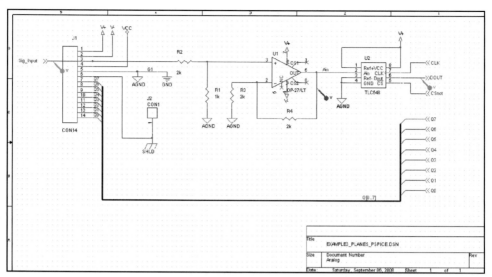

Figure 9.124 The analog schematic page.

schematic (e.g., Digital) and click OK. Add another schematic page to the SCHEMATIC1 folder and name it PSpice. The final schematic page structure is shown in Fig. 9.123B.

Begin by adding parts to the analog page. To display the analog page, double click the Analog page icon in the Project Manager. The analog page is shown in Fig. 9.124 and includes the passive components (R1-R4), the analog components (U1, U2) and the connector (J1). The U1 part has PSpice model assigned to it which will allow the simulation. The digital components are placed on the digital schematic page (see Fig. 9.127). New items in this project include off-page connectors and multiple ground symbols.

Using off-Page connectors with wires

Generally speaking, off-page connectors are used to continue signal nets across page boundaries. Off-page connectors are used in this example to connect signal lines between the ADC (on the analog page) and the microcontroller (on the digital page) and from the shift register (on the digital page) back to the connector (on the analog page). The off-page connectors are also used here to inject a PSpice signal originating from the PSpice page onto the analog input line (the Sig_Input net is shown in Fig. 9.124 and described later).

To place off-page connectors, select the Place Off-Page Connector tool button, , on the schematic page toolbar or select Off-Page Connector... from the Place menu. In the Place Off-Page Connector... dialog box, select one of the off-page connector symbols and enter the name to which the connector will be attached in the Name: text box, click OK, and place the off-page connector on the schematic page. You can change the name of an off-page connector after it has been placed on the schematic. *To change the name of an off-page connector*, double click the name, and enter the new name in the `Display Properties` dialog box.

Note that the off-page connector to pin 5 on U2 (chip select) does not contain the overbar (e.g., $\overline{CS}$, indicating an active low line), as does the pin name. Overbars can be generated on Capture schematic parts for nonpower-type pins, but do not use overbars on power symbols, as this will produce invalid netlist names. With PCB Editor, you can use overbars on off-page connectors not used for power nets without producing invalid netlist names. However, overbars on nets do not display the same as pin names. For example, when you make a Capture part, you can put an overbar above CS by typing the pin name as C\S\, and it will be displayed as $\overline{CS}$. But if you type C\S\ for a net alias name or an off-page connector, it will simply be displayed as C\S\, which will follow through and be displayed the same way in PCB Editor.

Off-page connectors cannot be used with power symbols, but they are not required, since power symbols are already global and known by all pages within the design.

Using off-page connectors with buses

Both nets and buses can be connected across page boundaries with off-page connectors. If nets belonging to buses cross page boundaries by off-page connectors, net aliases (using the abc button) are not required on the nets because the off-page connectors produce the aliases, otherwise net aliases are required to connect nets to the bus. For example, the nets connected to board connector J1 require aliases to be connected to the bus.

Setting up multiple ground systems on the schematic

Another difference between this design example and the previous one is the way the ground connections are made. In the previous example, there were two ground symbols (AGND and GND) but only one actual ground net (GND). In that design the grounds were physically separated on the board using a split plane, even though they were still of the same net. In this example, there are three ground symbols and three distinct ground nets (AGND, GND, and SHLD). The shield ground is indicated by a GND_EARTH symbol (renamed to SHLD), it is connected to J1.6 by itself, and it will be the only connection to the plane layer called *SHLD*. In Fig. 9.124 AGND and GND appear to be jointly connected to J1.5 but are actually separated by the special Capture part symbol G1 (shown in Fig. 9.125 along with the pin properties), which uses the PCB Editor footprint cap300. The footprint can be any two-pin footprint, and cap300 is used for convenience. Capture part G1 is a custom part and is not included with the standard OrCAD libraries but is included as GNDJCT part in the CHAP9EXAMPLES.OLB on the website if you care to see it. Part G1 contains two

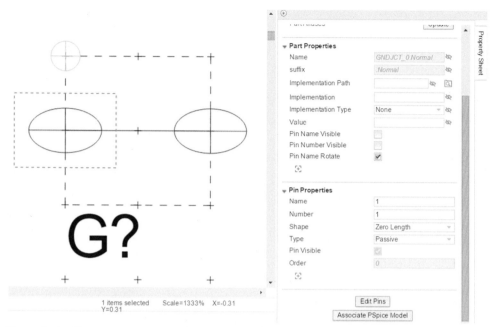

Figure 9.125 The Capture part G1 used as a multiground net connector.

pins that are graphically connected in the part but are not connected as far as the netlist is concerned. The purpose of G1 is to indicate on the schematic that the grounds are connected, while allowing the grounds to remain separate nets in the netlist.

After the ground connector part has been placed on the board and wired to the appropriate ground nets, you can turn off the part reference (G1) by selecting Do Not Display in the Display Properties dialog box (double click the part reference to show the dialog box).

You can also make the pin names invisible to make the part look like a wire. *To turn off the pin names*, select the part, right click, and select Edit Part from the pop-up menu. In the Part Properties pane (Fig. 9.126), clear the Pin Name (Number) Visible check box.

Note from Fig. 9.124 that a connector, J2, is used with the chassis ground symbol. J2 is a single pin connector with a single padstack footprint that is used to connect the buried shield to the chassis enclosure. When we begin working in PCB Editor, you will see how the three ground connections will be made on the board.

Fig. 9.127 shows the digital schematic page. Off-page connectors are used as described earlier. Unlike on the analog page, where each net in the bus had its own off-page connector, here the bus itself (and all the nets it contains) is attached to a single off-page connector that has the same name as the bus (e.g., Q[0..7]). Note also that off-page connectors are not used with the power and ground symbols, as they are global and known by all schematic pages in the design.

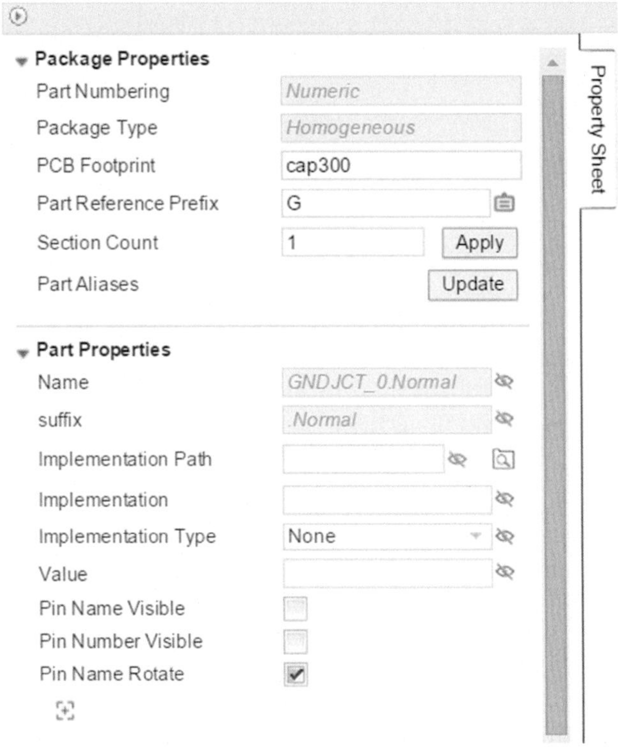

Figure 9.126 Making pin names and numbers invisible.

Setting up PSpice sources

Fig. 9.128 shows the PSpice page. Sources are VDC and VSIN, which can be found in the SOURCES.OLB library located in the Cadence `Tools/Capture/Library/PSpice` folder. Set the VDC and the VSIN source values as shown in the figure.

All PSpice simulations require a 0/GND symbol to which all sources can be referenced. The 0/GND symbol is included with the other GND symbols in the CAPSYM library. It is connected to both the analog and the digital grounds only during the simulation. After the circuit has been satisfactorily simulated, the 0/GND symbol must be deleted so that the different ground nets remain separate.

So that no footprints are added for the PSpice parts, make sure all PSpice parts are `PSpiceOnly = TRUE` and that the `PCB Footprint` cell is blank. To check these features double click a part to display the `Part Properties` spreadsheet. A partial spreadsheet is shown in Fig. 9.128.

Any parts that do not have PSpice templates will not be simulated. Parts U3 and U4 and all the connectors have no PSpice templates. When the simulation is run, they will be marked with green dots and ignored.

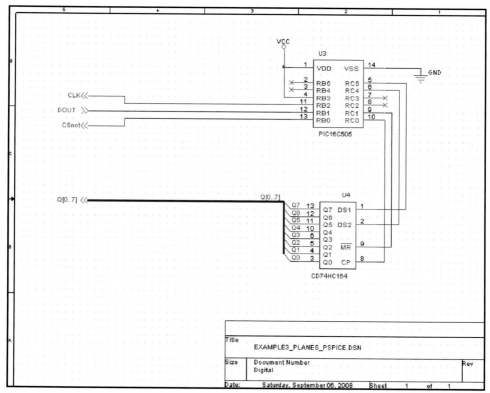

Figure 9.127 The digital schematic page.

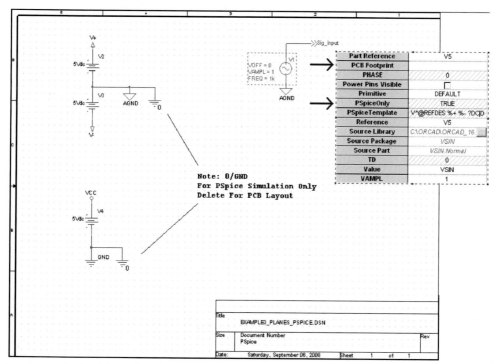

Figure 9.128 The PSpice simulation page.

Performing PSpice simulations

Once the circuit is made, a PSpice simulation profile needs to be established. A default simulation profile is included with the project, because the `PSpice Analog or Mixed A/D...` radio button was chosen during the project setup. All that needs to be done is to edit it for this design. *To edit the PSpice simulation profile*, select `Edit Simulation Profile` from the `PSpice` menu as shown in Fig. 9.129. If there is no simulation profile in your project, you can create the new one using `PSpice → New Simulation Profile`.

At the `Simulation Settings` dialog box (Fig. 9.130), select `Time Domain (Transient)` from the `Analysis type:` list. Enter a value in the `Run To Time:` box to display three or

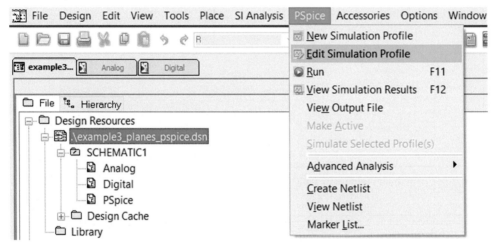

Figure 9.129 Editing the PSpice simulation profile.

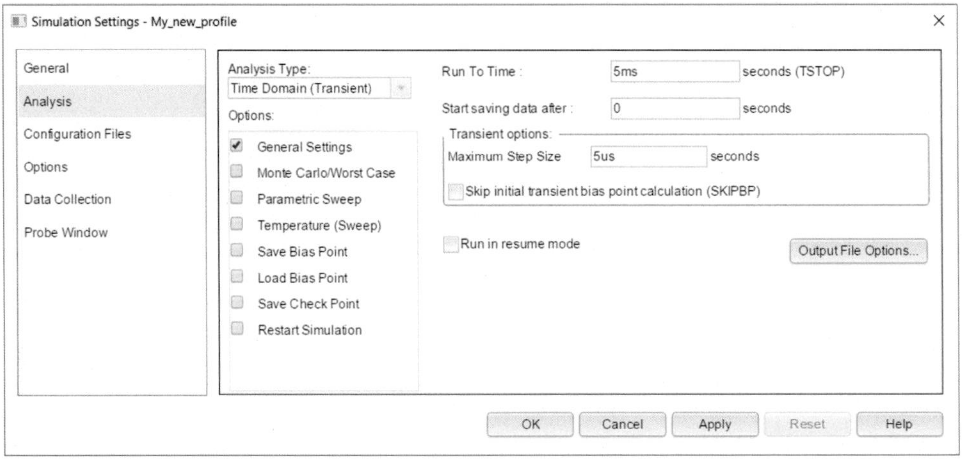

Figure 9.130 Setup for time domain analysis.

so complete cycles of the waveform (5 ms for a 1 kHz signal). You can specify a value in the Maximum step size: box also, but this is optional. A value that is about 1/1000 of the run time produces very smooth waveforms but takes longer to run. Click OK when you are finished.

Place the Voltage Markers on the Nets of interest in the Analog schematic page using PSpice → Markers (see Fig. 9.124 as an example with three markers placed). *To run the PSpice simulation*, click the Run PSpice button (*green triangle* button) located on the schematic page toolbar (shown in Fig. 9.131).

The PSpice results are shown in Fig. 9.132. Three voltage markers (probes) were placed on the design, but only two waveforms are displayed in the probe window,

Figure 9.131 Run PSpice button.

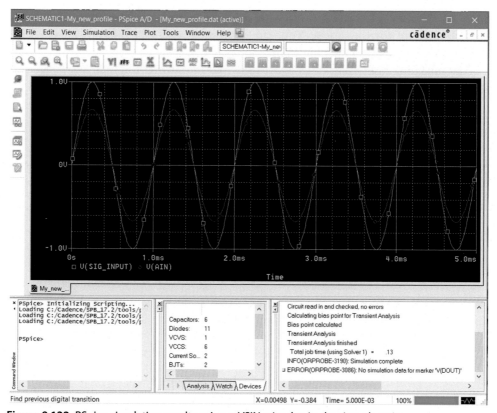

Figure 9.132 PSpice simulation results using a VSIN stimulus in the time domain.

because not all the parts in the design had PSpice models (PSpice templates) attached to them. PSpice will inform you that not all data were displayed by telling you, No simulation data for marker 'V(DOUT)', as is indicated in Fig. 9.132.

To perform time domain simulations, use the VSIN source; to perform frequency domain simulations, use the VAC source and select AC Sweep/Noise in the Simulation Settings dialog box (Fig. 9.130). Many other types of sources can be used to perform simulations. You can even create specialized stimulus files (including noise signals) using one of the VPWL_FILE sources. To see how to use these other sources, see *Help → PSpice Documentation* or *Help → Learning PSpice*.

Designing the board with PCB Editor

Once the PSpice simulations are complete and the circuit has been verified, the design is ready to be prepared for PCB Editor. One of the first tasks is to assign (or verify) footprint assignments for all parts. As described in the previous examples, a custom BOM can be generated to list the footprints to make it easier to identify missing or incorrect footprints (see the previous examples and Chapter 10: Artwork development and board fabrication, for details). A sample BOM for this design is shown in Table 9.6. Custom parts and footprints, such as CON1 and CONN14, are available from the website for this book. You can copy them to PCB Editor *symbols* folder before creating netlist and opening PCB Editor.

The remaining tasks were described in the preceding text or earlier examples and are listed here without details:
- Remove 0/GND symbols used for PSpice simulations.
- Perform an annotation to clean up part numbering (optional).
- Make sure that global power nets are properly utilized.

Table 9.6 Bill of materials footprint list for the dual-page plane example.

Reference	Part	Part library	Footprint
G1	GNDJCT	.../Ch9_CapturePartLib/CHAP9EXAMPLES.OLB	cap300
J1	CON14	ORCAD/.../CONNECTOR.OLB	conn14
J2	CON1	ORCAD/.../CONNECTOR.OLB	con1
R1	1 k	ORCAD/.../DISCRETE.OLB	smdres
R2	2 k	ORCAD/.../DISCRETE.OLB	smdres
R3	2 k	ORCAD/.../DISCRETE.OLB	smdres
R4	2 k	ORCAD/.../DISCRETE.OLB	smdres
U1	OP-27/LT	.../EXAMPLE3_PLANES_PSPICE.DSN	soic8
U2	TLC548	.../Ch9_CapturePartLib/CHAP9EXAMPLES.OLB	soic8
U3	PIC16C505	.../Ch9_CapturePartLib/CHAP9EXAMPLES.OLB	soic14
U4	CD74HC164	.../Ch9_CapturePartLib/CHAP9EXAMPLES.OLB	soic14

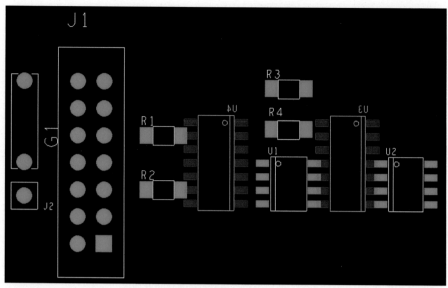

Figure 9.133 Initial part placement for Example 3.

- Perform a DRC in Capture to verify that the circuit design has no issues. Correct any errors and reperform the DRC as needed.
- Use Capture to generate the netlist and launch PCB Editor.

Create the board outline

As described in earlier examples, set the design size to a size A sheet (Setup → Design Parameters... → Design → Size), and draw a board outline (Outline → Design...).

A 3.00 × 2.00-in. or larger board is sufficient for this design. Use one of the Place Part tools to place parts inside the board outline. Place digital parts on the bottom side of the board, as shown in Fig. 9.133.

Placing parts on the bottom (back) of a board

To place parts on the bottom side of a board, select the General edit button, select Symbols in the Find filter tab, select the component, right click, and select Mirror from the pop-up menu. Left click off to the side to deselect the part. The part should now be on the bottom of the board. If you do not see the part, make sure the bottom Silk-Screen and associated layers are visible. Check the DRC for footprint and placement problems prior to doing anything else. Use Edit→Mirror to select and mirror several components at once.

Layer stack-up for a multiground system

The layer stack-up shown in Fig. 9.120 is used in this design. Set up Power and Ground planes, the Shield plane, and the analog and digital Routing layers as shown

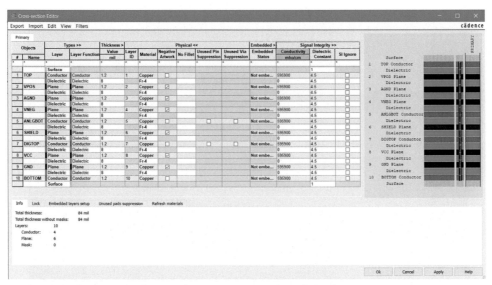

Figure 9.134 The board stack-up in the Cross-section Editor dialog box.

in the `Cross-section Editor` dialog box in Fig. 9.134 (click the `Xsection` button, ![icon]). To add a layer pair, right click and select `Add Layer Pair Above` (or `Below`) from the pop-up menu. Add 16 layers, then name and change the types as shown in the figure (eight Dielectric layers, two Routing layers, and six Plane layers). Select `Negative Artwork` for all the Plane layers, click the `Apply` button, then the `OK` button to complete the setup and dismiss the dialog box. Select distinctive colors to differentiate the planes using the `Color` tool as described in the earlier examples.

Once the parts are in place and the stack-up is defined, we can pour copper on the Plane layers, perform fan-outs, and begin routing traces on the board. We begin by pouring the copper on the planes.

Adding copper to the planes

As described in the previous examples, use the `Plane Outline` tool (`Outline → Plane...` from the menu bar) or one of the `Shape Add` tools (on the toolbar) to add the copper pours to the Plane layers. Remember that dynamic copper shapes are preferred over filled rectangles or static shapes and to assign each shape to its correct net.

Establishing net, plane, and constraint relationships

The next thing we want to do is to fan out power and ground, but before we can perform fan-outs or route traces, we need to set up blind via definitions and custom routing constraints to limit the layers on which certain nets can be routed. Table 9.7 shows a summary of which layers, vias, and physical constraint sets will be assigned to

Table 9.7 Relationships among nets, layers, vias and physical constraints.

Net	Routing	Plane	Via	Physical Cset
Analog nets	TOP, ANLGBOT	—	BBANLG	PCSanalog
V+	Top	VPOS	BBANLG	—
AGND	Top	AGND	BBANLG	—
V−	Top	VNEG	BBANLG	—
SHLD	—	SHIELD	VIA (modified)	—
VCC	BOTTOM	VCC	BBDIG-VCC-BOTTOM	—
GND	BOTTOM	GND	BBDIG-GND-BOTTOM	—
Digital bus nets	DIGTOP, BOTTOM	—	BBDIG-DIGTOP-BOTTOM	PCS1
Digital ADC nets	TOP, DIGTOP, BOTTOM	—	VIA	PCS2

each type of net. Blind and buried via definitions are described in the next section, and physical constraint definitions are described in the following section.

Defining blind vias

To use a blind or buried via in a design, an appropriate padstack must exist so that it can be used. If one does not exist, you need to make one. There are two ways to do this, as described next.

Padstacks used as vias exist as *library padstacks* and *board design (layout) padstacks*. A *library padstack* is a padstack definition contained in the *symbols* library. A *layout padstack* is a padstack definition associated with a pin or via in a board design. However, when a padstack is used in a board design, its definition (as used) is stored in the board layout file itself and not in any library. So there are two ways to work with via padstacks that are used as blind/buried vias, the first is through the Padstack Editor application and the second is through PCB Editor itself, which is the preferred method in most cases. Only a basic overview of the method using the Padstack Editor is given here. Using Padstack Editor is covered in detail in Chapter 8, Making and editing footprints. Via definitions for this design are made using the second method from within PCB Editor.

Basic overview of using Padstack Editor

To define and store a padstack in the PCB Editor library, you can create one from scratch or you can copy an existing padstack, modify it, then save it with a new name in the library using the Padstack Editor. You can launch it from within PCB Editor or independent of PCB Editor (called *stand-alone mode*).

To launch Padstack Editor from within PCB Editor select `Tools → Padstack` and select one of the `Modify padstack` options.

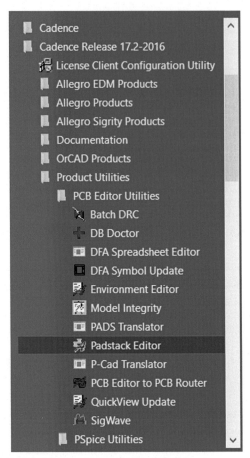

Figure 9.135 Launching Padstack Editor in stand-alone mode.

To make a padstack from scratch using the Padstack Editor in stand-alone mode, go to the Windows Start button on your tool tray and from the All Programs option, select Cadence Release 17.2-2016 → Padstack Editor, as shown in Fig. 9.135.

The Padstack Editor dialog box is shown in Fig. 9.136 opened in stand-alone mode. Using the Padstack Editor, you can construct through padstacks (for through-hole pins on components or for vias), blind/buried padstacks (for vias), or single-layer padstacks (for surface-mount component pins).

Padstack design is covered in detail in Chapter 8, Making and editing footprints, and mentioned only briefly here. In general, you set the drill diameter, pad shapes and sizes. In the Design Layers tab, you can specify layers in addition to the default ones. By selecting specific layers and certain types of connections, you can use this tool to construct blind/buried vias.

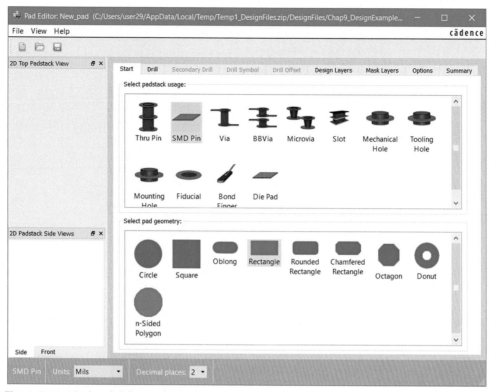

Figure 9.136 Padstack Editor.

You can also launch the Padstack Editor from PCB Editor by selecting `Tools →` `Padstack → Modify Padstack` from the menu. You can modify library padstacks or padstacks associated with the design only.

To use the Padstack Editor, you need to duplicate the layer stack-up in your board design and specify a specific layer/pad combination that satisfies the via requirements. The via padstack is saved to the library or the design and assigned to specific nets in the board design. This approach can be cumbersome, and the via is easily reusable in future designs only if they have an identical layer stack-up.

The other method is to set up vias interactively from the design using the `BBvia` tool. This is the easiest way to make a blind or buried via, because all the layer definitions are known, and set up, by the `BBvia` tool. Interactive design of blind and buried vias is initiated by selecting `Setup → Define B/B Vias`, as shown in Fig. 9.137, where you can select from two different methods of interactively designing custom blind and buried vias.

If you select `Define B/B Vias...`, you get the dialog box shown in Fig. 9.138. This method is used for the analog half of the board (see also Table 9.7). Give the via a

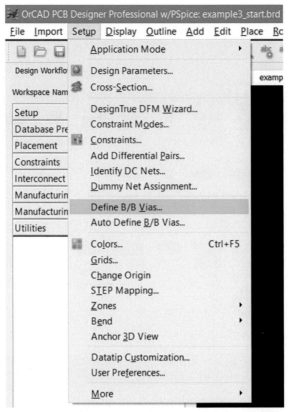

Figure 9.137 Two ways to interactively set up blind and buried vias in a design.

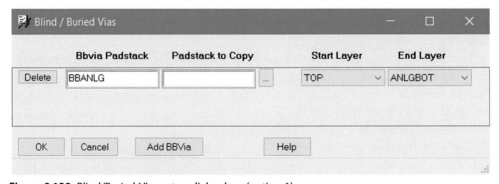

Figure 9.138 Blind/Buried Vias setup dialog box (option 1).

name (BBANLG in this example, for blind/buried analog) and select a padstack to copy from. This sets the basic definition of the new padstack (e.g., drill diameter and pad shapes and sizes). The Start and End layer entries define how "tall" the padstack will be (see Fig. 9.120). So in this example, if you choose VIA as the padstack to copy

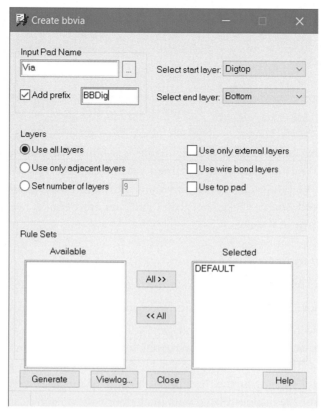

Figure 9.139 Blind/Buried Via setup dialog box (option 2).

(which of course is a through-hole padstack), the new padstack will be identical to it but will not go all the way through the board (it will go between only the Start and End layers). This method creates one padstack that goes from the `Start Layer` to the `End Layer` and includes all the layers between them. Click `OK` when finished or `AddBBVia` to make another one.

If you select `Auto Define B/B vias…`, then you get the dialog box shown in Fig. 9.139. We use this method to make the blind vias for the digital half of the board. Again select a padstack on which to base the new one. Check the `Add prefix` box and select a prefix name. Select the Start and End layers as before. You can select which layers will be used in setting up the via; we use all the layers in this example. Once the settings are entered, click `Generate` to start the process. Click `Close` when finished.

This method actually makes several vias. Essentially, it makes as many blind and buried vias as necessary to connect each of the adjacent planes (two layers at a time) and to connect the Start and End layers to each other and to each of the internal layers. You can view the `bbvia.log` file, and in this example, the following padstack vias were created:

- BBDIG-DIGTOP-VCC
- BBDIG-DIGTOP-GND
- BBDIG-DIGTOP-BOTTOM
- BBDIG-VCC-GND
- BBDIG-VCC-BOTTOM
- BBDIG-GND-BOTTOM

So any via that has BOTTOM in its name is a blind via, and the others where BOTTOM is not included are actually buried vias. One of the vias will never be used in this example, VCC-GND, because it would short the Power plane to GND and result in a scrapped board. Once the planes and vias are defined, we will assign the proper via(s) to each of the nets.

Assigning vias to nets

Use the Constraint Manager to assign vias to the nets. Open Constraint Manager by clicking the `Cmgr` button, [icon], on the toolbar. Select the `Physical` tab, and select `All Layers` under the `Net` folder. In the `Vias` column, you will see a listing of all vias in the design (including the default, `Via`). Select a net to modify by selecting the cell in the `Vias` column for that net. The `Edit via list` dialog box will be displayed, as shown in Fig. 9.140.

Note that the `BBANLG` is in the "available" list but not in the "used" list. Add the `BBANLG` via to all of the analog-related nets (including `V+`, `V-`, and `AGND`). Remove any vias that are prohibited or unnecessary. The final setup is shown in Fig. 9.141. Notice that the `CLK`, `CSNOT`, and `DOUT` nets use the default via, `VIA`, since it is a through-hole padstack, and these nets need to pass through the Shield layer (to connect the ADC on the analog side to the microcontroller on the digital side). Although the `SHLD` net does not need a via, the default via, `VIA`, is assigned to it because each net must have at least one via assigned to it, so the default via was assigned. Next we see how to restrict the routing of certain nets to specific layers.

Assigning nets to layers using custom, physical constraints

The next step is to assign nets to the proper Routing and Plane layers. The net layer assignments are shown in Table 9.7. To make net layer assignments, we need to set up new *physical constraint set* (CSet)—the custom set of rules which we can use later for some of our nets. *To create the new Physical CSet*, left click on the `All Layers` icon in the `Physical Constraint Set` folder of Constraint Manager, and select `Create →` `Physical Cset` from `Objects` menu (Figs. 9.142 and 9.143). Set the new name of your custom CSet and press OK. You will see the new row in the list of CSets, below the `DEFAULT` constraint set.

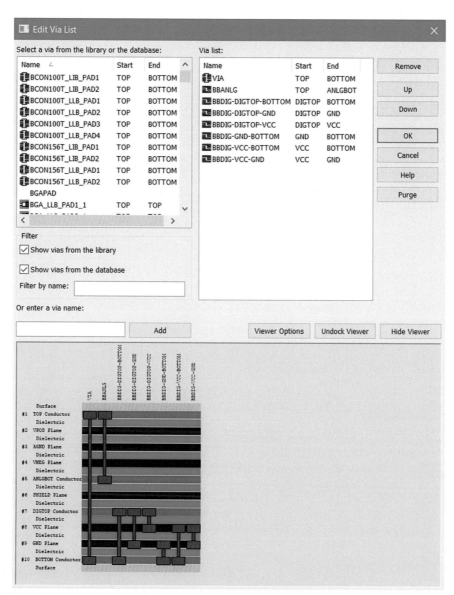

Figure 9.140 Using the Constraint Manager to assign vias to nets.

In this example, we use the constraint set to limit routing to specific layers, so the appropriate flag is set to FALSE under the Allow Etch column in the certain layers, as shown in Fig. 9.144.

Once the constraint set is defined, assign it to proper nets (see Table 9.7). *To assign a constraint set to a net*, select the All Layers icon under the Net folder (see Fig. 9.145).

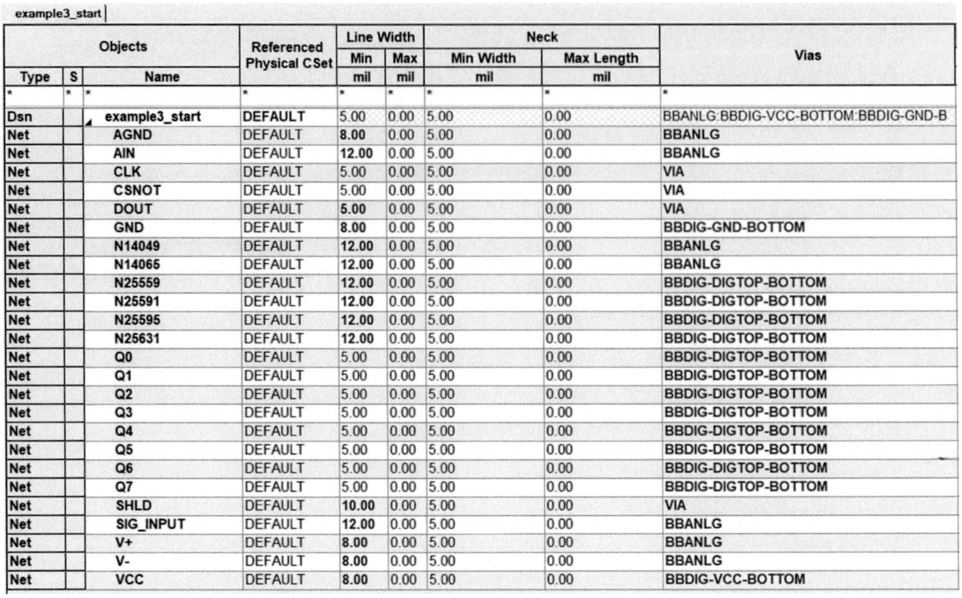

example3_start

Type	S	Name	Referenced Physical CSet	Line Width Min mil	Line Width Max mil	Neck Min Width mil	Neck Max Length mil	Vias
*	*	*	*	*	*	*	*	*
Dsn		example3_start	DEFAULT	5.00	0.00	5.00	0.00	BBANLG:BBDIG-VCC-BOTTOM:BBDIG-GND-B
Net		AGND	DEFAULT	8.00	0.00	5.00	0.00	BBANLG
Net		AIN	DEFAULT	12.00	0.00	5.00	0.00	BBANLG
Net		CLK	DEFAULT	5.00	0.00	5.00	0.00	VIA
Net		CSNOT	DEFAULT	5.00	0.00	5.00	0.00	VIA
Net		DOUT	DEFAULT	5.00	0.00	5.00	0.00	VIA
Net		GND	DEFAULT	8.00	0.00	5.00	0.00	BBDIG-GND-BOTTOM
Net		N14049	DEFAULT	12.00	0.00	5.00	0.00	BBANLG
Net		N14065	DEFAULT	12.00	0.00	5.00	0.00	BBANLG
Net		N25559	DEFAULT	12.00	0.00	5.00	0.00	BBDIG-DIGTOP-BOTTOM
Net		N25591	DEFAULT	12.00	0.00	5.00	0.00	BBDIG-DIGTOP-BOTTOM
Net		N25595	DEFAULT	12.00	0.00	5.00	0.00	BBDIG-DIGTOP-BOTTOM
Net		N25631	DEFAULT	12.00	0.00	5.00	0.00	BBDIG-DIGTOP-BOTTOM
Net		Q0	DEFAULT	5.00	0.00	5.00	0.00	BBDIG-DIGTOP-BOTTOM
Net		Q1	DEFAULT	5.00	0.00	5.00	0.00	BBDIG-DIGTOP-BOTTOM
Net		Q2	DEFAULT	5.00	0.00	5.00	0.00	BBDIG-DIGTOP-BOTTOM
Net		Q3	DEFAULT	5.00	0.00	5.00	0.00	BBDIG-DIGTOP-BOTTOM
Net		Q4	DEFAULT	5.00	0.00	5.00	0.00	BBDIG-DIGTOP-BOTTOM
Net		Q5	DEFAULT	5.00	0.00	5.00	0.00	BBDIG-DIGTOP-BOTTOM
Net		Q6	DEFAULT	5.00	0.00	5.00	0.00	BBDIG-DIGTOP-BOTTOM
Net		Q7	DEFAULT	5.00	0.00	5.00	0.00	BBDIG-DIGTOP-BOTTOM
Net		SHLD	DEFAULT	10.00	0.00	5.00	0.00	VIA
Net		SIG_INPUT	DEFAULT	12.00	0.00	5.00	0.00	BBANLG
Net		V+	DEFAULT	8.00	0.00	5.00	0.00	BBANLG
Net		V-	DEFAULT	8.00	0.00	5.00	0.00	BBANLG
Net		VCC	DEFAULT	8.00	0.00	5.00	0.00	BBDIG-VCC-BOTTOM

Figure 9.141 Via-net assignments.

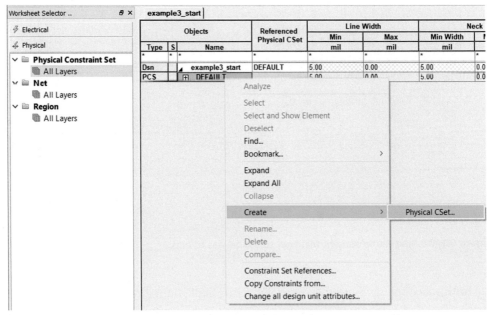

Figure 9.142 Setting up a new constraint with the Constraint Manager.

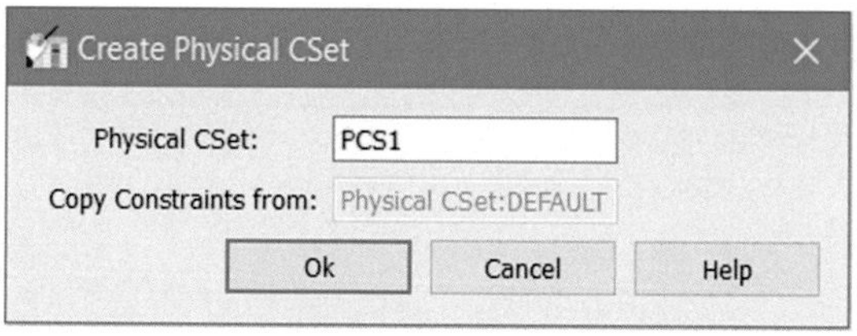

Figure 9.143 The Create Physical CSet dialog box.

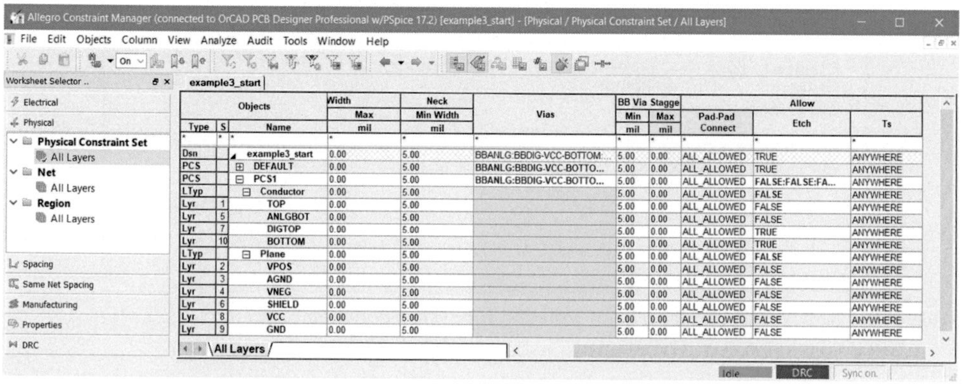

Figure 9.144 The new physical constraint listed in the Constraint Manager.

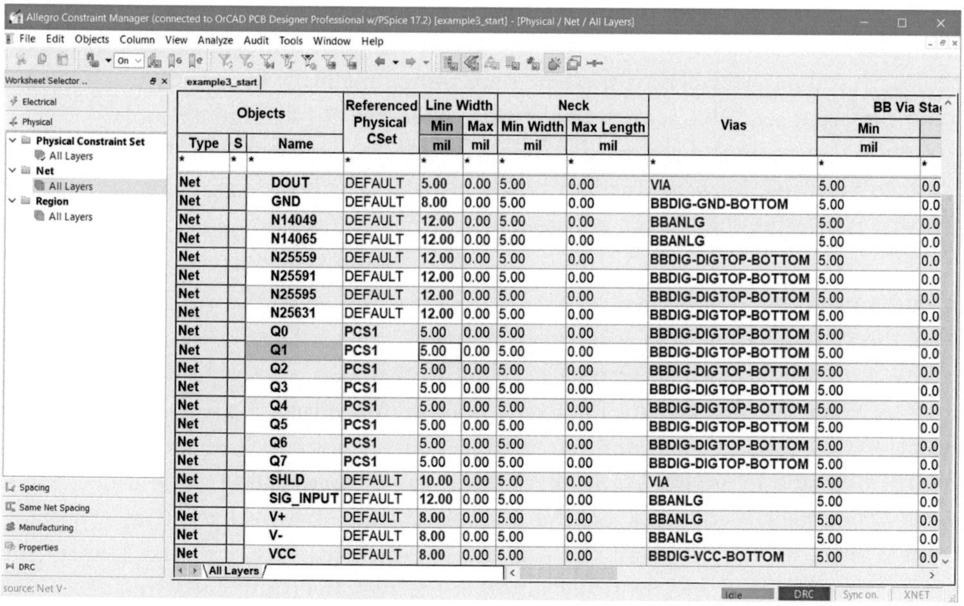

Figure 9.145 Assigning constraints to nets.

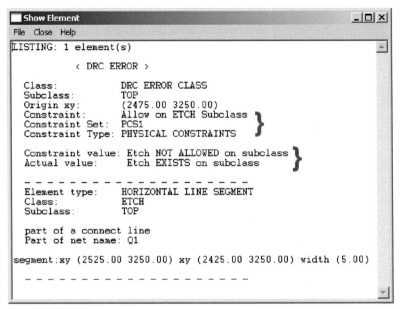

Figure 9.146 DRC errors resulting from violating a physical constraint set.

Select (using left mouse button) the cell or cells in the Referenced Physical CSet column for the nets to which you want to assign the new constraint. When you select the cell(s), a dropdown list will be displayed. Select the desired constraint set from the dropdown list. As shown in the figure, the new constraint, PCS1, was selected for the digital bus traces. With this constraint the autorouter will route these traces only on the DIGTOP and BOTTOM layers. If these nets are routed manually on a layer not included in the constraint set, DRC errors will be issued (see Fig. 9.146 as an example).

Continue naming and assigning constraint sets to the nets as outlined in Table 9.7.

Once vias have been assigned to nets and nets have been assigned to layers, the next step is to begin the fan-out process.

Fan-outs using blind vias

Once the vias are set up and assigned to the nets, we can begin performing the fan-outs. We do the first couple of ones by hand so that you can see how it works. We begin by doing the first fan-out for VCC. Make all the Plane layers and nets invisible except VCC. Set the Etch grid to 25 mil (All). Zoom to the area around U3 (PIC16C505) and locate pin U3.1 (check the Pins option in the Find tab if necessary). Select Add Connect... and click on the pin (make sure Pins is checked in Find tab). Route a small section out from the pad and left click to place a vertex. Display the Options tab (see Fig. 9.147) to make sure that the two layers involved are Bottom and Vcc and that the BBDIG-VCC-BOTTOM via is available. Then right click and select Add via from the pop-up menu. A bbvia should be inserted. Right click and select Done. Note

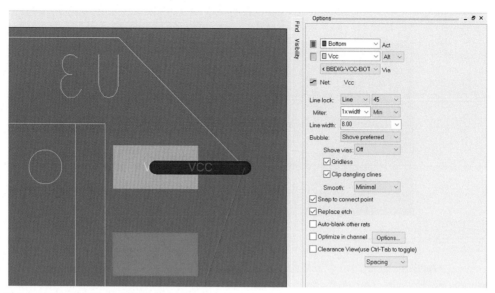

Figure 9.147 Routing a blind via to Vcc.

that the correct `Alt` layer has to be displayed in the `Options` tab or you may not be able to place a via if the via assigned in the Constraint Manager cannot physically make a connection from the `Act`. (active) layer to the `Alt`. (alternate) layer displayed in the `Options` tab.

If you toggle back and forth between the `Bottom`, `Vcc`, and `GND` layers, you should see that the via connects the trace and pad on the bottom to the VCC plane, and a clearance area around the via is on the GND layer. If you toggle through the analog planes, you should see no evidence of the trace or via at all.

Next we try one of the analog fan-outs. Locate resistor R3 and its pin R3.2. We will route a fan-out from this pin to the analog Ground plane. In the Constraint Manager, make sure that the `AGND` net (and layer) is visible and able to be routed. If you don't see AGND or GND net in a list, check if you removed the connections of AGND and GND to PSpice 0 net in the third page of schematic design which was made for simulation purpose. Select the `Add Connect...` tool, click on the R3 pad to start a trace, and left click a short distance (about 50 mil) to place a vertex. Check the `Options` tab and make sure that the active layer is `Top`, and the alternate layer is `AGND`, and that the `BBANLG` via is available. Right click in the work space and select `Add Via` from the pop-up menu.

Note that the process of using the blind via fan-outs is the same no matter which method was used to generate them. Now we use the autorouter to place the rest of the fan-outs.

We fan out the remaining `AGND` and `Vcc` nets and complete all the `V-`, `V+`, and `GND` fan-outs using the autorouter. Open the Constraint Manager, and make sure that all the power, and ground nets are visible, and no routing restrictions have been placed on them. Select `Route → PCB Router → Route Automatic...` from the menu. As described in detail in the

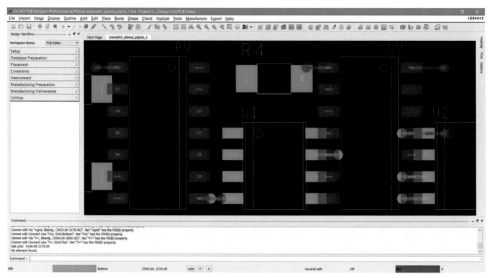

Figure 9.148 Fan-outs completed using blind vias.

previous examples, use the Automatic router dialog box to set the fan-out parameters. As a summary select the Router Setup tab, and in the Strategy group box, select the Specify routing passes radio button. Select the Routing Passes tab, and select only the Fanout option in the Pass Type column. Click the Params... button to display the router parameters dialog box. Select the desired options, such as Power Nets, and click OK to dismiss the dialog box. Click the Route button to perform the fan-out. You can also use the Create fanout tool to fan out pins and components.

An example of a completed fan-out is shown in Fig. 9.148. Notice where a via goes from Bottom to GND (digital GND) on U3 right under the corner of the pad on R4. Note also that this does not cause a DRC error, because the via under the pad of R4 does not go all the way through the board and therefore does not touch R4. Other occurrences of this type are shown on the pads for the ICs.

Once all the fan-outs are complete, the rest of the board can be routed. Set up the autorouter as described in the previous examples, and route the rest of the board. As a quick overview, select Route → PCB Router → Route Automatic... from the menu. At the Automatic router dialog box, select the Router Setup tab, and in the Strategy group box, select the Specify routing passes radio button. Select the Routing Passes tab, uncheck the Fanout option in the Pass Type column, and select the Route and Clean options. Click Params... and check the Signal Nets checkbox.

Click OK to dismiss the dialog box. Click the Route button to route the board.

Note: Designs like this can be tricky. If the autorouter has difficulty routing the board or performing the fan-outs, make sure that all the Plane layers and design constraints are set up properly, the proper vias are assigned to the correct nets, and the appropriate nets are enabled.

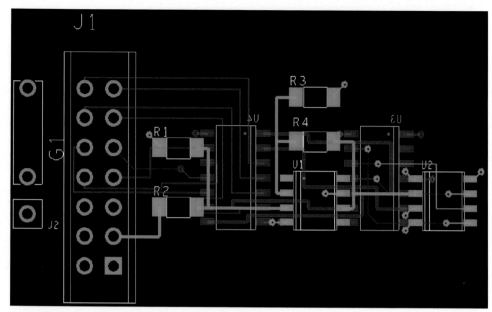

Figure 9.149 The final board design for Example 3.

An example of the fully routed board is shown in Fig. 9.149. The grid and all the Plane layers are turned off so that it is easier to see the traces.

Alternate methods of connecting separate Ground planes

In this example the two Ground planes are electrically separate and the component G1 in the schematic is used to connect the planes on the board by soldering a wire jumper, an inductor, or ferrite bead in the G1 footprint. In the following sections, alternatives to inserting and soldering the wires or components are discussed. The first alternative is to short the G1 padstacks with copper, and the second alternative is to short the planes with a shorting padstack.

Shorting the planes with copper etch

To reduce assembly complexity, a copper etch object can be used to short G1's padstacks rather than soldering a wire or installing a component into the footprint. The copper etch can be a trace or a copper area. *To add a thick trace* across the padstacks, select the Etch/Top classes in the Options pane. Select the Add Line tool, and from the Options pane, set the line width to 30 mil or so. Left click on the first padstack of G1 and draw a line (trace) to the second padstack. The copper etch line is shown in Fig. 9.150A.

Another way to make the connection is to use a copper area. To do so, select the Shape Add Rect tool. In the Options pane, select Static solid as the Shape Fill Type. You can leave the Dummy Net assigned to the shape. The static solid is shown in Fig. 9.150B.

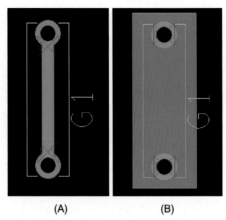

Figure 9.150 Adding copper etch, (A) a trace or (B) a shape, to short the planes.

When using either of these etch objects, DRC errors will occur because the shapes violate pad spacing rules. Since this is what we want, we can override the DRC errors. To do so, click the `Waive DRC` button on the toolbar or select `Check → Waive DRCs → Waive` from the menu. Make sure that the `DRC Errors` option is checked in the `Find` filter pane, then left click on the DRC markers to override them. Another way to avoid the DRC errors is to attach the `NET_SHORT` property to this shape. Select the shape, and right click, then choose `Net Short` from the pop-up menu, and then left click over each net to be connected together. Right click and choose `Complete Net Short`.

Shorting the planes with a Padstack

Rather than take up space on the board using a dedicated footprint, you can use a via attached to one of the Ground planes and modify it so that it is attached to both Ground planes at the same time. We place a `VIA` on the board, attach it to the `GND` plane, and modify it so that it is connected to the `AGND` plane too.

When using this approach, the special part (G1) in Capture is not used. Instead a small segment of a graphical line—using the `Place Line` tool instead of the `Place Wire` tool—is used to indicate that the two grounds are connected at the header pin. This eliminates the footprint in PCB Editor.

Go back the schematic page, and delete part G1 in Capture. Use the `Place Line` tool to make the `AGND` and `GND` nets look like they are connected as shown in Fig. 9.151. The `Line` tool is graphical only and does not create a connection in the netlist.

Perform an ECO to forward annotate the changes to PCB Editor as described in the earlier examples. In short, save the board design, and close PCB Editor. In Capture select the project icon, select `Tools → Create netlist...` from the menu. Select the `PCB Editor` tab, check the `Create or Update PCB Editor Board` box, and relaunch PCB Editor. When the board design is reopened, the footprint for G1 should be gone.

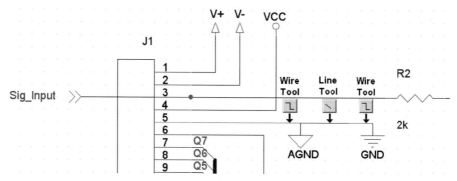

Figure 9.151 Placing a "virtual" ground connection using the Line tool.

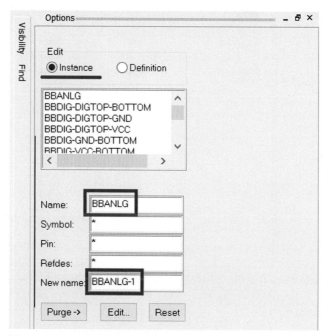

Figure 9.152 Modifying a via to short the two Ground planes.

The next step is to place a generic via on one of the planes and modify the via to connect it to the other plane. To do so, you should add VIA to the list of available vias for GND net in Constraint Manager. Then select the Add Connect tool, and from the Options pane, make the GND plane (class) as the active class and the Top layer as the alternate. Select Shapes in the Find pane. Left click on the GND plane near the connector J1 to place a vertex and start a trace. Immediately right click and select Add Via from the pop-up menu; right click again and select Done from the pop-up menu. The via (which should be padstack VIA) is now connected to the GND plane but nothing else.

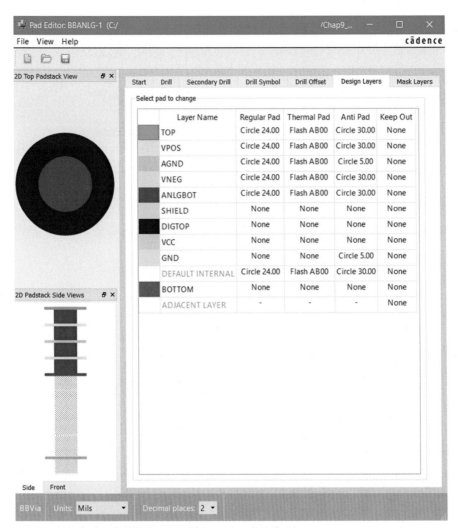

Figure 9.153 Use Padstack Editor to modify a padstack definition.

To modify the via, select Tools → Padstack → Modify Design Padstack... from the menu. Left click the via in the canvas to select it. In the Options pane, select the Instance radio button (see Fig. 9.152). The original VIA name will be listed, and a new via VIA-1 will also be listed. Click the Edit... button to display Padstack Editor. The Padstack Editor, Design Layers tab, is shown in Fig. 9.153 with the new padstack settings. Note that the Anti Pads are set to 5 mil in those layers we want to connect, which is much smaller than the drill diameter. So, when the holes are drilled into the board during the manufacturing process, the small clearances will be drilled away, and the copper on the planes will butt up against the drill hole. Then, when the hole is plated, the plating will short the planes

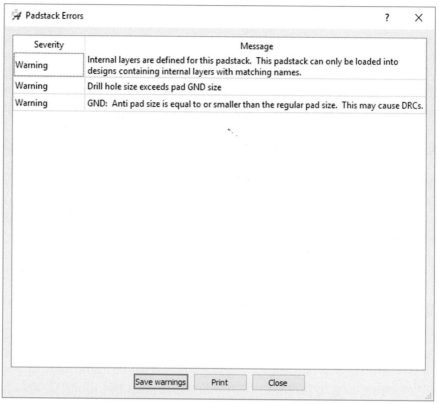

Figure 9.154 Pads drilled away warning.

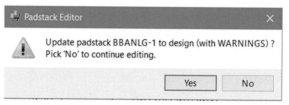

Figure 9.155 Save padstack modifications with warnings.

together. Remember to add clearances to the other plane layers (e.g., Vcc and SHLD), or they will be shorted to the planes too.

To save the changes, select File → Update to Design and Exit... from the menu. A warning box will be displayed telling you that the Anti Pad will be drilled away (see Fig. 9.154). That is what we are after, so close the warning box. Another warning will be displayed (see Fig. 9.155). Click Yes to complete saving the changes and quit editing.

To allow via to connect two nets together without creating DRC marker, switch to General Edit mode, check only Vias in the Find pane, select the via, right click and choose

`Net Short`. Then left click over each net to be connected together. Right click and choose `Complete Net Short`.

Note: This method is not necessarily recommended, since connection between the two Ground planes is not automatically "documented" by the software. The process should be manually documented on the schematic, and some type of marker should be placed in silk screen on the board, indicating which via is shorting the two planes together. In the event that some problem occurs and the planes need to be separated, it would be a simple matter to drill out the via. If the via is not marked somehow, it would be impossible for someone not familiar with the board design to know where or how the planes were shorted together. Even if a person were to look at the design in PCB Editor without some markings in the design, the only way to tell which via is shorting the planes is to look at all of the padstack definitions or create a Waived Design Rules Check Report (if the person happened to think of it) or look for `NET_SHORT` properties.

This concludes the third design example.

Example 4. High-speed digital design

This example demonstrates how to stack-up layers and design transmission lines for a high-speed digital PCB. The example also demonstrates how to create a moated ground area with a bridge around a high-frequency crystal oscillator, how to perform pin/gate swapping, and how to create a heat spreader using vias to the Ground plane. The example circuit is shown in Fig. 9.156.

The BOM for this example is shown in Table 9.8. The circuit consists of a (fictional) high-speed, low-pin-count microcontroller/digital-signal processor (uP-EXD10) driven

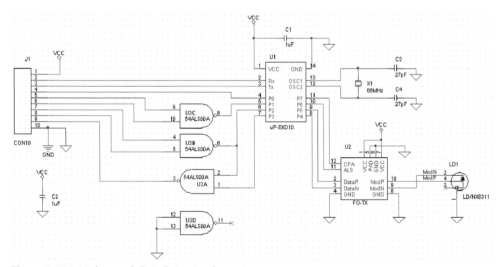

Figure 9.156 High-speed digital circuit schematic.

Table 9.8 Bill of materials for the digital design example.

Item	Quantity	Reference	Part	Nomenclature	Footprint
1	2	C1, C2	1 μF	Bypass capacitor	smdcap
2	2	C3, C4	27 pF	XTAL shunt capacitor	smdcap
3	1	J1	CON10	10-pin board connector	conn10xx100tr
4	1	LD1	LD/NX8311	Fiber optic laser diode	to18-4
5	1	U1	uP-EXD10	Microcontroller/DSP IC	soic14
6	1	U2	FO-TX	Digital-to-fiber interface IC	plcc12
7	1	U3	54ALS00A	NAND logic gate	soic14
8	1	X1	66 MHz	Crystal oscillator	XTAL2smd

IC, integrated circuit.

by a 66-MHz clock (X1), a digital-to-fiber optic interface IC (FO-TX, which mimics an ADN2530 but with fewer pins), a fiber-optic laser diode (LD1), and a couple of 54ALS00 NAND gates used for I/O decoding. The digital signals have rise and fall times from 200 ps to 1.9 ns and require controlled-impedance traces (see the Analog Devices ADN2530 data sheet for an example application). In a real design, more bypass capacitors would be used on the circuit, but the design is scaled down to keep the design simple. The parts and footprints are located on the website for this book.

Using the procedures described in the earlier examples, start a new Capture project, and place and connect the parts as shown in Fig. 9.156. After completing the schematic, make sure all footprints are assigned to the parts and create the netlist for and launch PCB Editor.

Start a new board project using the procedures described in the previous examples.

As in the previous examples, the first step is to make a board outline and place the parts inside the boundary. The initial board layout is shown in Fig. 9.157. Signal flow is from left to right, with the highest frequency components located close together near the laser diode connector on the right side of the board.

The next few steps were covered in detail in the previous examples. The following tables and figures show the design parameters for this example, but step-by-step instructions are not repeated here. The required steps are to (1) define the layer stack-up and enable the appropriate layers using the Cross Section Editor dialog box, (2) define two vias in addition to the default VIA using the Padstack Editor and Constraint Manager, and (3) fan out power and ground for the surface-mounted components.

Layer setup for microstrip transmission lines

Since there are so few parts, a simple four-layer board design is used. The layer stack-up and net assignments are shown in Fig. 9.158. The layer thicknesses depend on the board manufacturer; the values (units in mil) shown in the figure are typical. The Top layer and Ground plane will be used to route surface-type microstrip transmission lines

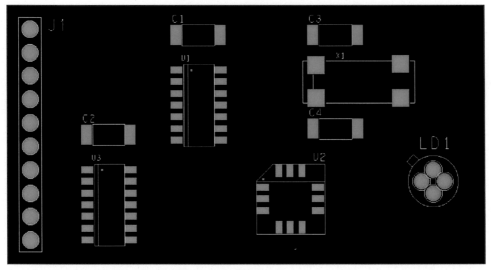

Figure 9.157 Initial board layout for the digital design example.

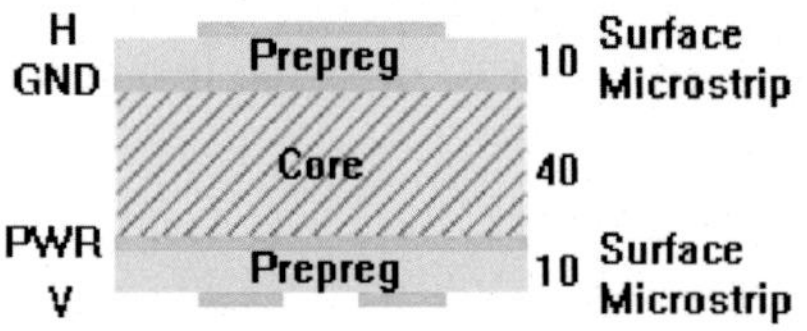

Layer Name	Layer type	Used by nets:
Top	Routing	Any
Ground	Plane	GND
Power	Plane	PWR
Bottom	Routing	Any

Figure 9.158 Layer stack-up for the digital example.

and most of the lower-speed digital traces. Only low-speed traces that cannot be routed on the top will be routed on the bottom layer.

Fig. 9.159 shows the layer stack-up as defined in the `Cross Section Editor` dialog box. Notice that the material thicknesses have been added.

Three via types are used in this example (see Table 9.9). `VIA` is the default via included in the *symbols* library folder, while `VIATENT` and `VIAHEAT` are custom vias (included in the design folder on the book's website). `VIATENT` is similar to the default `VIA` but is tented (i.e., the padstack contains no soldermask opening definitions) and used here for the fan-outs as a demonstration. Multiple copies of `VIAHEAT` will be used as heat pipes to connect a thermal pad (a copper pour area) beneath U2 to the Ground plane and function as a heat spreader. `VIAHEAT` has smaller dimensions so that they can be placed close together to provide low resistance to heat flow to the ground plane.

Constructing a heat spreader with copper pours and vias

The design of heat spreaders on PCBs depends significantly on the type of device and how it is attached to the board. For design examples and thermal management

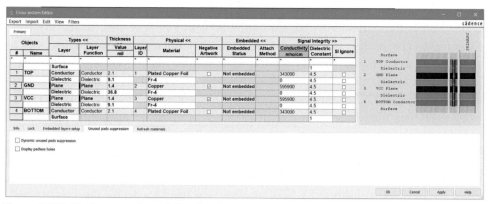

Figure 9.159 The layer stack-up defined in the Layout Cross Section dialog box.

Table 9.9 List of vias used in the high-speed digital example.

Via name	Function	Drill dia.	Pad dia.	Clearance dia.	Connection to plane	Soldermask opening
Via	Default	13	24	30	Full	Yes
ViaTent	Fan-outs	13	24	30	Full	No
ViaHeat	Heat pipes	10	20	26	Full	Yes

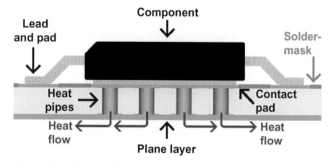

Figure 9.160 Functional diagram of a heat spreader.

calculation, see the application note references listed in Appendix E. The heat spreader demonstrated here is based on design suggestions described in the ADN2530 data sheet.

Before the board is fanned out or any traces are routed, the heat spreader is put into place so that the router avoids that area, thereby preventing having to rip up and reroute fan-outs or traces. A functional diagram of one type of heat spreader is shown in Fig. 9.160. The silicon die inside the component is thermally bonded to a metal pad on the bottom of a specially designed package. The pad is in turn thermally bonded (either by soldering or thermal compound) to a copper area on the top layer of the PCB. The copper area has an opening in the soldermask and multiple vias to connect it to a Plane layer (either ground or power depending on the chip design). The vias function as thermal conductors (heat pipes) that allow heat to flow away from the component. If a component dissipates excessive heat, the Plane layer can be mechanically (and thermally) connected to a larger heat sink or other mounting hardware to help dissipate the heat.

Via design for heat spreaders

Before we begin to construct the heat spreader, we need to define the VIAHEAT padstack. To make the heat pipe efficient at conducting heat, solid connections to the plane are used rather than thermal reliefs. To define the new via (VIAHEAT.pad), we begin with an existing padstack, save it with a new name, modify it to our specifications, then save it. If you have your board design open, launch the Padstack Editor by selecting Tools → Padstack → Modify Design Padstack... from the menu bar. Display the Options tab, select Via from the list, then click the Edit... button.

Before making any changes to the padstack choose File → Save as... from the menu bar. An information window will be displayed with the following warnings:

    Drill hole size exceeds pad size.

Click the Close button to dismiss the window (this will be explained shortly). When asked Save with warnings? click Yes. Save the padstack as VIAHEAT.pad in either the working directory or the *symbols* library.

Using the Parameters and Layers tabs as shown in Fig. 9.161 modify the padstack per the specifications in Table 9.9.

Since the top and bottom pads in this padstack are smaller than the drill diameter, they will be drilled out. Normally, we would not want this, but in this example, many vias will be placed side by side, and a copper rectangle will be placed over the whole group and will act as one big pad for the entire group of vias. Top and bottom pads can be left in place, but it is visually more appealing in the design if the pads are not shown.

Note also that the soldermask is left rather large but could have been removed because, as will be shown later, one large soldermask opening will be placed on the board to expose the heat spreader and heat pipes and will overwrite the individual soldermask openings.

After the VIAHEAT padstack is finished and saved, repeat the process to create the tented via, VIATENT, which will be used as the default fan-out via. The VIATENT padstack is identical to the default VIA padstack, except that no soldermask openings are defined, as shown in Fig. 9.162.

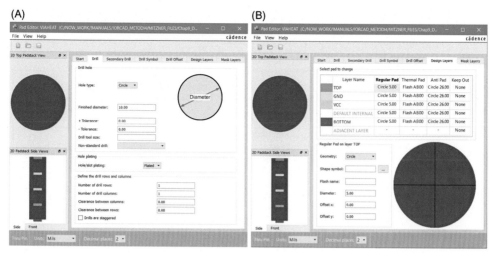

Figure 9.161 Use the Padstack Editor to define the heat pipes: (A) Drill tab and (B) Design Layers tab.

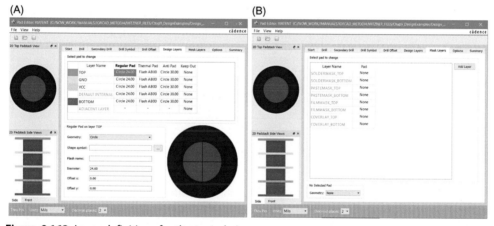

Figure 9.162 Layer definitions for the tented via.

Once the VIAHEAT padstack is finished, use the Constraint Manager to assign it to the GND net, as shown in Fig. 9.163. *To assign a via to a net*, left click in the cell to display the Edit Via List dialog box, as shown in Fig. 9.164. Left click a via twice in the left-hand box to add it to the Via list box.

You can also remove vias from the Via list by left clicking the one you want to remove, then clicking the Remove button. The via at the top of the list will be the default via. To change the priority of use, select the desired via and click the Up or Down button to move the via within the list. When you are finished, click OK. The ground layer will need the VIATENT via for the fan-outs and the VIAHEAT via for the

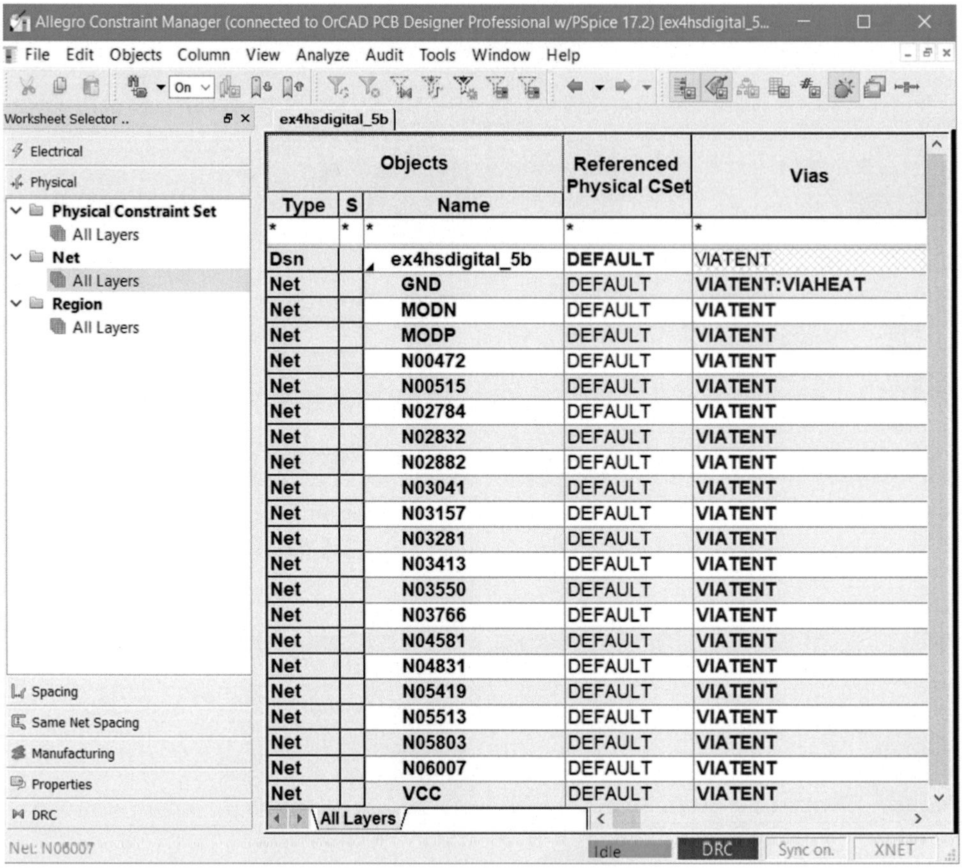

Figure 9.163 Use the Constraint Manager to assign vias to nets.

heat spreader, so assign both vias to the GND net. All other nets get only the VIATENT via, and no nets should be allowed to use the VIA via (select VIA and click the Remove button).

Now that the heat pipes have been defined, we can build the heat spreader. We begin by placing a copper plate on the Top layer to define the boundaries of the heat spreader. To do this click the Shape Add Rect button on the toolbar, display the Options pane, and select the options shown in Fig. 9.165.

Draw a copper pad in the center of the footprint, as shown in Fig. 9.166. *Note*: You may need to adjust the grid settings (Setup → Grids) to achieve the necessary drawing resolution.

The next step is to place the heat pipes. To do this, begin by clicking the Add Connect button. Display the Options pane, and select GND as the active layer (Act) and Top as the alternate layer (Alt). Left click on the design at the insertion point. Display

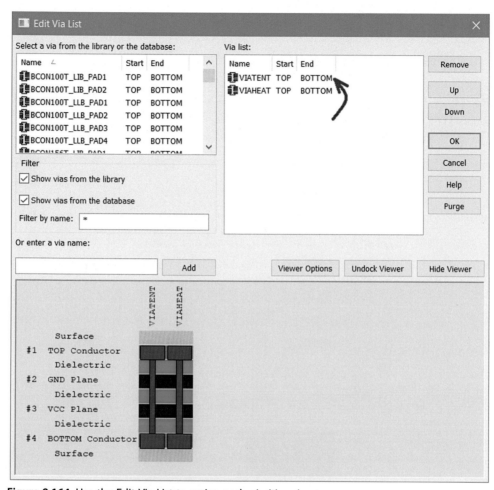

Figure 9.164 Use the Edit Via List to assign and prioritize via usage.

the `Options` pane again, and select `VIAHEAT` from the via list (see Fig. 9.167). Move your mouse back over to the design area, right click and select `Add via` from the pop-up menu. A `VIAHEAT` via should be placed. Repeat this process for each pipe. See Fig. 9.168 for reference.

Note that you can easily place many vias at once using the via array generator. As an alternative to placing the heat pipes in the board design, you can make a custom package that includes the heat pipes and copper plate.

For a good thermal connection to occur between the package and the heat spreader, an opening needs to be made in the soldermask. *To make an opening in the soldermask*, select the `Shape Add Rect` button on the toolbar, then select the `Package`

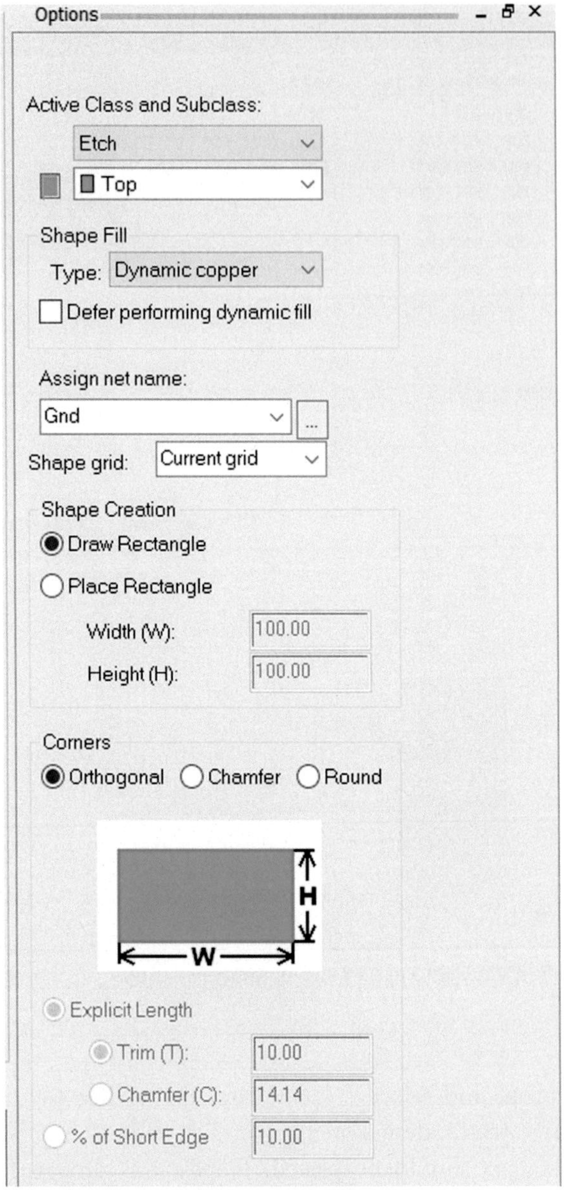

Figure 9.165 Options settings for the heat spreader plate.

Geometry class and Soldermask_Top subclass from the Options pane. Draw a filled rectangle over the copper pad that is 5 mil larger than the pad on all sides. If the thermal bond will be made by SMT soldering, repeat the preceding steps to place a filled rectangle on the Package Geometry class and Pastemask_Top subclass.

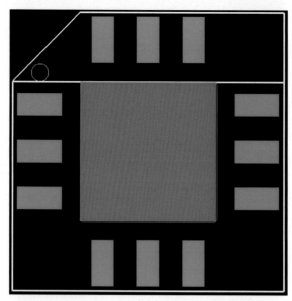

Figure 9.166 The heat spreader placed within the component footprint.

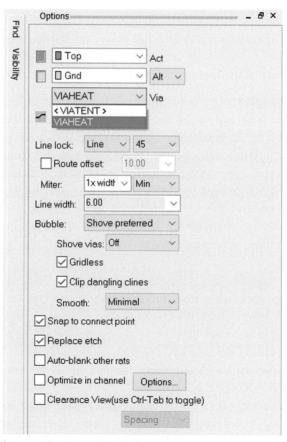

Figure 9.167 Select the tented via with the Options pane.

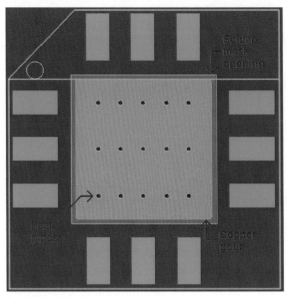

Figure 9.168 The final heat spreader design.

Note: Many of the footprints included in the library do not have a pastemask defined in the `Package Geometry` class. So, if you are planning to use pick and place assembly, you must modify some of the footprints to include the pastemask definitions.

The finished heat spreader is shown in Fig. 9.168. At this point click the `Fix` button (the *green thumb tack*), and fix all the vias and the dynamic copper plate so that they are not removed by the autorouter or any routing or cleanup processes.

The next step is to fan out the board. Earlier, we used the Constraint Manager to establish *net* default vias, but when the autorouter is used to fan out a board, it uses the *design* default vias. *To change the default design via*, open the Constraint Manager, select the `Physical` tab and the `All Layers` icon under the `Physical Constraint Set` folder. Select the `DEFAULT` constraint set. Left click the cell in the `Vias` column. In the `Edit Via List` dialog box, select the `VIATENT` via to add it to the list, and remove the `VIA` via, and click `OK`.

Now you can set the fan-out parameters using `VIATENT`. To set fan-out parameters, select `Setup → Design Parameters` from the menu bar. In the `Design Parameter Editor`, select the `Route` tab, then click the `Create Fanout Parameters` button, as shown in Fig. 9.169. In the `Create Fanout Parameters` dialog box, you can now select `VIATENT` whereas it would normally list only `VIA`.

Once these settings are correct, perform the fan-out as described in detail in the previous examples. In summary, select `Route → PCB Router → Route Automatic...` from the menu bar. At the `Automatic Router` dialog box, select the `Router Setup` tab, and select the `Specify routing passes` radio button. Next, select the `Routing Passes` tab,

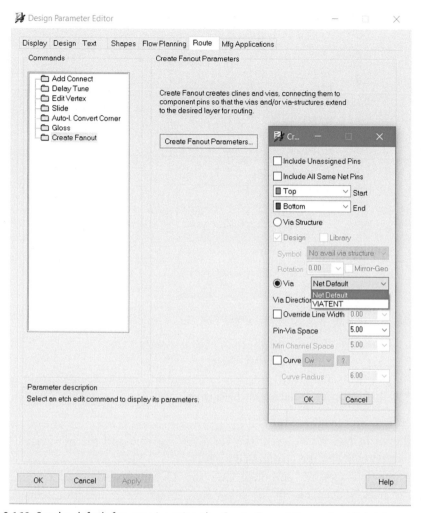

Figure 9.169 Set the default fan-out vias using the Create Fanout Parameters dialog box.

select the Fanout and deselect the Route and Clean options, click the Params... button. In the SPECCTRA Automatic Router Parameters dialog box, select the Fanout tab. In the Pin Types section, select the Specified: button, select the Power Nets, and deselect all others. Click OK, then click the Route button.

Once the router has finished, it is a good idea to check the vias to make sure PCB Editor did what you asked and performed the fan-out using the right via. To check the fan-out vias, display the Find pane and make sure that the Vias box is checked. Hover your pointer over a via, and an information box will be displayed, which will tell you which via was used.

If for some reason the wrong via was used, you can change it. *To change placed via types*, select Tools → Padstack → Replace from the menu bar. Display the Options tab

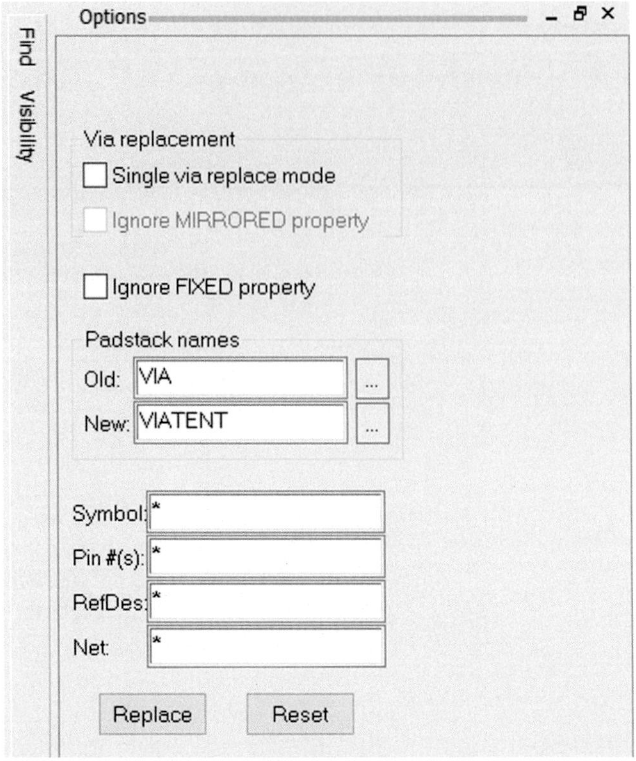

Figure 9.170 Use the Options pane to switch via types.

and, as shown in Fig. 9.170, select the via to be replaced and the new via. Click the Replace button to make the changes take effect.

Once the fan-outs are complete, use the Constraint Manager to fix ground and power nets, and enable all the other nets. Next we route critical traces before autorouting the board.

Determining critical trace length of transmission lines

Since the controlled-impedance traces are critical, they are routed next. The first step is to determine which traces need to be handled as transmission lines and which ones do not. As mentioned previously the digital-to-fiber interface IC, FO-TX (U2), was modeled after the Analog Devices ADN2530. In the data sheet the digital-signal lines going to the part and the modulation signals leaving the part (going to the laser diode) are to be handled as transmission lines. The digital control lines going to U2 need not be handled as transmission lines. The only traces left to consider are the ones related to the crystal oscillator and the NAND gates.

The literature states that the propagation time, PT, should be less than one half of the rise time, RT (or fall time, FT); that is, PT < 1/2 RT. If possible it is better if PT < 1/4 RT (see Chapter 6: Printed circuit board design for signal integrity, for more details). So we need to calculate PT for this board layout and look up RT and FT for the oscillator and the NAND gates. Since the crystal is a fictional part here, let us assume that RT = FT = pulse width = 1/4 the total period of a 66-MHz square wave. Under that assumption RT = 3.8 ns for the oscillator. The typical RT for ALS (Advanced Low-power Schottky) family logic is 1.9 ns.

The critical maximum length can be calculated using the following equation:

$$\text{Length}_{\text{trace}} < \frac{RT}{k \times t_{\text{PD}}} \tag{9.1}$$

where $\text{Length}_{\text{trace}}$ is the maximum allowed trace length in inches, RT is the signal rise time in picoseconds, k is the safety factor ($k = 2$ minimum), and t_{PD} is the propagation delay of the board material in picoseconds per inch.

The propagation delay for the surface microstrip (see Table 6.6 in Chapter 6: Printed circuit board design for signal integrity) is

$$t_{\text{PD}} = 85\sqrt{0.457\varepsilon_r + 0.67} \tag{9.2}$$

using $\varepsilon_r = 4.2$ for FR-4, $t_{\text{PD}} = 137$ ps/in., and the critical trace lengths are given in Table 9.10 for various values of k. As indicated, there is no way that the ADN2530 traces can be treated other than as transmission lines, but as long as none of the other traces is longer than 3.5 in., they need not be treated as transmission lines. Note that we are neglecting the length of the cables leaving the board through connector J1, but that is beyond the scope of the example.

Routing controlled impedance traces

The objective is to design surface microstrip transmission lines with a characteristic impedance of $Z_0 = 50\ \Omega$. Using the design equations from Chapter 6, Printed circuit

Table 9.10 Maximum safe trace lengths.

	RT (ns)	Maximum Length$_{\text{trace}}$ (in.)		
		$k = 2$	$k = 3$	$k = 4$
66-MHz OSC	3.8	13.9	9.26	6.95
ALS logic	1.9	6.95	4.63	3.47
ADN2530	0.026	0.095	0.063	0.048

RT, rise time.

board design for signal integrity, repeated here in the following equation, the width of the trace is calculated as

$$w = 7.47h \times e^{(-Z_0\sqrt{\varepsilon_r+1.41})/k} - 1.25t \qquad (9.3)$$

where, from Fig. 9.158, $t = 1.35$ mil (1 oz copper), $h = 10$ mil, $k = 87$ for $15 < w < 25$ mil (most references use this number—87 is used here), or $k = 79$ for $5 < w < 15$ mil (Montrose offers this option), $Z_0 = 50\ \Omega$ (the design goal), and the desired trace width in mil $w = 17.5$ mil (17 mil $= 50.9\ \Omega$).

To specify the width of a net, open the Constraint Manager. Select the Physical tab and select the All Layers icon under the Nets folder. Set the Min trace width to 6 mil and the Max value to 17.5 mil in the Line Width columns for the ModN and ModP nets.

With these settings the traces will be 6 mil by default (good for connections to the small surface-mount pads) but can be as wide as 17.5 mil (which is needed for the transmission lines).

Next the transmission lines are routed manually. Choose the Add Connect tool. Select a net on U2 at a point close to the pad to begin routing. Place a vertex just outside the place outline by left clicking once (this short, narrow trace allows for a thermal relief during reflow but is too short to interfere with the trace impedance).

With the vertex in place, we now want the 17.5 mil trace. *To change the trace width*, display the Options pane and enter 17.5 in the Line width: box, as shown in Fig. 9.171. The trace attached to your cursor should now be the correct width, and you can continue routing to the laser diode pin. Repeat this process for the other net. Fig. 9.172 shows the completed transmission line.

Note that, because of the pin-out of the component and the lead spacing of the diode, the lengths of the traces may not be equal (which is recommended in the data sheet). You can use the Show Element tool to measure the lengths of the traces. Fig. 9.173 shows a comparison of the two traces and reveals that the top trace (MODP) is about 58 mil longer than the bottom trace (MODN).

If we want the trace lengths to be within a certain tolerance, then we need to reroute the shorter traces to include extra length (using trombones, accordions, or sawtooths) or change the position or orientation of either or both components. As an example Fig. 9.174 shows U2 rotated 45 degrees and relocated. The trace lengths in this configuration are equal to within 0.01 mil.

To rotate a part 45 degrees, select the Move tool, display the Options pane, and select 45 in the Angle: selection list. Select the part (make sure the Symbols box is checked in the Find filter pane), right click, and select Rotate from the pop-up menu. Move the cursor around to rotate the part. When the part is in the correct rotation, left click, move the part to the correct place, and left click again to place the part.

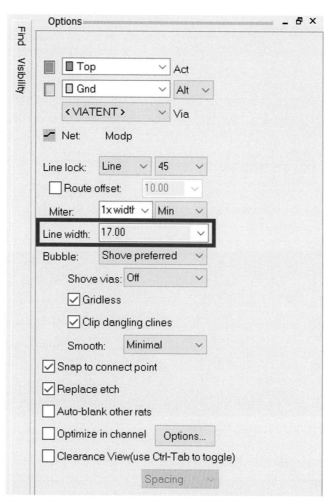

Figure 9.171 Use the Options pane to change the trace width during manual routing.

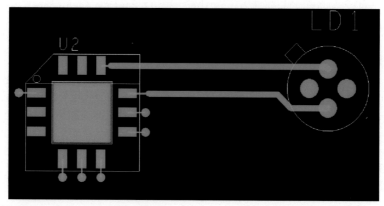

Figure 9.172 The routed transmission line.

(A)

(B)

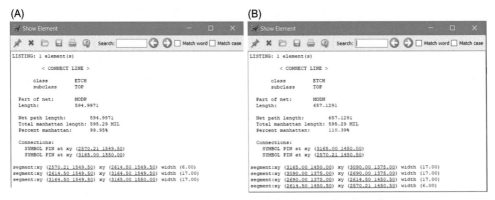

Figure 9.173 Use the Show Element tool to compare transmission line lengths.

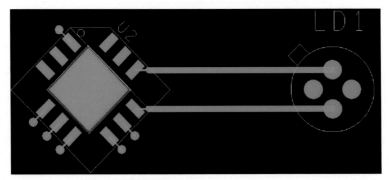

Figure 9.174 A component rotated 45 degrees.

Maximum neck length

When routing traces with varying widths, DRC errors may result because a trace has been "necked down." To clear maximum-neck-length DRC errors, open the Constraint Manager, select the Physical tab, and select the All Layers icon under the Physical Constraint Set folder. Change the DEFAULT Maximum Neck Length as necessary (e.g., 40 mil) to eliminate the DRC errors.

Moated ground areas for clock circuits

The oscillator is routed next. In many applications a moated ground plane around the clock circuitry is recommended to prevent stray ground currents from affecting other circuits. Before adding the moated ground area around the oscillator, the traces should be routed so that the size of the required ground area is known. Begin by enabling all the nets associated with the clock circuitry (set Rats on). Route the traces manually. Fig. 9.175 shows the routed, curved clock traces (along with the moat, which is described later).

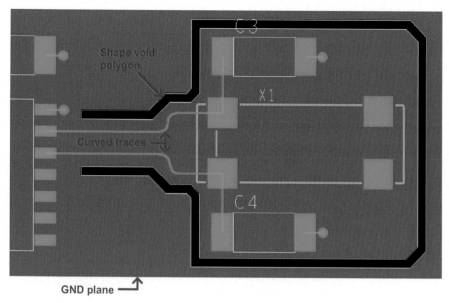

Figure 9.175 Clock circuitry with curved traces and a moated plane area.

Routing curved traces

Notice that the traces between the crystal (X1) and U1 are curved. While they are not necessary, the curved corners are used here as a demonstration. *To route curved traces*, select the Add Connect tool, left click the net near one of the pads on X1, route the trace straight out from the pad about 100 mil or so, and place a vertex by left clicking. Display the Options pane and select Arc from the Line lock: list (Fig. 9.176). As you move your mouse around, you will see that the trace is a curve instead of an angle. Place vertices to define the curves. You can switch between curves and lines as needed using the Options pane.

The next step is to etch a moat into the GND plane around the clock circuitry (see Fig. 9.175). *To etch a moat into a plane*, make the GND plane visible, and choose the Shape → Manual Isolation/Cavity → Polygon tool.

In the command window, PCB Editor will ask you to Pick shape or void to edit. Left click on the ground plane to select it (it will become highlighted). Create the void by picking void coordinates with the left mouse button. To define the shape shown in Fig. 9.175, 32 pick points were required. When you get to the last pick point, right click and select Complete from the pop-up menu.

Make sure to leave a "bridge" attached to the main ground plane. The bridge should be wide enough to include the ground pin and the area under the clock traces on U1. The local ground area under the clock circuitry must be attached electrically to the rest of the ground system, but the moat is used to "corral" the ground currents

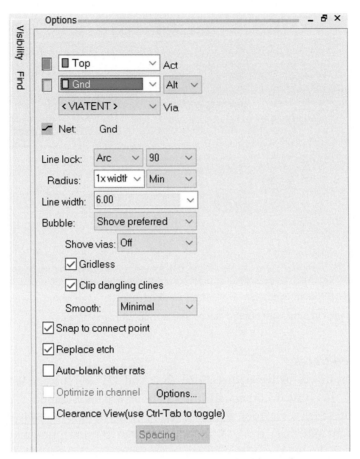

Figure 9.176 Use the Options pane to select between curved and straight traces.

back to the ground pin on the IC. Ground areas (and moats) will be placed on the Top and Bottom layers, and ground stitching will be used on all Ground planes; but before that is performed, the rest of the board needs to be routed.

Notice that you can also add the void in all layers at once by creating the shape in Route Keepout/All subclass.

Disable and lock all routed traces (use the Constraint Manager to turn on the Fixed parameter). Enable the remaining unrouted nets and set the routing grid to 25 mil (Setup → Grids, All Etch Spacing:). Autoroute the board (Route → PCB Router → Route Automatic..., Routing Passes: Route and Clean on, Fanout off). Fig. 9.177 shows the result. Many of the traces have wandered around due to poor usage of the gates (as assigned on the schematic), particularly around the areas marked 1 and 2. Note also that, at area 3, two traces were routed over gaps (the moat) in the Ground plane. As described in Chapter 6, Printed circuit board design for signal integrity, we do not

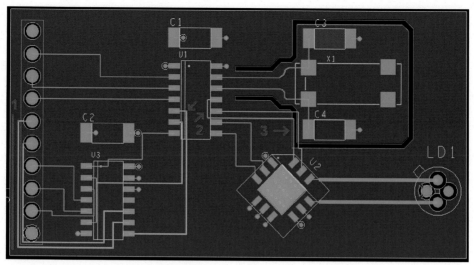

Figure 9.177 The design after autorouting.

want to allow this, as it will increase the loop inductance and introduce EMI issues. To fix these two problems, we now look at how to perform pin and gate swapping and how to define a route keep-out area to prevent the autorouter from routing traces over the moat.

Gate and pin swapping

Two methods can be used to swap gates and pins. The first is to swap the gates (or pins on a gate) on the schematic page and run an ECO to PCB Editor. The second method is to swap pins in PCB Editor and back annotate the changes to Capture to update the schematic.

The second method described is demonstrated here. Before doing the swap, save the design (or save as a new `.brd` file) so that the preswap design can be recovered if something goes wrong.

The first task is to unroute the gates. Make sure all gate nets are enabled, and all other nets are disabled and fixed. Select the `Delete` tool and drag a box across multiple nets to rip up more than one trace at a time.

Before a swap can be made, you need make sure pins are swappable in Capture. To verify that they are swappable, go to the schematic page, double click on a pin, select `<Current Properties>` from the filter list; its `Swap ID` will be −1 if it is *not* swappable, and it will be 0 or a positive number if it *is* swappable. If pins are not swappable, you need to change the property of the pins in the part definition in Capture.

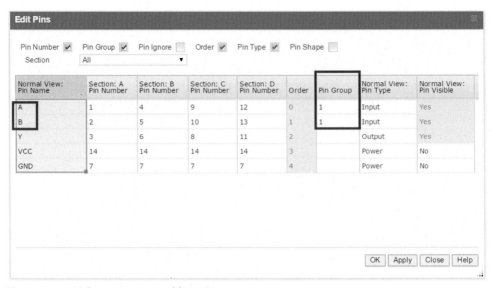

Figure 9.178 Making pins swappable in Capture.

To make pins swappable, go to the schematic in Capture, left click on the logic gate you want to make swappable, go to the `Edit` menu, and select `Part`. The part editor will be displayed. In the part editor, open the `Property Sheet`, press `Edit Pins` button, and in the `Pin Group` column, check that the swappable pins have identical numbers (see Fig. 9.178). So far any pin within the gate is available for swapping with any other pin of the same gate if it has the same `Pin Group` number. Click `OK` to dismiss the spreadsheet.

In the homogeneous part, all gates are identical, and Capture will automatically put the same integer number in the same input pin on all of the gates. Notice that in any homogeneous part all gates will be equally available for swapping with any other gate in that IC. But if the part is created as heterogeneous then every section differs from others. If you want to make some pins of one section of heterogeneous part to be swappable to some pins of other section, you can add the `SWAP_INFO` property to such part defining which sections may swap the pins which have the same `Pin Group` number. For example, `SWAP_INFO = (S1), (S2 + S3)` will mean that Sections 2 and 3 may have some pins which you can swap not only within the section but also between these sections.

To save the changes, click the `X` in the upper right-hand corner to close the part editor. When Capture asks if you want to change only the one or all of them, choose `Update All`. Choosing `Update Current` will cause problems with the netlister, because when you change a part with the part editor, the link between the part and the library from which it came will be broken, and the part will be given a modified name with a suffix (e.g., 54ALS00A will be renamed to 54ALS00A_1). Every time you modify a gate, the suffix is incremented. So if you modify a gate and choose Update Current, only that gate will be modified, and its suffix will be different from the other gates within the same IC package. You can check this by selecting the parts, right clicking, and selecting Edit Properties from the pop-up menu. Select <Current Properties> from the filter list and look at the Source Package cells for the individual gates. If gates from the same IC have different names (suffixes), the netlister will fail during an ECO (check the netlist.log file for details if this happens). If you choose Update All, then all gates in that IC will be given the same suffix. Gates contained in different ICs will not be changed (or renamed) by Update All, so if you want them to be swappable, you will need to modify them as well. If the parts within a package end up with different names, you cannot modify a single part to change its name to match the others, because Capture will increment its suffix to one greater than the highest suffix in the package, so they will always end up being different. The good way to fix it is to use the Replace Cache command in the Design Cache folder of the Project Manager pane.

Repeat the steps to make pins swappable for pins 10 and 11 on U1.

If you have had to modify parts a couple of times, a record of each change is maintained in the Design Cache folder in the Project Manager pane. This is not a problem, but you can remove the outdated parts from the Design Cache by selecting (left click on) the Design Cache folder then selecting Design → Cleanup Cache from the menu bar.

After you have made gates/pins swappable, you need to perform an ECO to let PCB Editor know that the pins can be swapped.

Once the ECO is complete and the board design is open, swapping is accomplished through the Place menu. The Place menu offers two swap options, Swap and Autoswap, and in each of the two options are additional options. In short the difference between the two options is that the Swap option gives *you* specific control over pin, function, and component swapping actions; while the Autoswap option allows *PCB Editor* control over swapping actions (with some input from you). We first look at the Swap option.

Using swap options
Pin swapping

In this example, we use the Swap option to manually uncross the two nets between U1 (pins 10 and 11) and U2 (pins 11 and 12). *To perform a manual pin swap*, select Place → Swap → Pins from the menu bar. In the command box, you will see Pick a pin you wish to swap. Pick pin 10 on U1. Then in the command box, you will see

```
last pick: 3500.00 3200.00
Pick a pin you wish to swap from those that are highlighted.
```

The package to which the pin belongs will have the other pins highlighted that were set in Capture with the same pin group number. Select pin 11 on U1. Then you should see

```
last pick: 3500.00 3200.00
Pick DONE/NEXT or pick the first pin of the next swap.
```

Right click and select Done from the pop-up menu. The nets should now be swapped between the two pins and no longer be crossed. A back annotation will be performed at the end of this section to show the effects of the pin swap in Capture.

You can perform pin swaps on logic gates too. When doing so, PCB Editor will allow you to swap pins only within a single gate and only pins that are of the same type. For example, if you pick pin A on a NAND gate (an input pin), it will allow you to swap it only with pin B on the same gate. You cannot swap pin A with any pin on another gate (even if it is on the same IC), and you cannot swap it with pin Y (the output) on the same gate, because pins A and Y are different types of pins.

Function (gate) swapping

If you need to swap an entire gate (e.g., U3A for U3C) and not just pins within a gate, then you use the Swap → Functions option. *To swap two gates* (and therefore their pins), select Place → Swap → Functions from the menu bar. In the command window, PCB Editor will instruct you to Pick a function you wish to swap.

In this example, select pin 1 on U3 (this is input pin A on gate A). All the pins on U3 will become highlighted and the command window will display

```
last pick: 2400.00 2700.00
Pick a function to swap with from those highlighted ...
```

Select pin U3.13 (this is input pin B on gate D); the command window displays:

```
last pick: 2600.00 2700.00
W-(SPMHA2-12): Pick NEXT/DONE or pick the first function of the next swap.
```

Right click and select Done from the pop-up menu. The rat's nest lines that were connected to gate A will now be connected to gate D, and the pins will be utilized in the best way to minimize length of the rat's nest (referred to as *virtual wire length* by the Autoswap tool).

Swap → Components does not perform gate or pin swapping, it just physically swaps (trades) the locations of two components on the board. The components do not even have to be the same type or be "swappable." *To perform a component swap*, select Place → Swap → Components from the menu bar. The command window will not ask you for anything, but left click one of the parts in the design (it will become highlighted) then left click another part. The two parts will immediately trade places. To undo the component swap, right click and select Oops or Cancel from the pop-up menu.

Back Annotating the swap operations to Capture

If you are working through these examples on your own board design, you can perform a back annotation to see the effects of the gate and pin swapping on the schematic. *To perform a back annotation, the steps are to be followed*:

- Save your PCB design and exit from PCB Editor.
- In Capture schematic project select the design icon in the Project Manager window, select Tools → Back Annotate from menu and choose the proper file using PCB Editor Board File: browse button.
- Check the path and name of the Output File where the result of back annotation will be stored.
- Check the path where the PCB Netlist is located and press OK.

When you go back to your schematic, it should be updated with the new information, as shown in Fig. 9.179 (compare this to Fig. 9.156).

In the case of the pin swap on the microcontroller (U1), pins 10 and 11 were actually moved on the part definition (and a suffix was added to the parts name, as described previously) while the nets were physically unaltered (see Fig. 9.180). This is

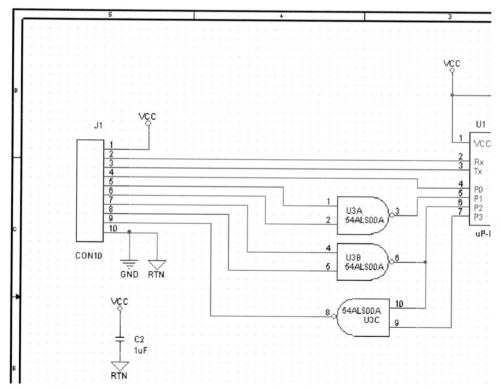

Figure 9.179 Logic gate connections in the schematic design in Capture after the pin swap.

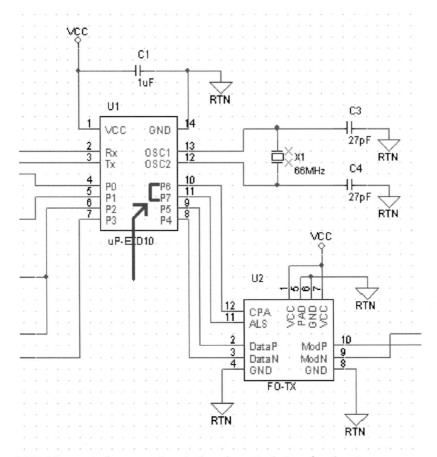

Figure 9.180 Microcontroller connections in the schematic design after the pin swap.

obviously different from the way pin swaps look in PCB Editor and can be easily missed. If you or another designer replaces this modified part with the original from the library and an ECO is performed, the pin swap will be effectively undone. Another point to consider in an example like this is that any software written for the microcontroller would have to be changed to accommodate the pin swap.

Using the Autoswap option

In using the Autoswap option, you basically let PCB Editor figure out the best gate usage and pin connections then back annotate the results to the schematic. You do have some control over the swapping action. From the Autoswap menu, you can have PCB Editor work the whole board design, a room, or a selection window. After you select the type of autoswap you want to do, you select Place → Autoswap → Parameters to execute the swap. Examples of each follow. Once you perform an autoswap, you

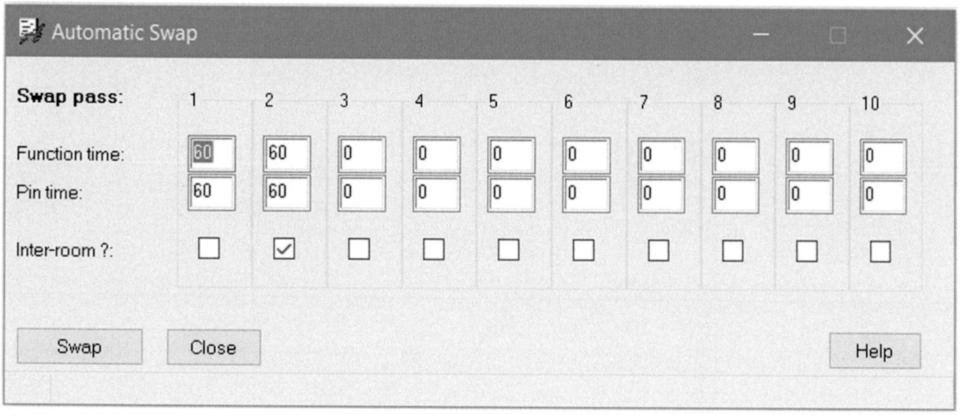

Figure 9.181 Setting up the Automatic Swap tool.

cannot Undo it, so to try all the autoswap types on this small design, you can try a swap and close the PCB design without saving it (and without doing a back annotation) then reopen the board design to try a different one.

Autoswapping an entire design

To have PCB Editor review and autoswap an entire board design, select Place → Autoswap → Design from the menu bar. Then select Place → Autoswap → Parameters... to display the Automatic Swap dialog box shown in Fig. 9.181.

The numbers represent the time limit (in minutes) for each pass. After a pass PCB Editor records the virtual wire length for each net and compares the lengths to the previous pass. PCB Editor will continue trying to shorten the wire lengths until all passes have been completed or the wire lengths cannot be made any shorter. By default only two passes are enabled. Passes with a 0 time limit are disabled.

The Inter-room option allows or prohibits PCB Editor from attempting to shorten the wire lengths by swapping gates between different rooms.

You can specify more or fewer passes and change the time limit depending on the complexity of your design. You can see what PCB Editor attempted to do during each pass by reading the swap.log file that PCB Editor generates after a swap operation. The swap.log file is located in the same file folder as the board design.

To begin the autoswap, click the Swap button. PCB Editor will look over all the components and nets and automatically swap all gates and pins in a way that will result in the shortest traces. Note that gates that were not placed somewhere on the design will not be considered for swapping and neither will gates with "NC" markers on pins or pins connected to nets that are fixed or nets that have a No Ripup property assigned in Constraint Manager. If you want unused gates that were not placed on the schematic to be considered for gate swapping, place them on the schematic and leave

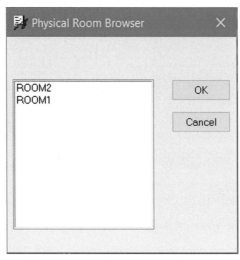

Figure 9.182 Use the Physical Room Browser for swap operations on rooms.

the pins floating. After the swap and subsequent back annotation, you can apply NC pins and delete unused gates.

Autoswapping a Room

You can restrict how much the design PCB Editor is allowed to modify by defining rooms around a component or groups of components and telling PCB Editor to work on only the rooms you want swapped. To use this option, you must have rooms defined (review Design Example 1 to see how to set up rooms). Once rooms are defined, select Place → Autoswap → Room from the menu bar. The Physical Room Browser will be displayed, as shown in Fig. 9.182. Select the room you want PCB Editor to review and autoswap, and click OK. Then select Place → Autoswap → Parameters... to display the Automatic Swap dialog box, and click the Swap button. PCB Editor will automatically swap all gates and pins in the selected room but will leave the rest of the design alone.

Autoswapping a window selection

You can also restrict how much of the design PCB Editor is allowed to modify by defining a window around a component or groups of components. To use the Autoswap Window function, select Place → Autoswap → Window from the menu. Define the window by clicking your left mouse on the design to define a window corner then drag a box (window) around the components you want swapped, and left click again to complete the window. If you want to define more windows, repeat these steps. When you are finished, choose Done from the pop-up menu. Then select Place → Autoswap → Parameters... to display the Automatic Swap dialog box, and click the Swap button. PCB Editor will automatically swap all gates and pins within the selected window (if swapping results in an improvement) but will leave the rest of the design alone.

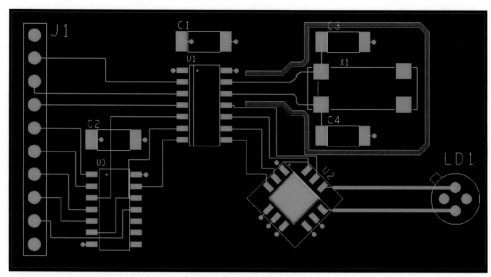

Figure 9.183 The board design after autoswap and autoroute.

Viewing the swap list and the swap log

Once you have made a swap type selection, you can review what PCB Editor will be looking at with a swap list before you actually perform the swap. To display a swap list, select Place → Autoswap → List from the menu. The swap list shows you where PCB Editor will focus during the swapping operation, but it does not give you information about the swap itself. You can obtain detailed swap information by opening the swap.log file generated after a swap is performed. The swap.log file is located in the same file folder as the board design.

Once the swaps have been completed you can rerun the autorouter. Fig. 9.183 shows the results of the autoswap and autoroute. The figure also shows the route keep-out area placed over the ground moat. Notice how the autorouter hugs the keep-out area but does not cross it. Defining a route keep-out area is described next.

Defining a route keep-out area

Shape Add button

To define a route keep-out area on your board, begin by making the Route Keepout class active on the Options pane. In the subclass list, you can choose any of the layers. In this example, select the All subclass to place the keep-out on all layers simultaneously. Next, select the Shape Add button, , on the toolbar. Draw a polygon on the Route Keepout / All class/subclass that matches the moat in the Ground plane. This will keep the router from routing any traces on Routing (conductor) layers and will also create a void in the VCC plane layer similar to what the void shape did on the Ground plane (see Fig. 9.175). The keep-out also works if you pour a copper ground area on the Top and Bottom layers, such as the one demonstrated in Example 3.

This completes the Example 4 design.

Positive planes

As we saw in the previous design examples, keeping track of the necessary details to use negative Plane layers can be a bit of a task (making flash symbols, modifying padstacks, etc.). Negative planes are used because they have been used historically. One of the reported advantages of using negative planes is that the artwork files are smaller than for positive artwork files. With the current PC speed and memory capacity, file size is less of an issue. Another argument for negative planes is a shorter processing time for negative planes at the manufacturer's end. That certainly was true for vector plotting with the flash lamps, but for current photoplotting equipment it has become less of an issue. Yet another argument for negative planes is that, during the board design, the positive planes can be distracting and result in slow drawing regeneration. But PCB Editor displays the negative planes in the positive view ("what you see is what you get" mode) anyway, and you can easily turn off the Plane layers. In addition, the author is aware of some designers who use positive planes for most of their work. That being the case and for the sake of completeness, we look at a very simple design to demonstrate how to use positive planes and take care of the corresponding details.

The circuit for this example is shown in Fig. 9.184. As shown, the circuit consists of only two resistors. We focus on the Plane layers that will be used by the VCC and GND nets, so this simple circuit suffices.

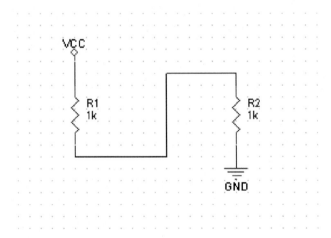

Figure 9.184 Simple circuit to test positive planes.

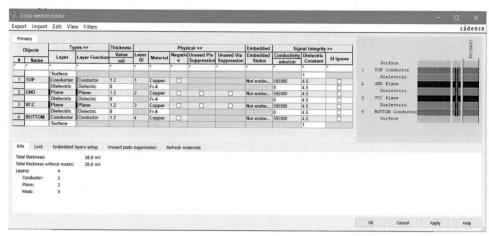

Figure 9.185 Layer stack-up for the positive plane circuit.

As with the previous design examples, the netlist is created and PCB Editor is launched. A board outline is drawn and the parts placed. The next step is to define the layer stack-up.

In the previous design examples the Ground and Power planes were defined as negative Plane layers. In this example, we use positive planes. The layer stack-up is shown in Fig. 9.185. As indicated the layer stack-up is similar to the other design examples except that the Negative Artwork boxes are unchecked.

A positive plane layer should still be specified as PLANE rather than CONDUCTOR so that the autorouter does not try to route traces on that layer. Routing traces on Plane layers (particularly GND planes) results in slots in the plane, which disrupts the return paths for signals on Routing layers and can lead to signal integrity issues. Setting the type as Plane prevents this.

Once the layer stack-up is specified, the next step is to draw the copper areas on the planes. The process is identical to the previous examples. Select the Shape Add tool, select the GND or VCC subclass in the Etch class, and draw a dynamic copper area attached to the appropriate net.

Unlike with negative planes, you do *not* have to enable the Thermal Pads setting in the Design Parameters dialog box to see the thermals. Fig. 9.186 shows the two resistors and the VCC plane. Note that R1 is connected to the plane with a thermal relief that was automatically generated by PCB Editor. The padstacks used in this design are native to the *symbols* library and have no thermals assigned to them. This result is similar to the Ground plane placed on the Top layer in Example 3.

A close-up inspection of the padstack reveals that the inner pad is indeed 55 mil, but the diameter of the clearances (outer diameter of the thermal relief) is only 65 mil (leaving a space of only 5 mil between the pad and the copper area). So, on a positive plane,

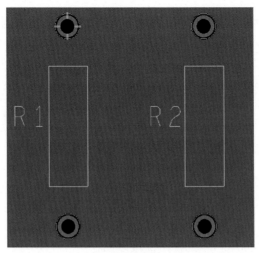

Figure 9.186 Thermal reliefs shown on a positive Plane layer with default settings.

neither the thermal relief diameter nor the Anti Pad diameter in the padstack definition determines the diameter of a clearance or the outer diameter of a thermal relief.

Since PCB Editor automatically generates the thermal reliefs, a question arises as to how it determines the thermal relief dimensions. The answer was described briefly in Example 2. The space between the copper area and the padstack is determined by the spacing constraints set in the Constraint Manager, and the spoke width is determined by the trace width setting in the Constraint Manager. If we change the shape-to-pin spacing constraint to 10 mil (see Fig. 9.187), the thermal and clearance areas should increase. And if we change the trace width constraint to 10 mil (see Fig. 9.188), the spoke width should increase.

Fig. 9.189 shows the difference between (A) the default constraints and (B) the result of changing the trace width and spacing to 10 mil—the spoke widths and the void width are wider.

Besides the constraint settings in the Constraint Manager, you also have control over thermal reliefs on positive planes using the Design Parameter Editor. As shown in Fig. 9.190, additional parameters can be set by selecting the Shapes tab and clicking the Edit global dynamic shape parameters button. At the Global Dynamic Shape Parameters dialog box, you can control the rotation (Orthogonal, Diagonal, or Full contact, etc.), the number of connects (spokes), and set fixed or scaled spoke widths (relative to the trace width constraint).

One thing that is different about this plane and a typical negative plane is that the pads not connected to the planes are present, whereas with the negative planes, they are absent. Typically, it is desired that the unused pads on Plane layers are removed. Although the pads cannot be removed in the design mode, they can be removed during the artwork production.

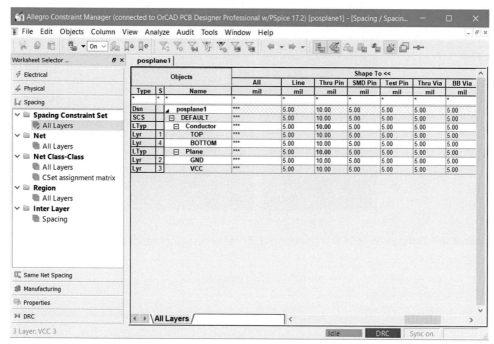

Figure 9.187 Changing the trace (and void) spacing constraint.

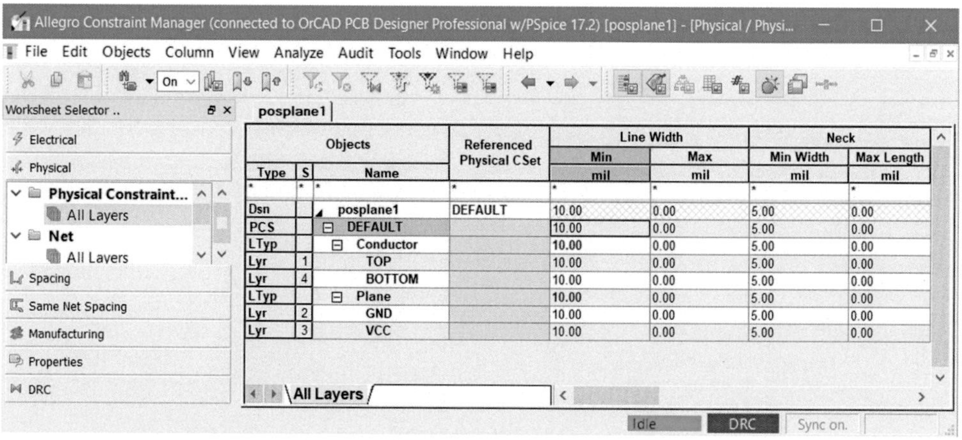

Figure 9.188 Setting the trace (and spoke) width constraint.

Positive plane artwork production

Artwork production is covered in detail in Chapter 10, Artwork development and board fabrication, but we touch on it here. *To create Gerber files* select Export → Gerber... menu. The inner pads can be removed on positive planes during the artwork

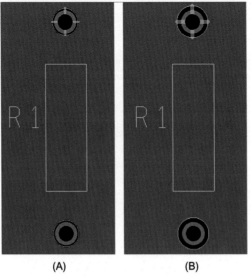

(A)   (B)

Figure 9.189 Comparison of VCC (A) before and (B) after constraint changes.

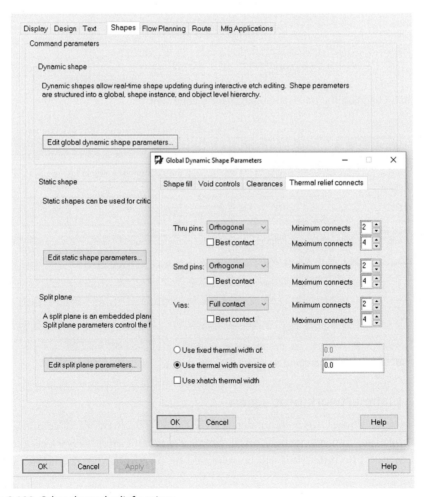

Figure 9.190 Other thermal relief settings.

production, provided that three conditions are met: (1) the `Suppress unconnected internal pads` option needs to be selected in the Padstack Editor's `Options` tab (see Fig. 9.191), (2) the `Suppress unconnected pads` option needs to be checked in the `Artwork Control Form`, and (3) the film `Plot mode:` needs to be processed as `Positive` artwork (also in the `Artwork Control Form`, as shown in Fig. 9.192).

The IPC-2221B (2012) standard specifies that unconnected pads on routing layers should be maintained, but unconnected pads on Plane layers can be removed, so when using positive planes, suppress unconnected pads on only the Plane layers and leave the option unchecked for Routing layers.

Fig. 9.193 shows the VCC layer photoplot artwork for this design (silk-screen and other layers not shown). The figure shows that the unused pads were removed as

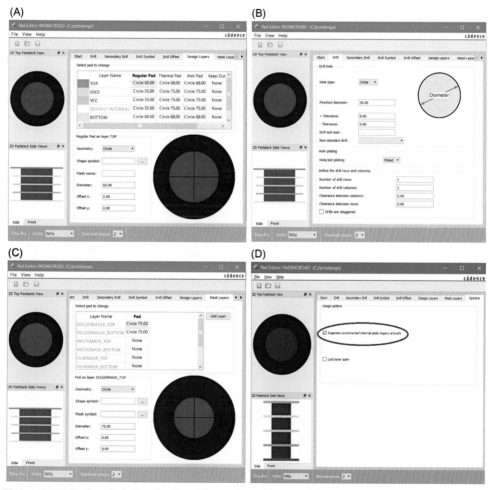

Figure 9.191 Padstack design for the positive Plane layers example.

Figure 9.192 Artwork Control Form for positive planes.

intended. Artwork verification is described in detail in Chapter 10, Artwork development and board fabrication.

Positive versus negative plane file sizes

It was mentioned at the beginning of this section that it is commonly held that the positive plane file sizes are larger than the negative ones. Let us take a look at the file sizes of the resulting artwork for this example. The GND plane artwork for this design in the positive image is 54 lines of Gerber code at 1.15 kB, while the negative artwork file is

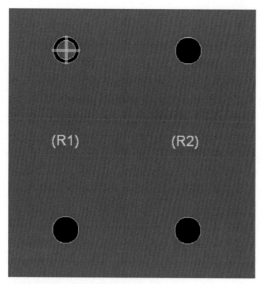

Figure 9.193 Gerber artwork for the positive VCC plane.

128 lines of code at 2.34 kB. So, at least for this simple design, the positive artwork files are actually *smaller* than the negative files (as generated by PCB Editor anyway).

Pros and cons of using positive versus negative planes

In general, there are two drawbacks to using positive planes. The first is that, in the padstack definition, the Suppress unconnected internal pads option must be set for every padstack in the design for the artwork to be produced correctly. However, it is much simpler to do that than to make and assign thermal flashes for the padstacks. The second potential drawback is that constraints set for trace width and spacing affect the thermal relief geometry (possibly adversely), whereas thermal flashes on negative planes are fixed, and once they are engineered, they are unaffected by changes made in the Constraint Manager. But the latter potential drawback can be mitigated to some extent by using the Global Shape Parameters settings as described previously.

The bottom line though is that, when using OrCAD PCB Editor, positive planes are easier to implement than negative planes as long as the layer stack-up and thermal relief design do not have to be highly engineered.

Design templates

Making a custom Capture template

If you design a lot of projects that are similar to each other, setting up a project template in Capture can be a real time saver and can help eliminate errors by reusing known good project setups. Project templates can be used only when setting up a project through the `Analog or Mixed A/D` option from the `New Project` dialog box. However, since you can make a PCB design from either the `Analog or Mixed A/D` option or the `PC Board Wizard` option, you can still take advantage of creating your own Capture templates for PCB design projects.

To make a custom Capture project template, start a new Analog or Mixed A/D project; set it up with power supply and ground symbols, connectors, or whatever you want; then save it in the OrCAD `tools/capture/templates/pspice` folder with the other templates. It will automatically be added to the templates list, so that you can select it the next time you start a new project.

Making a custom PCB Editor board template

As with the Capture templates, if you design a lot of PCBs that are similar to each other, setting up a custom PCB Editor board file (`.brd`) can save a lot of time and eliminate many manufacturing errors by reusing known good PCB setups.

To make a custom PCB template start PCB Editor, begin with a new board design by selecting `File → New` from the menu. At the `New Drawing` dialog box, enter a name for the template and select `Board` from the `Drawing Type:` list, as shown in Fig. 9.194. The Board Wizard will be demonstrated later, after a board template and technology file has been developed.

If you see `Create a New Design` dialog box, select the desired `Units`, `Sheet Size`, `Accuracy` and `Extents`, and click `OK`. When the board is opened, use the `Design Parameters` dialog box to set up the basic design parameters, including

```
Display tab:
    Display plated holes
    Display thermal pads
    DRC marker size
```

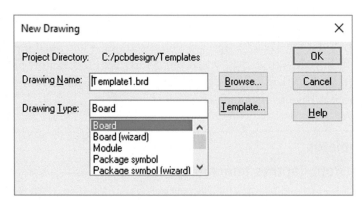

Figure 9.194 Open a new, blank board to start a board template.

```
Design tab:
    Design size
    Design extents
Shapes tab:
    Edit Global Dynamic shape parameters (RS274X for artwork format)
Route tab:
    Add connect defaults (trace width and angles)
    Create Fanout parameters
Manufacturing Applications tab:
    Edit silkscreen parameters
```

When the parameters have been set, click `Apply` then `OK` to dismiss the dialog box.

In addition to setting up previous parameters, the following items can also be set up:

- Default grid settings (`Setup → Grids`).
- A default layer stack-up using the `Cross-section Editor` dialog box.
- A board outline (select `Outline → Design` from the menu).
- Default artwork folders using the `Artwork Control Form`.
- A default, custom color scheme and layer visibility using the `Color Dialog` box.
- Design constraints using the Constraint Manager. Default settings might include the default trace width (`Physical` tab) and the default trace spacing (`Spacing` and `Same Net Spacing` tabs).

Once all of the design parameters and constraints have been established, save the board.

You can also save the Constraint Manager settings in a separate technology file that can be imported into any board. This is described in the next section.

To reuse the board template, assign it as the `Input Board File:` during the netlist creation in Capture, as shown in Fig. 9.195. The new board will inherit all the template board properties.

Making a custom PCB Editor technology template

If you often design boards with similar design constraints, you can save them in a board template. You can also save the Constraint Manager settings to a stand-alone technology file, so that, if you start a new design that is not based on the board template, you can still reuse the design constraints by importing the technology file.

To create a technology file, start from a board template or an active board design that has the constraints you want to be able to reuse. Open the Constraint Manager and select `File → Export → Technology File...` from the menu. Save the `filename.tcf` file in a folder that you can easily access for future designs.

When you start a new board and want to load the technology file, select `File →` `Import → Technology File...` from the `Constraint Manager` menu. The technology file

Figure 9.195 Use the board template during the initial netlist creation.

will replace the existing constraints and layer stack-up with the setup saved in the technology file.

Using the board wizard

Over time you will likely develop several board characteristics that you have used but not every board uses all of them. For example, you might have two or three board outlines you often use, but the layer stack-ups are different from one board to the next. Perhaps you have different technology files you use and have several mechanical symbols or drawing symbols that you need to select from, but they also vary from board to board. Rather than make a board template for every possible combination of these characteristics, you can save each of the characteristics and symbols in a library and use the Board Wizard to select specific characteristics and combine them into one new board design. The Board Wizard allows you to reuse your favorite board characteristics while keeping your template library more easily organized.

When you start a new board, the wizard will look for templates, technology and parameter files in the current working directory where you start the process but will look in the *symbols* library for mechanical and drawing symbols. Before you begin, make sure that any mechanical and drawing symbols you plan on using are in the *symbols* library and any board templates, technology and parameter files you will use are copied to the file path where you are setting up the new board. The board parameter file may contain the design settings, color settings, text sizes and/or other PCB Editor parameters. To save the parameter file select Export → More → Color/ Board Parameters.

To begin a new board using the wizard, start PCB Editor and select File → New from the menu. At the New Drawing dialog box, enter a name and path for the new board design and select Board (wizard) from the Drawing Type: list, as shown in Fig. 9.196, and click OK.

The Board Wizard dialog box will be displayed, which provides an introduction to the board wizard. Click the Next button to continue.

The next step is to load a board template (*name*.brd) if you have one. To load the board template, select the Yes radio button then click the Browse... button to locate the file (see Fig. 9.197). Remember that the board template needs to be located in the OrCAD \symbols\template directory path, or the wizard will not be able to find it. At the Board Wizard Template Browser dialog box, select the board template on which

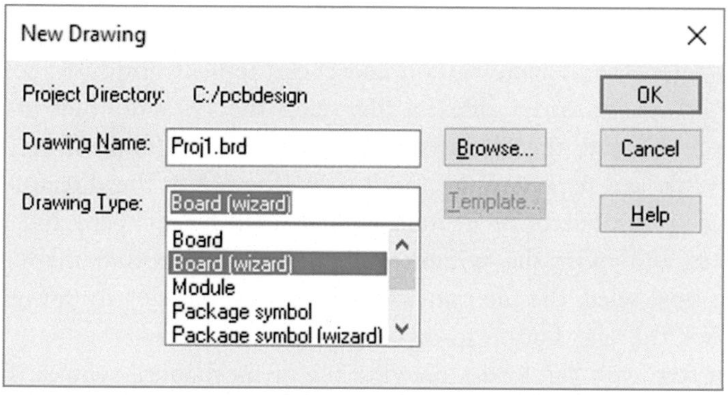

Figure 9.196 Select the Board (wizard) from the New Drawing dialog box.

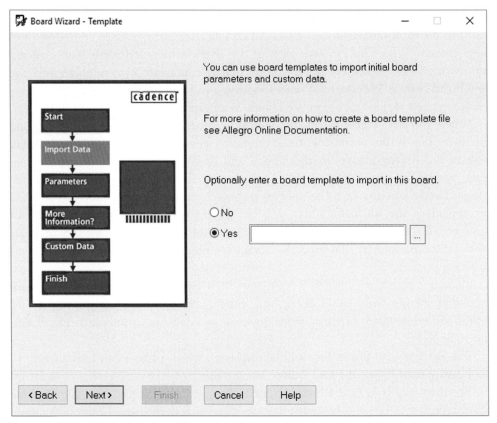

Figure 9.197 Loading a board template.

you want to base the new board, then click OK to dismiss the Browser dialog box. Click the Next button to continue to the next step.

Next you can load a technology file (*name*.tcf) and a parameter file (*name*.prm). A technology and/or parameter file loaded by the wizard will override the file originally loaded with the board template. If you want to keep the original technology or parameter file, select the No radio button and click the Next button to continue to the next step. Otherwise, to load a different file, select the Yes radio button and click the Browse... button to locate the file, as shown in Fig. 9.198. Remember that copies of the technology file and parameter file need to be located in the directory path where you are setting up the board, or in the folder defined by techpath and parampath in user preferences, otherwise the wizard will not be able to find them. At the File Browser dialog box, select the file you want to use and click OK to dismiss the Browser dialog box. Click the Next button to continue to the next step.

In the next step, you can load a drawing file or mechanical symbol (Fig. 9.199). If you have no drawing or mechanical symbols you want imported into the board, select

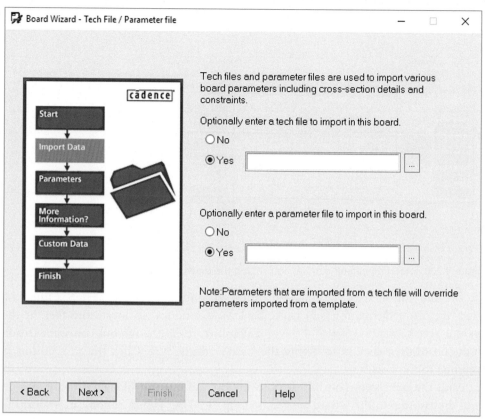

Figure 9.198 Loading a technology file.

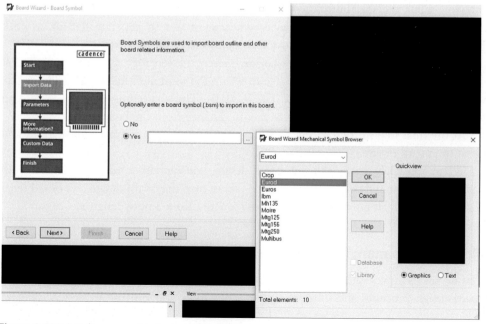

Figure 9.199 Loading a mechanical symbol drawing.

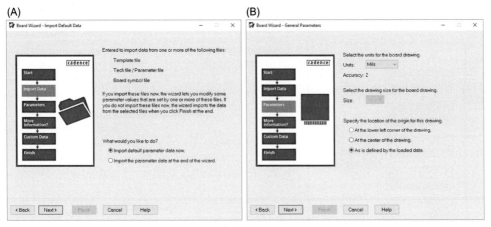

Figure 9.200 Importing default data (A) and setting the drawing origin (B).

the No radio button and click the Next button to continue to the next step. Otherwise select the Yes radio button and click the Browse... button to locate the file. At the Board Wizard Mechanical Symbol Browser dialog box, select the symbol you want to add to the board then click OK to dismiss the Browser dialog box. Click the Next button to continue to the next step.

With the next step, you can elect to import the selected files right away or wait until the last step in the wizard process, see Fig. 9.200A. Typically you can import the data right away. Click the Next button to continue to the next step.

At the following step, Fig. 9.200B, you can specify the origin of the new board drawing. Click the Next button to continue to the next step.

The next step allows you to specify whether or not to generate default artwork films, Fig. 9.201A. Click the Next button to continue to the next step.

At the next step you can generate negative internal layers for power planes, and after clicking the Next button you will be able to set up the minimum line width and spacing, and to select the default via padstack (Fig. 9.201B). The succeeding step will allow you to specify the Route Keepin and Package Keepin distance from the board edge.

At the final step in the process click the Finish button to complete the wizard and open the board design.

When you finally get to the new board design, you can either import logic data from a previously generated netlist (Import → Netlist) or save the board design as a board template and use it as an input file when creating a netlist from a new design in Capture.

Moving on to manufacturing

The design examples included here are to provide an overview of the basic steps in the board design processes and introduce the various tools you can use to design different

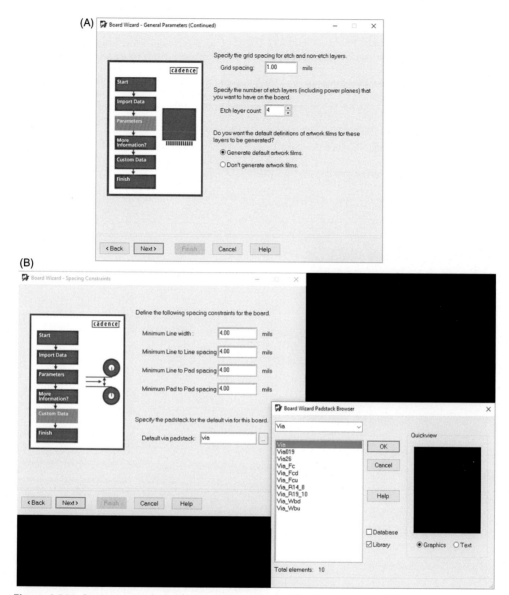

Figure 9.201 Setting artwork design parameters (A) and finishing the wizard (B).

types of boards. These examples were *not* designed with manufacturability in mind, as that is another subject, which encompasses additional things to consider and would have made the examples overly cumbersome. Chapter 10, Artwork development and board fabrication, combines design for manufacturing topics introduced in Chapter 5, Introduction to design for manufacturing, with design processes introduced here.

References

OrCAD Capture User's Guide. Product Version 17.2-2016. Cadence Design Systems Inc.

IPC-2221B. (2012). *Generic standard on printed board design*. Northbrook, IL: IPC/Association Connecting Electronic Industries.

Ott, H. W. (1988). *Noise reduction techniques in electronic systems* (2nd ed., p. 129). New York: Wiley.

Further reading

OrCAD PSpice User's Guide. Product Version 17.2-2016. Cadence Design Systems Inc.

CHAPTER 10

Artwork development and board fabrication

Contents

Developing artwork for the board design was briefly introduced in Chapter 2, Introduction to the printed circuit board design flow, by example. Here we look at this step in greater detail. To illustrate the process, we will create and route a new simple PCB project, place mounting holes and assembly fiducials. Then we will setup the artwork (Gerber) and numeric control (NC) drill files. We will see how the PCB Editor auto-silkscreen tool is working, to prepare the high quality silkscreen photoplot data. We will generate Gerber, Drill and Mill manufacturing files, as well as an assembly Pick & Place program.

Schematic design in Capture

As with the previous examples, we use a simple circuit to minimize those details and allow us to focus on the board layout and manufacturing steps. The circuit (shown in Fig. 10.1) consists of a 10-pin printed circuit board connector (J1), a bypass capacitor (C1), a logic IC (U1), load resistors (R1 and R2), and two plated mounting holes that electrically connect the ground plane to the PCB's mounting bracket. Once we move the

Complete PCB Design Using OrCAD® Capture and PCB Editor
DOI: https://doi.org/10.1016/B978-0-12-817684-9.00010-2

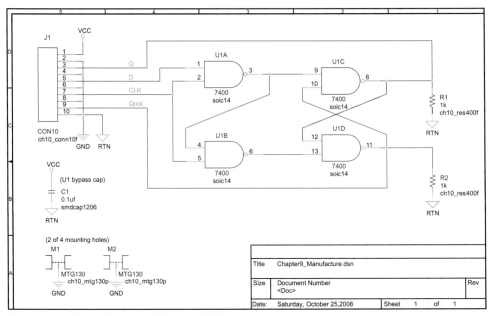

Figure 10.1 Circuit design used for the fabrication process.

Table 10.1 Bill of materials for the manufacturing design example.

Item	Qty.	Ref.	Part	Symbol	Padstack	Flash
1	1	C1	0.1 μF	smdcap1206	smd45rec71	N/A
2	1	J1	CON10	CH10_Conn10f	CH10_pad60cir36f	TR_90_60
3	2	M1, M2	MTG130	CH10_mtg130p	CH10_pad400cir130f	AB00
4	2	R1, R2	1k	CH10_res400f	CH10_pad72cir42f	TR_102_72
5	1	U1	7400	soic14	smd50_25	N/A
6	2		(NP mtg hole)	CH10_mtg130np	CH10_Hole130np	Circle 160
7			(VIA)		VIA	AB00

design to PCB Editor, two more mounting holes will be added, which are nonplated and electrically isolated from the board. The two types are used so that, when we generate the drill files, you will be able to see the differences between the two types of drill holes.

Table 10.1 shows the bill of materials for this design. Only the footprint symbol for U1 (soic14) is from the native PCB Editor symbols library. The footprint for C1 was made in the footprint design example from Chapter 8, Making and editing footprints; the electrically connected mounting hole, mtg130p, is a custom symbol; and the remaining symbols are modifications of existing footprints. The footprint symbols are on the website for this book, and you need to copy them (and the associated padstacks and flash symbols) into the PCB Editor's *symbols* folder. You also need the mtg130np mechanical symbol, which is included on the website in the same folder.

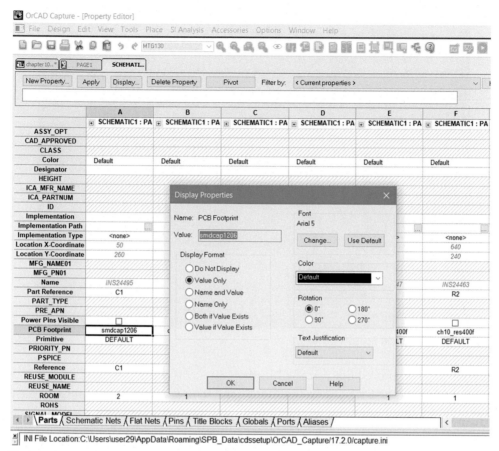

Figure 10.2 Use the Property Editor to display part information on the schematic.

The footprint names for each of the components are shown on the schematic. *To display component information on the schematic,* double click the part to display the `Property Editor` spreadsheet (see Fig. 10.2). Select the cell (row or column) that contains the information you want displayed. Click the `Display...` button at the top of the spreadsheet to show the `Display Properties` dialog box. Select one of the display formats and click `OK`. Click the `Apply` button at the top of the spreadsheet and close the spreadsheet.

Once the schematic is completed, generate a PCB Editor netlist as described in Chapter 9, PCB design examples.

The board design with PCB Editor
Routing the board

Fig. 10.3 shows the routed board. The stack-up consists of the top and bottom routing layers and the VCC and GND plane layers. The stack-up was defined and the traces

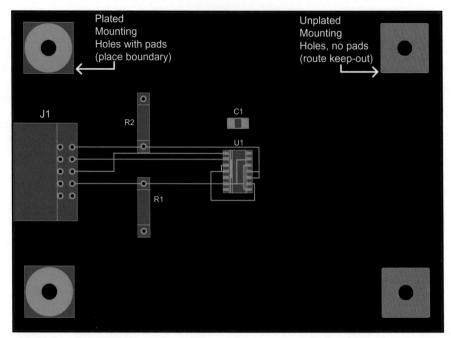

Figure 10.3 The routed board.

routed using the procedures described in the Chapter 9, PCB design examples. The padstacks used in the footprint symbols for J1 and the two resistors (R1 and R2) have flash symbols already defined, so they should automatically show their thermal reliefs when the plane layers are visible.

Along with the components and routed traces, the board contains four mounting holes. The two on the left side of the board are the ones included on the schematic and are component symbols (CH10_mtg130p.psm) with plated padstack holes (CH10_pad400cir130f. pad). The f indicates that the flash symbols have been predefined. They are placed on the board using the same procedure used for placing the other components.

The two on the right side of the figure are mechanical symbols (CH10_mtg130np. bsm) with nonplated padstack holes (CH10_Hole130np.pad) and are added to the board from within PCB Editor using the following steps. Make sure you have copied CH10_mtg130np.dra, CH10_mtg130np.bsm and CH10_Hole130np.pad into your PCB Editor *symbols* library.

Placing mechanical symbols

Mechanical symbols for mounting holes

To place the mechanical symbol mounting holes, select Place → Mechanical Symbols from the menu bar or select the Place Manual button on the toolbar. At the Placement dialog box (see Fig. 10.4), select Mechanical symbols from the Placement List. Click the Advanced Settings tab and make sure that the Library option is checked. Select the

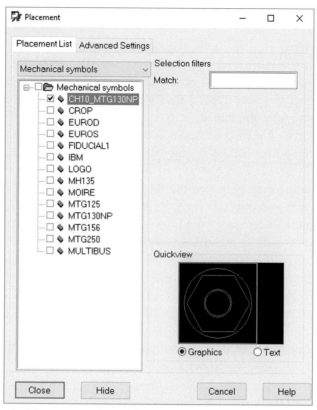

Figure 10.4 Using the Placement tool to add mounting holes.

Placement List tab again, and check the box beside the CH10_MTG130NP icon. Move your pointer to the design area; the symbol will be attached to it. Left click to place the mounting hole. Recheck the box for each hole you want to add to the board design. Click OK to dismiss the dialog box.

Since the plated holes are footprints, they contain a place boundary like any other component, but on this footprint place boundaries were added to both the top and bottom layers. The symbol was designed that way so no components are allowed to be located near the holes and cannot contact the mounting hardware (standoffs, washers, or nuts, say).

The unplated holes are mechanical symbols. They contain place boundaries on the top and bottom (not shown in the figure) for the same reason, but they also contain route keep-out areas (Route keepout/All). The keep-out areas are needed so that traces cannot be routed under mounting hardware or across the unplated hole (and subsequently drilled away). The design rule checker (DRC) may not detect such a situation, because the Constraint Manager does not know about the mounting hardware that will be put into the hole, and there are no pads against which to maintain spacing rules.

Mechanical symbols for fiducials

Fiducials are marks or objects used for aligning the board with *solderpaste stencils* and *pick and place machines*.

Stencils are thin stainless steel plates or sheets through which solderpaste is squeegeed onto the board. The thickness of the stencil establishes the thickness (and therefore the amount) of the paste applied to the pads. The stencil manufacturer uses the pastemask file to fabricate the stencil.

Fiducial objects are placed at particular coordinates on both the board and the stencil, and optical equipment is used to align the stencil to the board during application of solderpaste.

During pick and place operations the fiducials on the board are used to precisely identify the location and orientation of the board so that the components are accurately placed into the solderpaste.

On the circuit board the fiducial is typically a copper mark produced during the etch process. Because the fiducials are relied on optically, the soldermask is usually removed at the fiducial location. Since the fiducial needs to be visible through the stencil, an exact copy of the fiducial needs to be included with the pastemask pattern. A fiducial object then requires etch, soldermask, and pastemask elements.

Note: Many of the native PCB Editor footprint symbols do not contain pastemask elements; so you need to add them to the footprints if you plan to fabricate a stencil and use pick and place processes to populate your board.

We next look at *how to make a fiducial*. Ideally we want the fiducial to be a symbol that is an element of the board rather than a component. So we want to construct the symbol on the classes and subclasses shown in Table 10.2. Unfortunately a pastemask subclass does not exist under the Board Geometry class. So, we will make one. Begin by opening a new mechanical drawing in PCB Editor as described in the mounting hole example.

To define a new subclass, select Setup → More → Subclasses... from the menu. At the Define Subclass dialog box shown in Fig. 10.5, select the class where you want to assign the new subclass. For this example, click the box next to Board Geometry.

At the Define Subclass dialog box (see Fig. 10.6), type PASTEMASK_TOP in the New Subclass: text box then hit the Enter key on your keyboard. The new class will be added to the list and displayed in the Options pane and the Color dialog box.

Table 10.2 The elements of a fiducial.

Function	Class	Subclass
Copper object	Etch	Top
Soldermask opening	Board geometry	Soldermask_Top
Pastemask opening	Board geometry	Pastemask_Top

Figure 10.5 Defining a new subclass for a board geometry pastemask.

The geometry of a fiducial is typically specified by the user. In this example the fiducial will be a 40×40 mil^2. The first step is to place the copper square. Change the All Etch grid setting (Setup → Grids) to 10 by 10 mil. Zoom in to the center of the drawing area (0, 0). Select the Etch/Top class and subclass from the Options pane. Select the Shape Place Rect tool and draw a 40×40 mil box centered at coordinate (0, 0).

Next we draw the soldermask opening. Change the Non-Etch grid setting to 5 by 5 mil. Select the Board Geometry/Soldermask_Top class and subclass from the Options pane. Select the Shape Place Rect tool and draw a 50×50 mil box (5 mil larger than the pad on all sides) centered at coordinate (0, 0).

Finally the pastemask is defined. Repeat these procedures using the Shape Place Rect tool on the Board Geometry/Pastemask_Top class and subclass and draw a 40×40-mil box centered at coordinate (0, 0) directly over the copper pad. Save the drawing to create the .bsm file.

To place the fiducial, use the procedure described for placing a mechanical symbol. Place two or more fiducials on the outer perimeter of the board.

Generating manufacturing data

Manufacturing data include two general types of files: artwork (Gerber) files and NC files (for drilling and milling). These files are made using two or more distinct processes. We begin with the artwork files.

Figure 10.6 Define non-etch subclass dialog box.

Generating the artwork

The artwork files describe elements of the design and include the Etch, Soldermask, SolderPaste, and Silk–Screen layers.

Most of the manufacturing data are basically present in the board design but not in a format that manufacturers use. The artwork creation process performs that function.

Before we begin the process, one more element needs to be added to the board, a photoplot outline.

Adding a photoplot outline

The board outline generated by the `Design Outline` dialog box serves as a guide for the designer during the design process, but maybe sometimes it does not contain manufacturing data for the manufacturer. That information may be contained in a photoplot outline.

To draw a photoplot outline, select the `Manufacturing` class and `Photoplot_Outline` subclass from the `Options` pane. From the toolbar, select the `Add Rect` tool (not the `Shape Add Rect` button). Draw a rectangle directly over the top of the board outline located on the `Board Geometry/Design_Outline` layer. A photoplot outline is shown in Fig. 10.7.

Optional artwork items

All the silk-screen elements in the design are associated with the individual components. You can add additional etch and silk-screen markings that are specific to the board. For example, you can use objects (mechanical symbols) on the top or bottom layers to add a company logo or add silk screen to display the board part number or manufacturing date, say, as shown in Fig. 10.7.

Once the board is ready to go, the next step is to generate the artwork files.

Figure 10.7 Photoplot outline and optional manufacturing objects.

Generating the artwork files

Artwork control form

The artwork files are generated using the Artwork Control Form dialog box (shown in Fig. 10.8). *To display the* Artwork Control Form *dialog box*, select Export → Gerber from the menu bar or click the Artwork button on the toolbar (if it is displayed). The Artwork Control Form contains two tabs, the Film Control tab and the General Parameters tab. The default settings may cause two warning boxes to be displayed. The procedure for clearing these warnings is given in the next section.

General parameters

The General Parameters tab is shown in Fig. 10.8. Use the tab to set the artwork format. Select the Gerber RS274X option if it is not already checked. In the Decimal places: box, enter the same value that is in the Integer places: box. This will handle

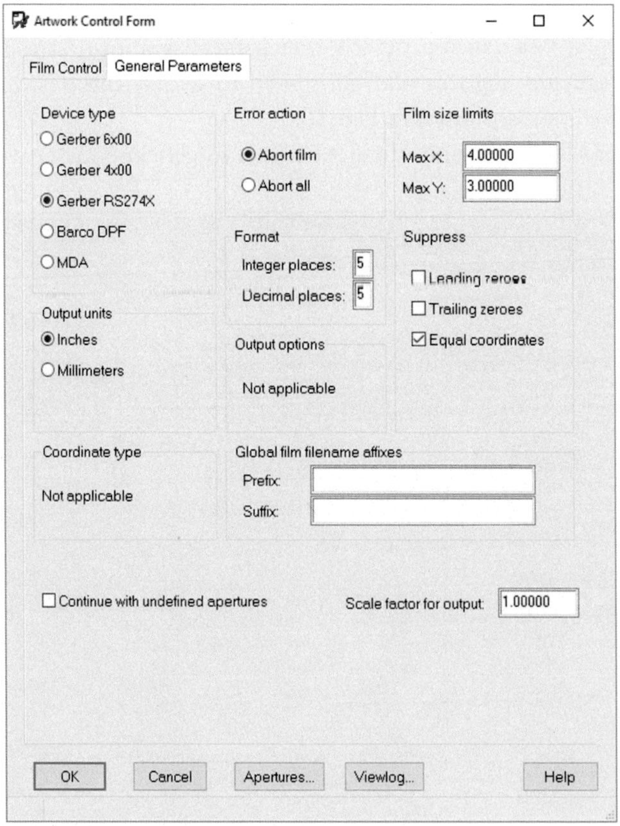

Figure 10.8 Selecting the artwork format.

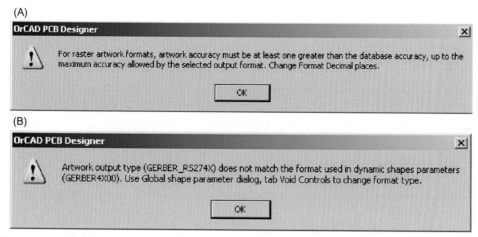

Figure 10.9 Artwork Control Form warnings: (A) decimal places, (B) artwork output type.

the warning shown in Fig. 10.9A. Click OK to close the Artwork Control Form dialog box.

To handle the second warning in Fig. 10.9, *change the* Artwork Format *in the* Design Parameter Editor *to match the* Device *type in the* Artwork Control Form (Fig. 10.8). To do this, select Setup → Design Parameters... from the menu. At the Design Parameter Editor dialog box, select the Shapes tab then click the Edit global dynamic shape parameters... button to display the dialog box (see Fig. 10.10). Select the Void controls tab and choose Gerber RS274X from the Artwork format: list. Click OK twice to dismiss both dialog boxes.

Film control

The Film Control tab shown in Fig. 10.11 contains folders that in turn contain subclass items. To see the subclass items, click on the + symbol next to the folder you want to view. Initially the form contains folders related to the layer stack-up only. Folders for the other items (e.g., silk screen and soldermask) must be added.

A folder contains related items. For example, recall that there are several types of silk-screen items [component reference designators (REFDES), component outlines, board geometry text, etc.], which belong to different classes and subclasses. These items are collected and organized in folders, and a single silk-screen Gerber file is made from the collective data in that folder.

The first step then is to take an account of all the types of elements you want manufactured with your board. Table 10.3 shows the conductor type artwork elements that need to be generated; these are autogenerated and already exist in the Artwork Control Form. Table 10.4 lists some of the most common nonconductor elements that can be placed on the board. Solderpaste (Pastemask) class elements can

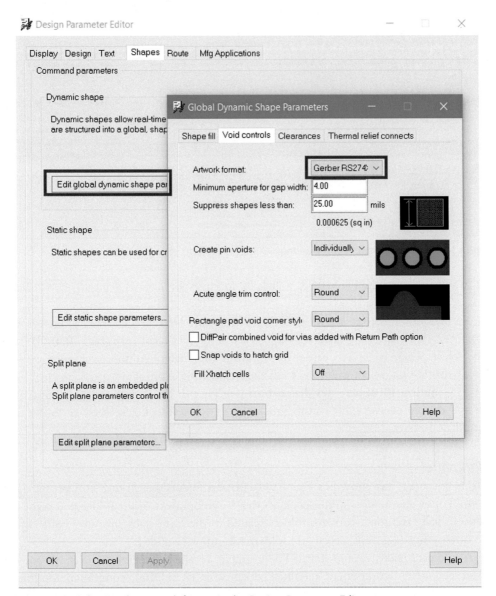

Figure 10.10 Selecting the artwork format in the Design Parameter Editor.

also be generated if you are planning on using pick and place machines to populate your board. Three classes contain pick and place elements, as described in the section *Generating Pick and Place files*. They are added to the Artwork Control Form the same way the soldermask and silk-screen elements are added, which is described next.

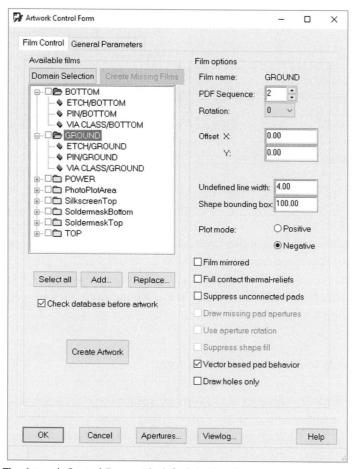

Figure 10.11 The Artwork Control Form with default layers.

Table 10.3 Conductor artwork elements.

Class	Subclass
Etch	Any layer in the stack–up
Pin	Any layer in the stack–up
Via	Any layer in the stack–up

For this design the top side of the board needs a soldermask and silk screen, while the bottom side of the board needs only the soldermask. Since there is no silk screen on the bottom, we have no bottom silk-screen folder. So from the table, we have three folders: a top-layer silk-screen folder, which includes the Board Geometry, Ref Des, and Package Geometry classes; a top-layer soldermask folder, which includes the

Table 10.4 Nonconductor artwork elements.

	Subclass (*_Top and *_Bottom)	
Class	**Silk screen**	**Soldermask**
Board geometry	X	X
Component value	X	
Component device type	X	
Component reference designator	X	
Component tolerance	X	
Component user part number	X	
Package geometry	X	X
Pin		X
Via		X

Pin and Via classes; and a bottom-layer soldermask folder, which contains the same classes as the top-layer soldermask (only on the bottom layer).

In addition, we need a Gerber file for the board outline. Other types of Gerber files can be made, but for the time being, these are the only ones we need. We begin by adding the top-layer soldermask folder.

Adding elements to the artwork control form

When you add a folder to the control form, all classes that are visible in the design are added. So we begin by turning off all layers except for the classes and subclasses related to the top-layer soldermask. Use the Color dialog box to control the visibility of the classes. *To turn off all the layers at once,* click the Off button (Global Visibility:) at the top of the dialog box, then click the Apply button at the bottom.

Next, check the box for the Soldermask_Top subclasses in each of the Package Geometry, Board Geometry, Pin, and Via classes. The Package and Board Geometry classes have nothing on them in this design, but it does not hurt to add them to get in the practice of including them for future designs.

Once these elements are the only ones visible, open the Artwork Control Form. We will add a new folder for the top-layer soldermask items. *To add a new folder to the* Artwork Control Form, right click on any one of the existing folders and select Add from the pop-up menu (do not use the Add button at the bottom of the form). Enter a name for the folder in the New Film dialog box (see Fig. 10.12). The name does not have to match any PCB Editor class or subclass name, but it can be helpful if it does or is similar. The new folder now will be listed in the form, and it will contain the subclasses visible in the design.

Using this procedure, make a new SoldermaskBottom folder and include the similar classes as the top-layer soldermask folder (VIA, PIN, PACKAGE GEOMETRY, etc.).

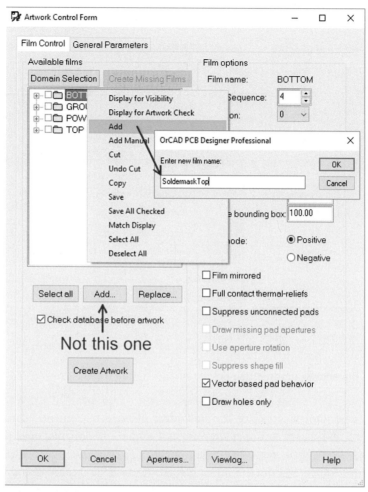

Figure 10.12 Adding a new folder to the artwork list.

Next we add a silk-screen object. As indicated in Table 10.4, many silk-screen objects can be added to the design. You can use the procedure just described to create the silk-screen films too. But there is another way to generate the silk-screen films using the `Autosilk` tool.

During the part placement steps, it is very possible that silk-screen objects can accidentally be removed (or be unable to be printed) on the soldermask because of openings in the mask, or they can become hidden under components or even printed on top of other silk-screen objects. Keeping the silk-screen objects under control can be a task. Fortunately PCB Editor has a tool that helps us keep track of them, `Autosilk`.

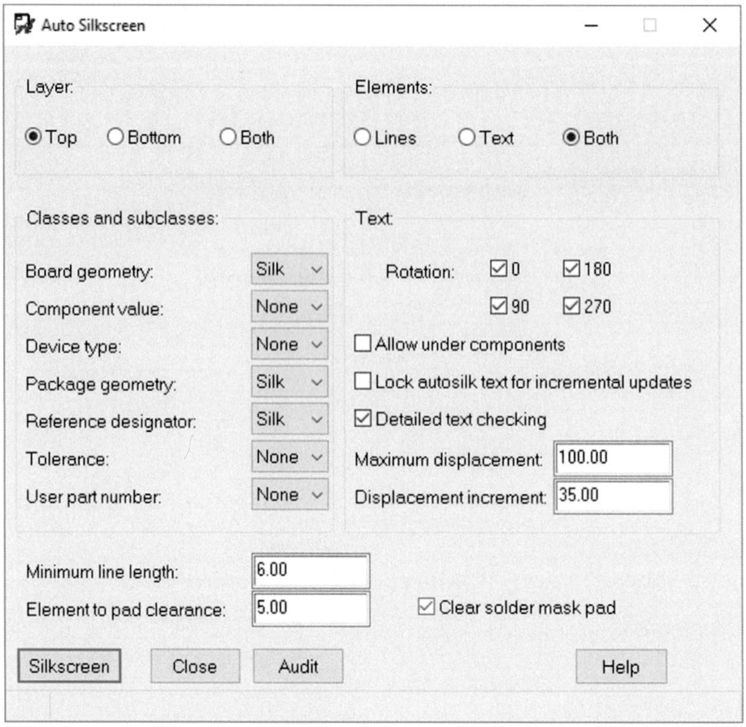

Figure 10.13 The Autosilk screen dialog box.

The Autosilk tool takes all of the individual silk-screen elements, combines them into one layer, and makes sure that they are not under components, over padstacks, or too far away from their respective components. *To start the Autosilk process*, select Manufacture → Silkscreen from the menu bar. At the Auto Silkscreen dialog box (shown in Fig. 10.13), you can select which elements you want added to the common silk-screen files (one for Autosilk_Top and one for Autosilk_Bottom). You can also set default values for some of the physical properties related to silk-screen objects. When the settings are complete, click the Silkscreen button. When the process is complete, the dialog box will dismiss itself and the results will be displayed in the command window. To view the new silk screen, make sure the Manufacturing/Autosilk_Top (or _Bottom) subclass is visible. It is also helpful to make the new silk screen a color different from the other ones.

If you do not notice a big difference, it could be that the new silk screen is directly on top of the original. This will happen if the Autosilk tool does not find anything wrong with the existing ones. To make sure the new silk screen is correct, turn off the original ones (all of them) and look at just the new one, which will be the sum of all the others. Otherwise, you should notice that some of the REFDES were moved away from fan-out vias and outside the place boundary shapes. Depending on how the autorouter routed your board, you may also notice that some of the component

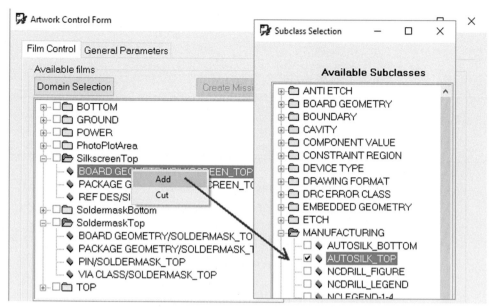

Figure 10.14 Adding an artwork item to an existing folder.

outlines are removed where vias are close to the component body. We now add the new silk screen to the artwork form.

If you previously made a silk-screen folder you can add the new autosilk data to the existing folder and remove the other, individual silk-screen elements.

To add an item to an existing folder, right click on any one of the existing items in that folder and select Add from the pop-up menu. At the Subclass Selection dialog box (see Fig. 10.14), check the boxes for the items you want to add and click the OK button. The items will be added to the folder. *To delete an item from a film folder,* right click on the item and select Cut from the pop-up menu.

The last item to add is the board outline item. The board geometry outline drawn at the beginning of the board design process does not contain Gerber data and is for the designer's benefit. Gerber data for a board or design outline is contained in the photoplot outline.

Using the previous procedures, add a photoplot outline folder that contains the Manufacturing/Photoplot_Outline class and subclass. An example of the completed artwork list is shown in Fig. 10.15.

We are almost ready to generate the Gerber files. Before we do, check to make sure that the ground and power layers are marked as Negative in the Plot Mode: area of the Artwork Control Form and you have entered a value in the Undefined Line Width: box (a number between 5 and 10 is usually good). If you forget to add a default line-width value, any lines or text that have a width of 0 will be ignored (resulting in large blank areas on the silk screen). And recall from Chapter 8, Making

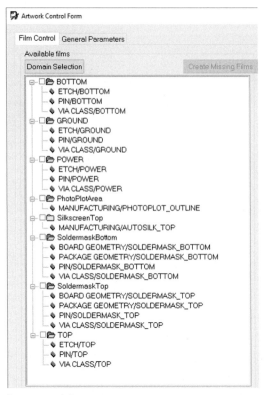

Figure 10.15 The complete artwork list.

and editing footprints, that many of the silk-screen outlines that come with the library use rectangles instead of lines, and rectangles always have a default width of zero. Also note that if you decide to plot the board outline using the shape created in `Design_Outline` instead of `Photoplot_Outline` drawing, you should add `Board Geometry/Design_Outline` subclass to the `PhotoPlotArea` folder and set the `Undefined line width` value to 10 for it as well.

Also, make sure that the `Vector based pad behavior` option is checked. This applies only to negative planes, but it affects how the unconnected pads on the plane layers will be represented in the artwork. Leaving this option checked removes unused pads from negative planes, which results in a larger (safer) clearance gap.

Next, populate the aperture list (skip this step if you've selected Gerber RS274X format). Click the `Apertures...` button (see Fig. 10.16). An `Edit Aperture Wheels` dialog box will be displayed with at least one entry. Click the `Edit` button to display the `Edit Aperture Stations` dialog box. If you do not see a list of entries, click the `Auto` button and select `Without Rotation` from the pop-up menu. After the aperture wheel list has been populated, click `OK` twice to get back to the `Artwork Control Form`.

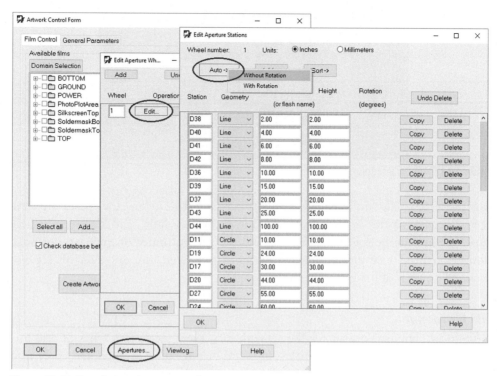

Figure 10.16 Populating the aperture list.

To generate the Gerber files, click the `Select All` button in the `Artwork Control Form` to check all the folders (or just check the ones you want—the others will not be written or overwritten if they already exist). Then click the `Create Artwork` button. The command window will report the results and you can read through the photoplot log to look for any errors. Check the working directory and make sure that the files are present. In this example, eight files with the `.art` extension should have been generated. These are the Gerber files that will be sent to the board manufacturer.

That completes the artwork process. Now we move onto making the drill files.

Generating drill files

Before generating the drill data, it is a good idea to take an inventory of the drill holes. You can do so using the drill legend and the `Padstack Definition Report`. The report provides all information about each padstack, but the drill legend is much easier to read.

To make a drill legend, zoom out a bit so you have room to place the drill legend. Select `Manufacture → Create Drill Table...` from the menu. Set the required options in the `Drill Legend` dialog box, press `OK`, and the legend will be attached to your pointer. Left click to place it (typically above the board outline or off to one side, but any

DRILL CHART · TOP to BOTTOM			
ALL UNITS ARE IN MILS			
FIGURE	SIZE	PLATED	QTY
·	14.0	PLATED	8
▫	36.0	PLATED	10
◦	42.0	PLATED	4
○	130.0	PLATED	2
+	130.0	NON-PLATED	2

Figure 10.17 The drill legend.

place is OK). The drill legend (shown in Fig. 10.17) lists the holes by drill figure (specified in the `.pad` file as designed in Padstack Editor), and it shows the hole sizes, whether the hole is plated or not, and the total quantity of each type of hole. Notice too that, once the drill legend has been generated, the drill symbols are also displayed on each of the padstacks in the design.

Drill file format

As there are different formats for the artwork files, there are several types of drill formats. You need to know the format your board manufacturer requires and apply it to your design. *To specify the drill format*, select the `Ncdrill Param` button on the toolbar or select `Export → NC Parameters...` from the menu to display the `NC Parameters` dialog box (see Fig. 10.18). At the `NC Parameters` dialog box, you can enter a header if your manufacturer asks for one, but PCB Editor does not care if you enter one. Enter your board manufacturer's specifications for the format (e.g., 2.4), zero suppression, Excellon format options, and the like. Once the values are set, click the `OK` button to save your settings.

The next step is to create the drill file. Select `Export → NC Drill...` from the menu to display the `NC Drill` dialog box shown in Fig. 10.19. If you have not scaled your design drawing in any way, enter a scale factor of 1.

If your board manufacturer prefers a particular tool sequence, you can select that here; otherwise, use the default setting.

Select the `Auto tool select` option. This option causes PCB Editor to look for a tool list. If one does not exist, it will create one automatically. It also adds drill tool header information to Excellon formatted files. If the `Auto tool select` option is not enabled, no tool list is created and the drill file contains only drill locations and manual stops where the computer numerical control machine operator is required to indicate the drill tool from a manually generated tool list.

If you know for sure that your board manufacturer can process drill files that contain both plated and nonplated hole information, you can leave the `Separate files for`

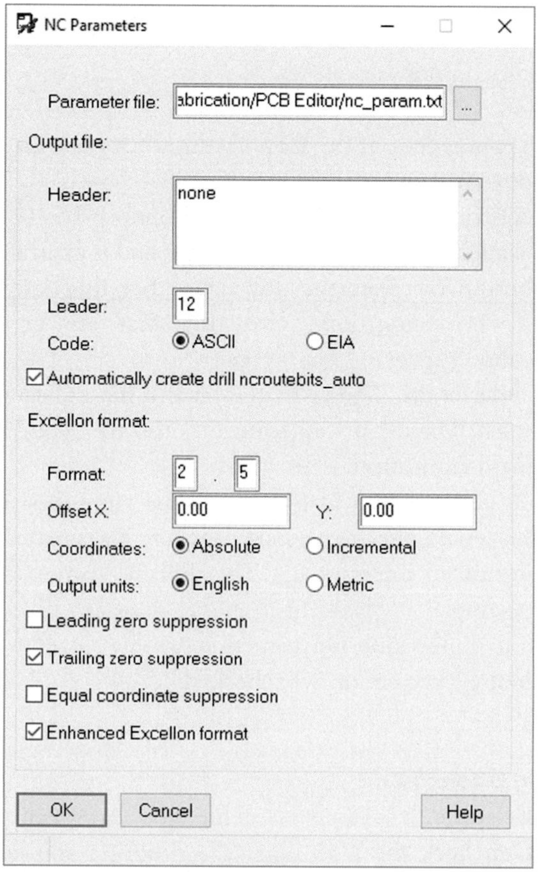

Figure 10.18 The NC (Drill) Parameters dialog box.

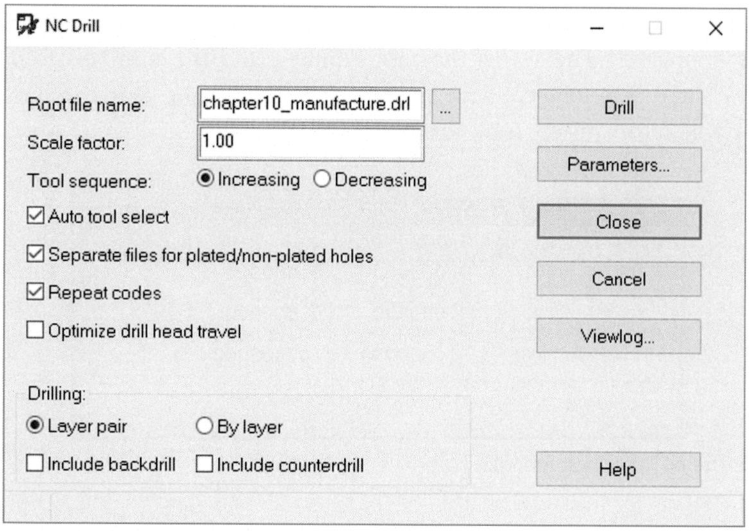

Figure 10.19 The NC Drill dialog box.

plated/non-plated holes option unchecked. Otherwise, check the box to make two distinct drill files. Nonplated hole drill files have np in the filename, as described later, whereas the plated holes do not.

The last two options (repeat codes and optimized head travel) also depend on your board manufacturer's preferences. If the manufacturer does not specify one way or the other, check them both.

In the Drilling: option box, select the Layer pair option. The By layer option is used for back-drilling and blind/buried-via drilling operations and is explained briefly later.

Click the Drill button to create the drill file(s). For this design with the settings chosen in the NC Drill dialog box, two drill files are generated, chapter10_manufacture-1-4.drl and chapter10_manufacture-1-4-np.drl. The naming convention is *filename*-startlayer-endlayer.drl. The np in the second file indicates that the file is for the nonplated drill holes. These *.drl files are the NC files your board manufacturer will use during the board fabrication.

If you chose the By layer option in Fig. 10.19, the same files would be generated, but additional drill files would also be generated, where a separate drill file is generated for each layer combination in the stack-up. The drill file names would then be, for a four-layer board for example, *filename*-bl-1-2.drl, *filename*-bl-2-3.drl..., *filename*-bl-3-4.drl, in addition to the one file named *filename*-1-4.drl. The bl in the file name indicates that the by layer option was checked.

Manually generating a tool list

Instead of using the default nc_tools_auto.txt file, you can specify your own tool list. To do so, you must uncheck the Auto tool select option and create a tool list file, which must have the file name nc_tools.txt. *To generate a tool list*, open a text editor such as Notepad and save it as nc_tools.txt. A tool list is shown in Fig. 10.20. Each line begins with the diameter of the drill hole. The P or N indicates whether the hole is plated or nonplated. The T*nn* is the tool number, and the numbers following that indicate the + and − tolerance. Save the tool list file into the working directory, and PCB Editor will automatically look for it there.

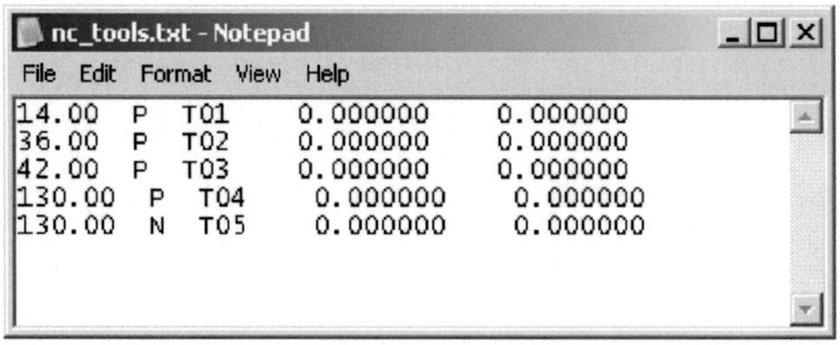


```
14.00   P   T01     0.000000     0.000000
36.00   P   T02     0.000000     0.000000
42.00   P   T03     0.000000     0.000000
130.00   P   T04     0.000000     0.000000
130.00   N   T05     0.000000     0.000000
```

Figure 10.20 The nc_tools drill tool file.

Generating route path files

For most designs and most manufacturers the artwork and drill files provide enough information for them to fabricate your board. But, under certain circumstances, you may also need to generate an *NC route path* file for the milling operation used to cut your board from a larger panel or for cutting slots or irregularly shaped drill holes in the board.

As described in Chapter 4, Introduction to industry standards, the PCB manufacturers use standard panel sizes to fabricate boards. To help reduce costs, several designs may be placed on a single panel during fabrication and cut apart by a milling machine. The cutting path that the machine takes is called the *NC route path*. Normally you need not worry about it, since you do not know how the board manufacturer will utilize the panel space. However, if, for example, you want to relay to your manufacturer specific cutting requirements or you buy your boards in panels for mass production pick and place processes and separate the pieces yourself, you need to generate an NC route file.

Generating an NC route file typically requires two steps, but additional steps can be employed, which will be discussed here for completeness. The first required step is to create an `ncroutebits.txt` tool file and the second is to place an NC route outline. The optional steps are placing cut marks at the board corners and placing cutting path direction indicators on the route path outline.

Establishing an NC route tool list file

As for the NC drill process, you can automatically create the routing tool list, or you can specify it manually. Setting up an NC route tool list is similar to making the drill tool list described previously. A sample is shown in Fig. 10.21. Use a simple text editor (such as Notepad) and open a blank file. The contents have a line that includes the bit diameter (in inches if your design is using English units or millimeters for metric units) and a tool number for each cutting tool that will be used in the design, as defined by the line widths on the `Ncroute_Path` subclass. For this example the bit diameter is 0.100 in. (100 mil). Save the file into your working directory with the filename `ncroutebits.txt` and then close the file.

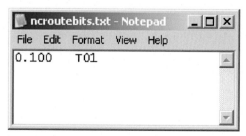

Figure 10.21 The ncroutebits tool file.

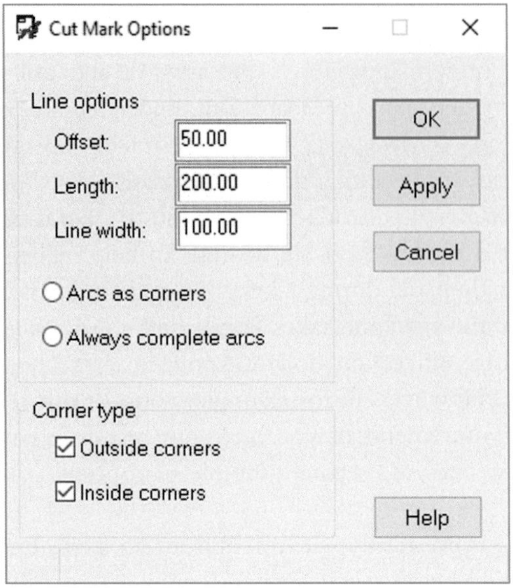

Figure 10.22 Configuring the cut marks.

Cut marks

The actual cutting information is contained in the route path outline described later and not in the cut marks. But we begin with this optional step because cut marks can be used as a guide in placing the route path outline, and it will help make sense of the later steps.

We specified that the bit used to cut around the board is 100 mil in diameter, so the cut marks and the placement of the route outline will be based on that. To place *cut marks on your board*, select Manufacture → Cut Marks from the menu bar. At the Cut Mark Options dialog box (see Fig. 10.22), enter 100 as the Line width:, because that is the diameter of the cutting bit. Next, assuming that the edge of the bit will cut right along the board edge, the center of the cut path has to be offset by one half of the bit width (50 mil) from the board so that the bit does not cut into the board itself. Therefore enter 50 in the Offset: option. The line length is not critical, as it simply determines the visual length of the mark. A line length of twice the width is a good rule of thumb. If your board is not a simple rectangle, you should select both inside and outside corners in the Corner type box, otherwise the outside only is adequate. When you are ready to place the marks, click the Apply button. PCB Editor automatically places cut marks around the board outline on the Board Geometry/Cut_Marks class and subclass. An example is shown in Fig. 10.23.

Drawing the cut path

The next step is to place an outline along the cutting path. Begin by selecting the Ncroute_Path subclass under the Board Geometry class in the Options pane. Select

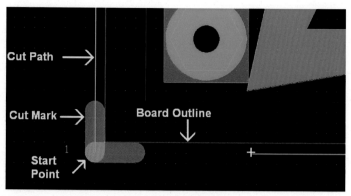

Figure 10.23 Drawing a cut path around the board outline.

the `Add Line` tool from the toolbar. Set the width of the cut in the `Line width:` box in the `Options` pane (100 mil). The center of the cut marks will serve as a drawing guide. Left click and release at the center of the cut mark to begin drawing the cutting path. Continue around the board, drawing the cut path through the center of the cut marks. When you return to the starting point, left click to end the last segment at the starting point and then right click and select `Done` from the pop-up menu.

Note: You must use the `Add Line` tool (not a rectangle) to draw the cut path so that you can specify the cut width. PCB Editor uses the width of the line to determine which cutting tool to use from the `ncroutebits` file when it generates the route file (`*.rou`). If you have only one cut path (and width) and only one tool in the `ncroutebits` file, PCB Editor will assume you mean to use that bit for that cut. But if you have multiple cuts (e.g., for edge slots) and multiple bits, then you have to specify the line width and have the correct bits listed in the bit file.

Fig. 10.23 shows a small `1` near the starting point. This is an indicator used to tell the NC machine where to start. A `2` at one of the adjacent corners defines the direction that the cut should be made (not the order the outline was drawn). The numbers are optional, but we will place them as a demonstration.

To place cutting order text objects, select the `Add Text` tool from the toolbar and the `Ncroute_Path` subclass under the `Board Geometry` class in the `Options` pane (same as for the cutting path outline). Left click near the corner you would like to have the cutting tool start (it does not have to be the corner you started the outline). Enter a `1` in the text block, right click, and select `Done` from the pop-up menu. That defines the starting point. Use the same procedure to place a `2` at the next corner. The second number defines the direction of the cut path. You can place starting and direction indicators on as many distinct cutting paths as there are on your board. You do not have to place a number at every corner of a closed path unless your board requires a special cutting process.

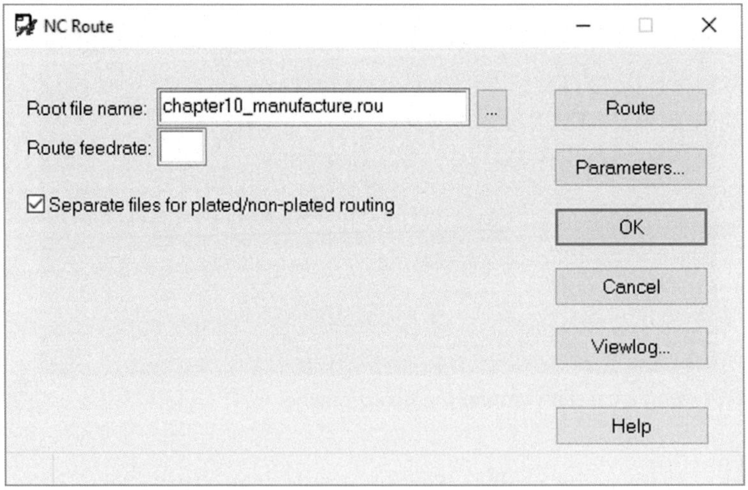

Figure 10.24 The NC Route dialog box for starting the route process.

Generating the route file

The last step in this example is to generate the NC route file. To do so, select Export → NC Route... from the menu. At the NC Route dialog box (see Fig. 10.24), enter a feed rate if you know what it is or leave it blank. You can change the NC Parameters if necessary; we will use the same ones specified during the drill setup. Click the Route button to continue. When the route process is complete, the route log will be displayed. The NC route file will be named *boardname*.rou.

Verifying the artwork

Before you send your artwork files to the board manufacturer, it is a good idea to review them.

To use PCB Editor to review your artwork files, open a new, blank board design in PCB Editor. To do so select File → New from the PCB Editor menu bar. At the New Drawing dialog box, select Board from the Drawing Type: selection list. Enter a name for the board and choose a destination using the Browse... button then click the OK button. A new, blank design will be opened.

Use the Design tab in the Design Parameter Editor dialog box (Setup → Design Parameters... from the menu) to set up the drawing size needed for the artwork files. Use the Cross Section dialog box (Xsection button on the toolbar) to set up etch layers for each conductor artwork file you intend to import (e.g., top, ground).

Next, select Import → More → Artwork... from the menu (see Fig. 10.25).

At the Load Cadence Artwork dialog box (see Fig. 10.26), select the class and subclass for the artwork you want to import then use the Filename: Browse... button to locate that artwork file. Click the Load file button.

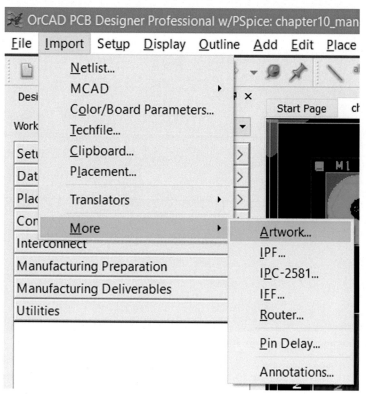

Figure 10.25 Importing artwork into PCB Editor.

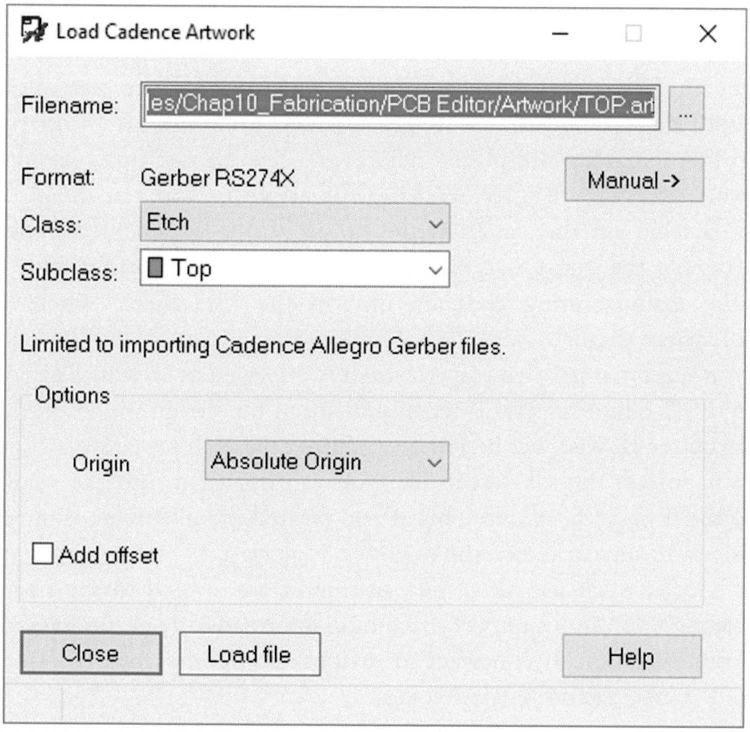

Figure 10.26 Loading an artwork file.

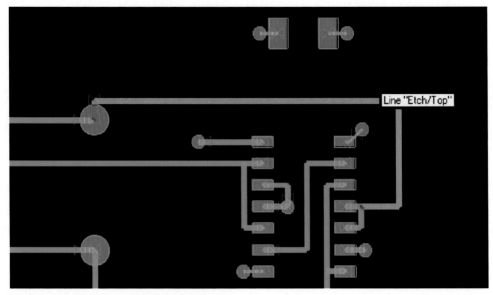

Figure 10.27 The imported top-layer artwork.

The artwork will be attached to your pointer as an outline with the artwork origin as the insertion point. Click the P button at the bottom of the design window to display the Pick dialog box. Enter 0 0 and click the Pick button. The origin of the artwork file will be placed at the origin of the current drawing.

Fig. 10.27 shows an example of what the top-layer looks like when importing the Top.art file. The traces (clines) are now just lines and no net information is associated with them. The pads are shapes and have a dummy net property attached to them, so DRC markers are placed wherever a line crosses a pad (even if it is not supposed to). Since we are just reviewing the artwork, you can disable the DRC markers by turning off that layer in the Color Dialog panel (go to the Stack-Up folder and disable the entire DRC column). Alternatively you can set up layer subclasses in the manufacturing class and import the files there, where the DRC checker will ignore them.

When you import a negative plane layer, it is displayed as an actual negative image, not the way PCB Editor displays it when you are in the design mode. In the negative view, what you see is what will be removed during the etch process.

When you import the silk-screen file, you can import it into the Board Geometry class or into the Manufacturing class, but if you created the silkscreen using the autosilk feature, it is best to import it into the Manufacturing class.

The types of artwork files that you can import are limited to the basic etch and nonetch objects. You cannot import machining information (e.g., drilling and milling) or photoplot information. If you want to review that information, you need to use a computer-aided manufacturing (CAM) tool.

Using CAD tools to 3D model the printed circuit board design

Before you send the Gerber files off to the board house to be manufactured, you might want to know if the board will be physically compatible with the house mounting hardware and enclosure (i.e., will it fit?). You can export the design information from PCB Editor to a .DXF file, which you can open with most popular CAD drawing packages.

Sincerely speaking the design has a tremendous amount of information, a lot of which does not need to be exported to a drawing file. To select layers to plot, use the Colors Dialog panel to turn on only the layers you want to include in the drawing.

To export a PCB design as a .DXF file, select Export → MCAD → DXF... from the menu. At the DXF Out dialog box (Fig. 10.28), enter a name in the file name box. Click the Layer conversion file entry Browse... button. At the dialog box, click the Open button to accept the default file name or enter a new name and then click Open.

The conversion file needs to be set up. To do so, click the Edit button to display the conversion file editing dialog box shown in Fig. 10.29. You can specify layers and names or map the PCB Editor classes directly to the new DXF layers. The options shown in the figure map them directly. To execute the conversion, click the Map button. The class and subclass names will be combined into one name in the DXF layer boxes. Click OK to dismiss the conversion dialog box and return to the DXF Out dialog box.

In the Data Configuration section, you can leave everything unchecked, but the objects in the DXF file will be of two dimensional. If you want a 3D representation, then at a minimum you need to check the Export symbols and padstacks as BLOCKS option. If you also check the Fill options, flat objects (traces and silk screens) will be filled in instead of just an outline. This will make the DXF drawing look richer, but it significantly increases the memory usage and drawing regeneration time.

Once the remaining options are set in the DXF Out dialog box, click the Export button.

When the conversion is complete, the DXF file can be opened with a CAD program.

Fig. 10.30 shows a 3D model of the design using a popular drawing program (the colors were modified with the CAD program). The CAD program can then be used to add the enclosure and mounting hardware to verify form, fit, and function. Dimensions and other manufacturing information can easily be added to the drawing file to fully describe manufacturing and assembly instructions.

The height information that the CAD program uses is derived from the PACKAGE_HEIGHT_MAX property of the component's place-bound outline in PCB Editor. You can view the property by selecting Edit → Object Properties... from the menu and left clicking on the component. The information shown in Fig. 10.31 is from the place boundary outline of J1. As mentioned earlier, place boundaries can be drawn

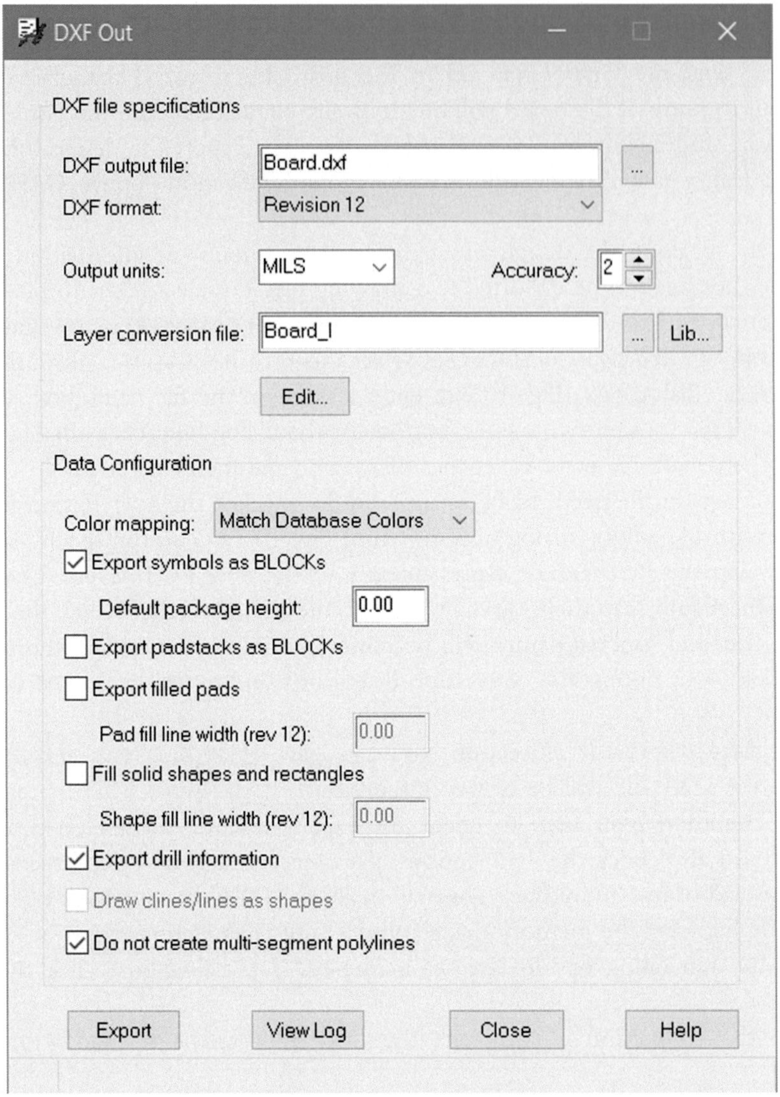

Figure 10.28 Setting up the DXF Out dialog box to generate data for a CAD drawing.

using filled rectangles or shapes. In some CAD programs, filled rectangles behave as solid objects whereas shapes behave as wire frame objects.

There are other options to export the information to CAD program. You can export IDF, IDX, or STEP model of your project. You can also view the 3D image in the PCB Editor using Display → 3D Canvas. Press Shift + Mouse Wheel and drag the mouse to rotate the view. In order to see the more realistic image you can attach the 3D STEP model to each component pattern using Setup → Step Mapping dialog box.

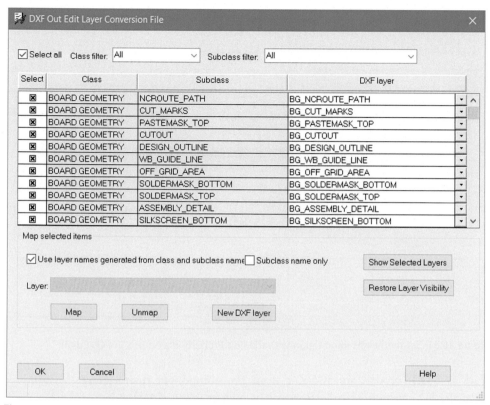

Figure 10.29 Mapping PCB Editor classes into DXF layers.

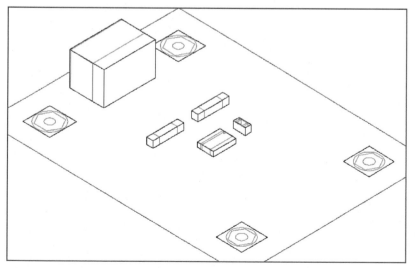

Figure 10.30 The 3D CAD model of the PCB.

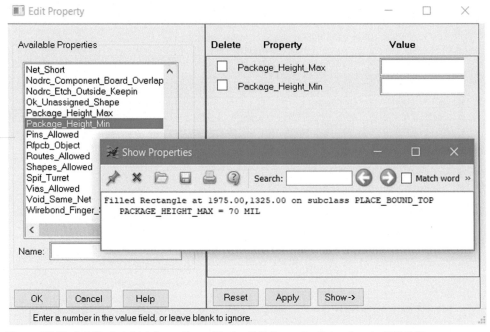

Figure 10.31 Set the height information with the Edit Property dialog box in PCB Editor.

Various STEP models are available from the component manufacturer websites or from the 3D component collection portals such as www.ultralibrarian.com where you can find the model by its part number. It's recommended that the component STEP models are changed in the MCAD program so that its orientation and origin were the same as for the PCB footprint in your design. This will simplify the mapping procedure as well as improve the ECAD—MCAD collaboration process.

Fabricating the board

Not all board houses are the same. While most follow certain fabrication standards, not all have the same fabrication capabilities (e.g., number of layers, copper thickness, and drill sizes). There are also differences in design submission policies and minimum order and billing practices.

Many board houses use an online file submission and quote process, but not all do. Some have you send the Gerber files by e-mail, and after someone looks at them with CAM software (such as GerbTool or CAM350), the company sends you a quote via e-mail.

An example of the generated Gerber files that would be sent to the board manufacturer is shown in Fig. 10.32.

Name ▲	Size	Type
BOTTOM.art	2 KB	ART File
chapter10_manufacture-1-4.drl	1 KB	DRL File
chapter10_manufacture-1-4-np.drl	1 KB	DRL File
chapter10_manufacture.rou	1 KB	ROU File
GROUND.art	7 KB	ART File
Outline.art	1 KB	ART File
POWER.art	4 KB	ART File
SilkscreenTop.art	7 KB	ART File
SoldermaskBottom.art	2 KB	ART File
SoldermaskTop.art	2 KB	ART File
TOP.art	3 KB	ART File

Figure 10.32 A list of artwork files.

Fabricated PCB inspection and testing

Regardless of which house fabricates your boards, you should perform an inspection of the boards, especially if it is a first run. You may also want to do some initial electrical tests (continuity and isolation), especially on the power and ground, before you place any components on the board. IPC-2515A and IPC-6011 outline many types of inspection and acceptance tests.

The next step is to populate the boards. The next section describes how to generate a file describing your board and parts that pick and place machines can use to populate the board.

Generating pick and place files

There are many types of pick and place machines, each with its own programming and setup requirements. As an example here, we just look at the basics most machines require. At a minimum the machine needs to know which parts are to be placed, where they are to be placed, and in what rotation. When placing large quantities of parts on a board, it is also often necessary to include part numbers, as there may be more than one type of component with the same value. For example, the only way to differentiate between a size 1206 smd, 0.1%, 1k resistor and a size 1206 smd, 1%, 1k resistor is by the part numbers. So we will customize the pick and place list to include the components' part numbers so that the assembler will know which part to place in which tray on the machine and how to program it.

The pick and place part list is a report. In general you generate reports by selecting Export → Reports from the menu or the Reports button on the toolbar if it is displayed. The Reports dialog box is shown in Fig. 10.33. There is no specific report for

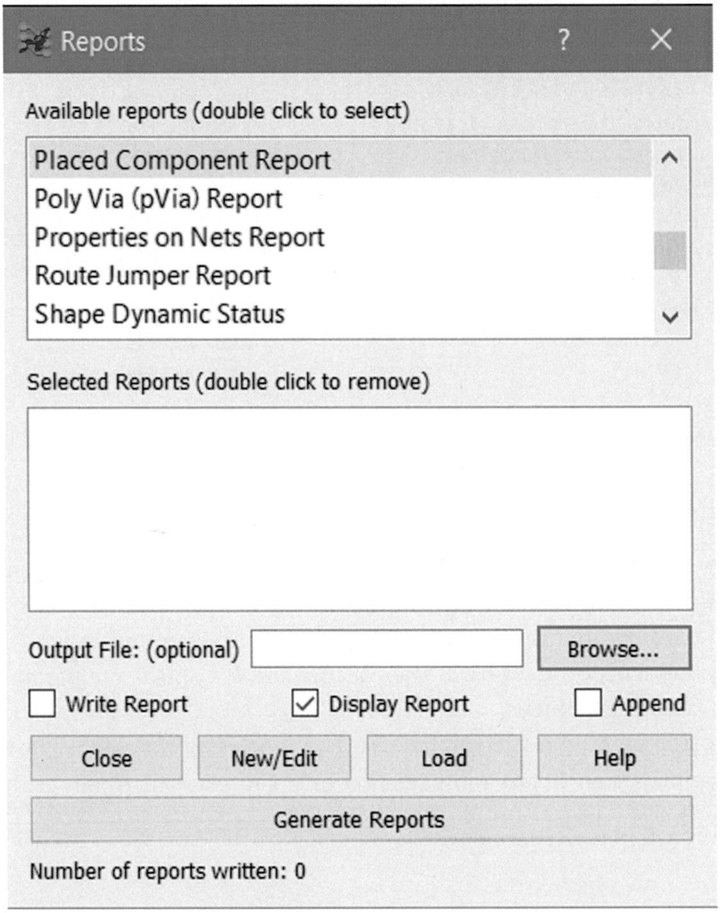

Figure 10.33 The reports selection dialog box.

pick and place machines, but the Placed Component Report comes close, so we begin with that.

The default Placed Component Report is shown in Fig. 10.34. This report provides the REFDES, the symbol X and Y coordinates, and the rotation, but no part numbers, and it contains other information that we do not care about.

We can create a report that will provide the part number, but this information must already be assigned to each part. An easy way to do that is to add the part number to the part in Capture. *To add a part number to a component*, select the component (or components) on the schematic, right click and select Edit Properties... from the pop-up menu. Enter the part numbers in the PART_NUMBER cells in the spreadsheet, as shown in Fig. 10.35. Perform an engineering change order to update the part number information to the board design in PCB Editor (if it has not already been done).

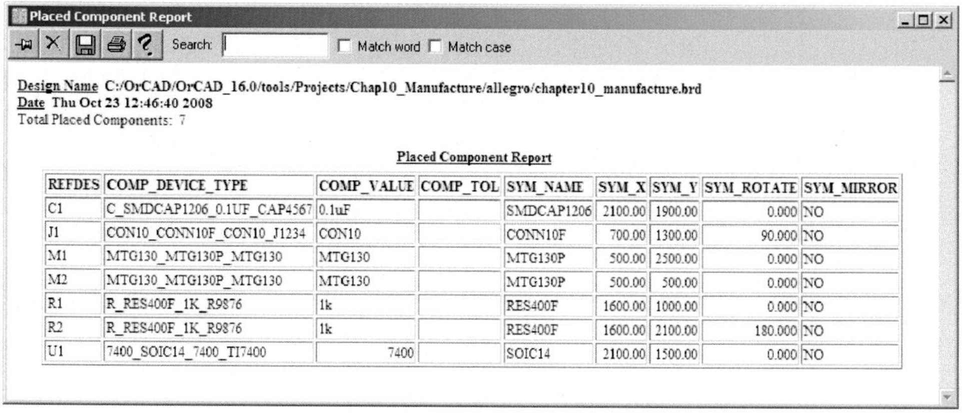
Placed Component Report

Search: Match word Match case

Design Name C:/OrCAD/OrCAD_16.0/tools/Projects/Chap10_Manufacture/allegro/chapter10_manufacture.brd
Date Thu Oct 23 12:46:40 2008
Total Placed Components: 7

Placed Component Report

REFDES	COMP_DEVICE_TYPE	COMP_VALUE	COMP_TOL	SYM_NAME	SYM_X	SYM_Y	SYM_ROTATE	SYM_MIRROR
C1	C_SMDCAP1206_0.1UF_CAP4567	0.1uF		SMDCAP1206	2100.00	1900.00	0.000	NO
J1	CON10_CONN10F_CON10_J1234	CON10		CONN10F	700.00	1300.00	90.000	NO
M1	MTG130_MTG130P_MTG130	MTG130		MTG130P	500.00	2500.00	0.000	NO
M2	MTG130_MTG130P_MTG130	MTG130		MTG130P	500.00	500.00	0.000	NO
R1	R_RES400F_1K_R9876	1k		RES400F	1600.00	1000.00	0.000	NO
R2	R_RES400F_1K_R9876	1k		RES400F	1600.00	2100.00	180.000	NO
U1	7400_SOIC14_7400_TI7400	7400		SOIC14	2100.00	1500.00	0.000	NO

Figure 10.34 The default placed component report.

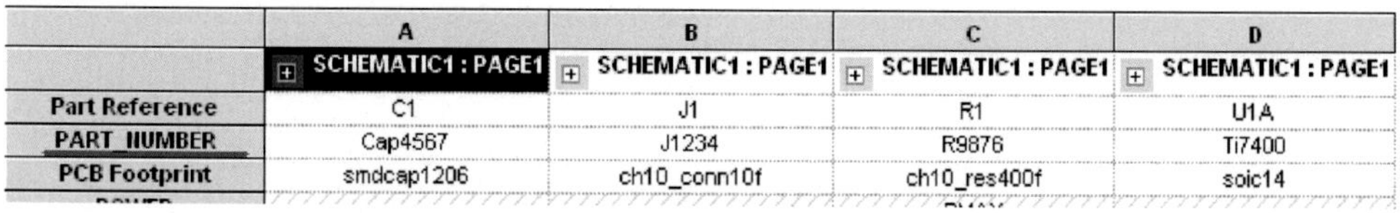

	A	B	C	D
	SCHEMATIC1 : PAGE1	SCHEMATIC1 : PAGE1	SCHEMATIC1 : PAGE1	SCHEMATIC1 : PAGE1
Part Reference	C1	J1	R1	U1A
PART_NUMBER	Cap4567	J1234	R9876	Ti7400
PCB Footprint	smdcap1206	ch10_conn10f	ch10_res400f	soic14

Figure 10.35 Using Capture to add part numbers to components.

Make sure that the parts have the part number property attached to them. To check, select the `Show Element` tool from the toolbar and select one of the parts. The `PART_NUMBER` property should be visible in the `Show Element` text box that pops up.

The next step is to create a custom report. *To make a new report*, display the `Reports` dialog box (see Fig. 10.33). Click the `New/Edit` button at the bottom. At the `Extract UI` dialog box (see Fig. 10.36), select the `Data Fields` tab, select `COMPONENT` from the database list, and double-click `REFDES` in the `Available Fields` list to add it to the `Current Configuration` window. Next, select the `Properties` tab, scroll down the `Available Properties` window until you find `PART_NUMBER`, then double-click it to add it to the list. Return to the `Data Fields` tab and look for `SYM_CENTER_X` and `SYM_CENTER_Y` and add them to the list. Also add `SYM_ROTATE` to the list. You may also add the `SYM_MIRROR` to the list, to show the component's side for 2-sided SMT assembly.

The final list is shown in Fig. 10.36. The order that the properties are listed in the dialog box is the order that they will appear in the report. If you need change the order, use the up and down arrows to the right of the list to move a selected item.

To save this report, click the `Save` button and enter a name in the dialog box that pops up. Click the `OK` button to dismiss the `Extract UI` dialog box.

To generate the new report, scroll all the way down to the bottom of the `Available Reports` list in the `Reports` dialog box (Fig. 10.37), and your report will be listed.

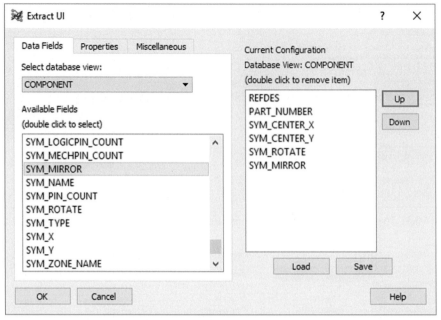

Figure 10.36 Making a custom report in PCB Editor.

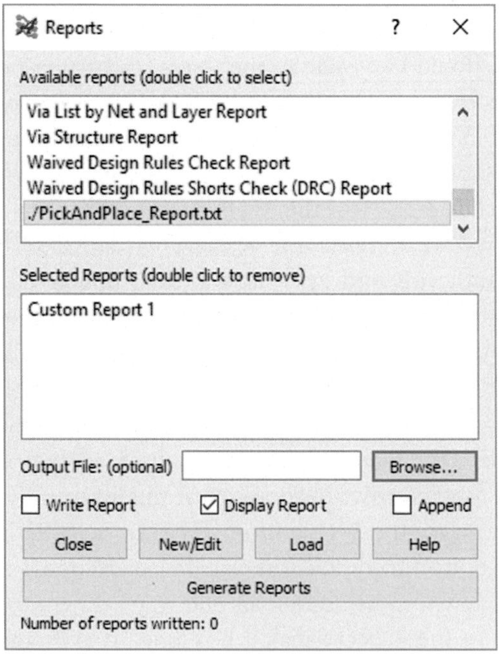

Figure 10.37 The Custom report shown in the Available Reports list.

./PickAndPlace_Report.txt

Search: Match word Match case

Design Name C:/pcbdesign/DesignFiles/Chap10_Fabrication/PCB Editor/chapter10_manufacture.brd
Date Wed May 29 13:27:08 2019

./PickAndPlace_Report.txt

REFDES	PART_NUMBER	SYM_CENTER_X	SYM_CENTER_Y	SYM_ROTATE	SYM_MIRROR
C1	Cap4567	2100.00	1900.00	0.000	NO
J1	J1234	475.00	1500.00	90.000	NO
R1	R9876	1600.00	1200.00	0.000	NO
R2	R9876	1600.00	1900.00	180.000	NO
U1	Ti7400	2100.00	1500.00	0.000	NO
M1		500.00	2500.00	0.000	NO
M2		500.00	500.00	0.000	NO

Figure 10.38 A custom report for pick and place machines.

Double click the new report to select it then click the `Generate Reports` button. The report is shown in Fig. 10.38. You can save the report as an `html` file or select and copy the data from the report and paste it into a text or Excel file, which the pick and place machine programmer can use to set up the machine to populate your board.

References

IPC-2515A (2000). Sectional Requirements for Implementation of Bare Board Product Electrical Testing Data Description [BDTST]. Northbrook, IL: IPC/Association Connecting Electronic Industries.

IPC-6011 (1996). *Generic performance specification for printed boards*. Northbrook, IL: IPC/Association Connecting Electronic Industries.

CHAPTER 11

Component information system

Contents

Introduction

This chapter introduces component information system (CIS) as a part management system that is available as an option for use with OrCAD® Capture. It helps manage part properties, including part properties required at each step in the printed circuit board (PCB) design process, from implementation through manufacturing, and to integrate into a flow for schematic design, simulation tests, and PCB layout.

CIS provides access to local and remote part databases that contain all relevant property information for the parts used in your designs. This information may include company part numbers, part descriptions, PCB layout footprints, technical parameters,

Complete PCB Design Using OrCAD® Capture and PCB Editor
DOI: https://doi.org/10.1016/B978-0-12-817684-9.00011-4

such as speed, tolerance, and ratings, PSpice simulation parameters, and purchasing information. The examples of simulation with PSpice were given in Chapter 9, Printed circuit board design examples, but here we will clarify some database and part property settings that will help us to use the simulation more efficiently.

You can configure CIS so that it will transfer any, or all, of the properties associated with the part to the schematic when you place the part. Typically, only the part number, value, and PCB footprint properties are transferred to the schematic. CIS maintains a link to the engineering database part, so that you can retrieve other part properties at any time. By linking placed parts to your preferred parts, database gives you access to complete part information during the schematic design process. In case you need a part for your design that is not yet in the parts database, you can create that in the design and add the same to your part database immediately or at a later time. You can also link a nondatabase part you have created before to a database at any time.

A simplified overview about the data structure of CIS is shown in Fig. 11.1. It shows how all the different data sources, files, and tools work together. There are

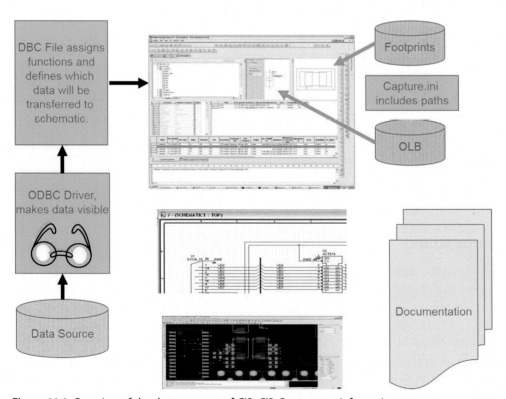

Figure 11.1 Overview of the data structure of CIS. *CIS*, Component information system.

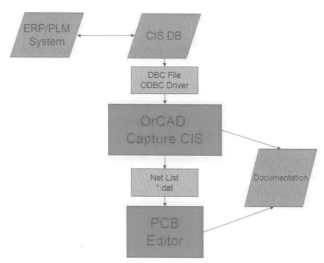

Figure 11.2 One of the possible configurations of the database with CIS. *CIS*, Component information system.

two different main flows, one for the properties and the other for the symbols. All properties (metadata) are stored in a data source (database). In this structure, the open database connectivity (ODBC) driver and the database configuration (DBC) file format the data and forward it to the CIS user interfaces. The `capture.ini` file creates the connection between symbols and footprint values and the binary data, as it will be shown in the following sections.

CIS may use a local or shared standalone database. This may be based upon an original extract of data from enterprise resource planning (ERP)/product lifecycle management (PLM) system; may have some periodic, typically overnight, update mechanism; or may be based upon database views created within the ERP/PLM database specifically for use by CIS to provide access to "live" data. The exact implementation details of the database will be determined by what data the company has and what use they wish to make of it (see Fig. 11.2 for example).

Properties in component information system

The previous section discussed about metadata, sources, properties, etc. Before starting to give information about how to do the different connections, and generate the ODBC driver and the DBC file, it is necessary to know the properties that must be included in a database. As mentioned earlier, it is company dependent, which properties are necessary to add to a database depending on the bill of materials (BOM), documentation, etc. However, in terms of CIS there are five mandatory properties that all part tables in your database must include (even when they are empty fields). These properties are as follows:

- *Part number*: unique number CIS uses for communication with database. Never change this number, otherwise you will lose the connection to the database and need to link every single part manually to rebuild the connection.
- *Part type*: creates the category tree view in CIS Explorer browser window.
- *Schematic part*: name of the binary part description (`OLB/schematic part`). The `OLB` (OrCAD library) name is not mandatory, but it makes the search faster.
- *PCB footprint*: used for preview in CIS Explorer and is mandatory for netlist generation.
- *Value*: the shown value or type on the schematic page. Value is an intelligent property and it understands units and metric prefixes of resistance, capacitance, and inductance (such as `Ohm`, `pF`, and `uH`), because it is one of the major properties for PSpice.

If you want to use your schematic part for PSpice simulations, there are three other mandatory properties:

- *Implementation:* usually the same as the part name. This should correspond to the MODEL declaration in the PSpice model file that represents this part.
- *Implementation type:* for schematic parts to be simulated, its value should be `PSpice model`.
- *PSpice template:* a long string of variables and pin name notations that link the pins on the schematic symbol with the PSpice model and is used for the *PSpice netlist* generation.

Note: The recommendation is to define these three properties in the schematic part, because with the procedure that we will see for the association of PSpice models in the section *CIS and PSpice*, such properties will be added automatically. It means that these properties shouldn't be added into the database.

In addition to these properties, there are other optional properties depending in a part on the needs of your flow. The first step in creating a part database is to determine the properties to be included for each part. Typical properties in a part database are *part description*, *tolerance*, *rating*, *speed*, *timing parameters*, *manufacturer*, and *cost*. CIS supports an unlimited number of properties, so you can include as much information in your part database as you need. Define as much as you need but keep it simple as possible.

The names you use in the database can be different to the property names you assign to the placed parts. For example, you can name the `Part Number` property "`My Company Part Number`." Also, you may call a property `Tolerance` in the database and `Tol` on the placed part. Database property types and placed-part property names are defined by separate `DBC` file that maps the fields in the database to the properties in the schematic.

Note: Do not use the same property name more than once. For example, if you have two manufacturer columns in your database, call them `Manufacturer1` and `Manufacturer2`.

When you transfer a property, this property is included in the schematic as an attribute of the placed part. Normally, you transfer properties that are required by CIS (such as Part Number and Schematic Part), used in the design process (such as Value, Tolerance, and Rating), or needed for use by other software products (such as PCB Footprint). Properties that aren't transferred can still be included in a BOM report.

Reserved properties

There are many property names that are reserved and cannot be used, such as
- *Power:* system component property for power dissipation
- *Voltage:* system net property for voltage level
- *Color:* defines the use display color for all elements in capture canvas

Transfer component information system properties to PCB Editor

It is not the main focus of this chapter; however, it is necessary to explain how to transfer the schematic part properties to PCB Editor in order to understand, why the properties issue is so important not only for PCB Editor. but also for bill of materials, documentation, etc.

To generate the netlist for PCB Editor in OrCAD Capture CIS, select the design icon in the project manager window and click on Tools → Create Netlist. A window will pop up. Then click on the setup button like in Fig. 11.3.

Use the Browse button (...) to choose the correct allegro.cfg. The standard file is stored in your OrCAD install directory\tools\capture folder. Select Edit to open the file (Fig. 11.4). Add the missing properties you want to transfer to allegro in appropriate sections. The syntax is "Property Name = Yes".

We recommend to store the changed allegro.cfg not in the install directory. Move it to a user-defined folder outside installation.

All properties in Capture can be transferred to PCB Editor. The allegro.cfg has five sections for the different types of properties as you can see in Fig. 11.5.
- *ComponentDefinitionProps*—properties that are used to build the *device type* for the PCB Editor netlist. Don't change this definition after you have started with Capture or DE CIS.
- *ComponentInstanceProps*—all component properties that you want to transfer to PCB Editor in addition to the component definition properties.
- *Netprops*—net properties for *spacing, physical,* and *electrical rules.*
- *Functionprops*—a *function* is a part of a component. These properties are function based.
- *pinprops*—pin properties.

    You can customize the allegro.cfg to your specific flow and transfer *user properties* to PCB Editor.

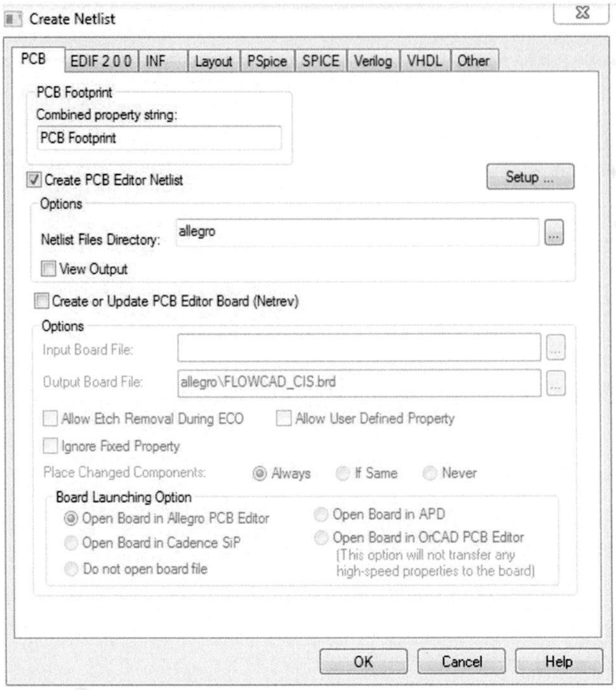

Figure 11.3 Create Netlist dialog box.

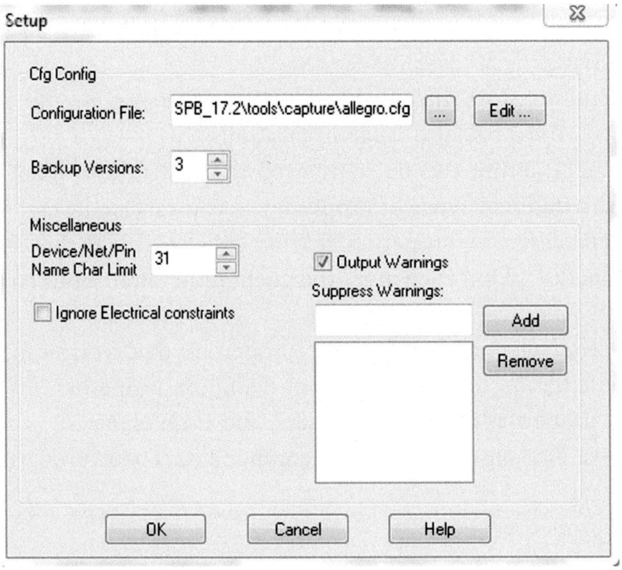

Figure 11.4 Setup configuration for the allegro.cfg file selection.

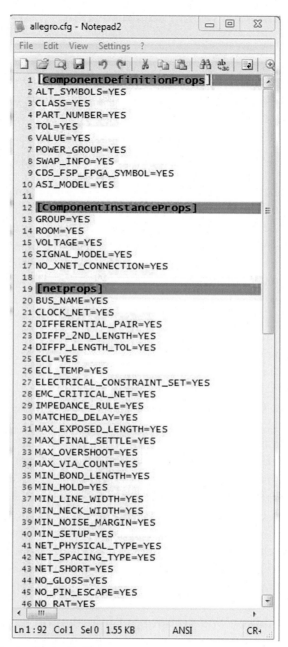

Figure 11.5 Properties in allegro.cfg file.

Component information system administration

This section corresponds to an administrative task, which should be done just once in a company. It explains how to configure the ODBC driver with the database and how to generate the DBC file that is responsible for the management of the parts' properties. It is also explained what the capture.ini is and what is its role in the CIS option.

Database systems

OrCAD allows us to use different kinds of databases. In the default installation, there is a configured SQLite database, which is free database software and easy to manage. However, it is possible to use text files (ASCII), Excel spreadsheets, Microsoft Access, and SQL database such as Microsoft SQL Server and Oracle. The first two ones may be used for testing or individuals. Access may be suitable for small workgroups, but database applications offer many advantages: comprehensive data management features, better performance for part searching, form-based entry for entering part information, safeguard against duplicate part numbers, and so on.

Before you create your database, you need to choose which database format you want to use. Then, when you create your database, you need to carefully set up its structure and organization including:

- Part property assignments
- Field formats for properties
- Number of tables used in the database implementation

On the other side, CIS can work with database field formats for text and numbers. CIS converts different database field formats to text field format when properties are transferred to placed parts. From that perspective a good strategy is to use *database views* to change field format for the fields passed from existing data sources. This *view* can also be used to change Property Names to make them conform to CIS requirements. For example, in PLM or ERP systems the price information often may be with *"currency"* field format and with *"gl-abc"* property name, where GL means *general ledgers*, and ABC means *activity-based costing. To get the OrCAD CIS flow working and more understandable for the users*, create the "database view," and change this field format to char or varchar and name it Price.

Note for SQLite: Use only ANSI SQL-92 compliant data types for your field formats. If you use noncompliant data types, CIS may misinterpret property values.

Connection of database and component information system

We are right now in the point that you already have defined a database with all required fields and records. Before you can start the CIS configuration, you need to define the ODBC data source for your database. This ODBC driver needs to be defined on every client system and the used data source configured must be consistent on all of the machines.

A data source consists of a database filename and an associated ODBC driver with which the database can be accessed from CIS. Because this driver needs to be defined on every client machine, it can be very time-consuming in larger teams. It can be configured by IT to manage ODBC driver as part of the user account.

If you are setting up a client-server database, or you want to link different databases, the ODBC driver also manages the connections between server and client or linked databases.

To set the ODBC driver for OrCAD 17.2, CIS needs a 64-bit driver. For that, click on `Start → Settings → Control Panel → Administrative Tools → Data Sources (ODBC)`.

The settings file CAPTURE.INI

Capture.ini is a text file that every user needs to store in their OrCAD Capture settings. Per default, it is located in `$HOME\cdssetup\OrCAD_Capture\17.2.0`, where `$HOME` is defined by user variable `HOME`.

In this file we can find the most important information for OrCAD Capture. For CIS, we have to define

- *Part Management*—location of `DBC` database configuration file
- *Footprint Viewer Type*—Allegro per default
- *Allegro Footprints*—directories where the footprints are located
- *Part Library Directories*—directories where the part libraries are located
- *CIS Browse Directories*—directories where documents referenced in CIS are located (e.g., PDFs)
  In this file you can also include the global settings such as
- CIS settings
- Design template—titleblock, page sizes, fonts, etc.
- Electrical Rules Check (ERC) matrix
- Printer settings

The database configuration file

CIS requires a database configuration (`.DBC`) file to make use of your part database. That is why, the first step you have to do is to create a new *DBC file*, which goes to administrate the transferring data from your database (Fig. 11.6) . This configuration file

- identifies the ODBC data source to use as part database and specifies the tables to use within that database;
- identifies part properties that are transferred to your design, when you place or link a database part;
- sets the visibility for each of the transferred properties;
- contains the part type associations;
- maps the property names.

Note: Keep the configuration (.DBC) file in a read-only directory that is accessible to all CIS users. You should make the directory read-only to prevent users from accidentally changing the configuration.

In OrCAD Capture CIS, you can access to this file by clicking on Options → CIS Configuration. Please consider that this option is available only when a design is opened. Immediately after clicking, a window, as shown in Fig. 11.7, pops up with many options:

- Setup—to open the current file
- Browse—to browse for a different DBC file
- New—to create a new DBC file. This opens the DBC wizard
- Save As—to save the current file with a different filename

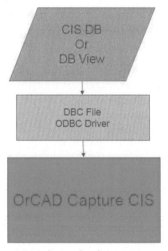

Figure 11.6 ODBC file connection. *ODBC,* Open database connectivity.

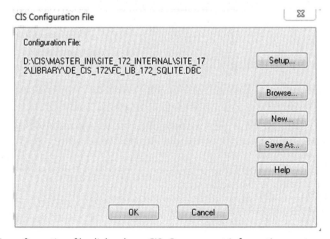

Figure 11.7 CIS configuration file dialog box. *CIS,* Component information system.

Database configuration file—Part Database tab

After selecting `Setup` (it is considered that you have already defined a new `dbc` file) next window pops up (Fig. 11.8).

In the upper left corner, you find the name of the data source from the ODBC driver and a list of all available tables. You can choose which tables should be used by CIS by marking them.

The two grayed out columns are the data read from database. You can't change this in `DBC` file setup. To change this data you should open the database.

On the right side of the picture, you can see `Select Mapping Table`. It can be set to add *mechanical parts* to your electrical parts. The mechanical parts are virtual parts (such as screws and nuts), which are only added to the BOM to create a complete export.

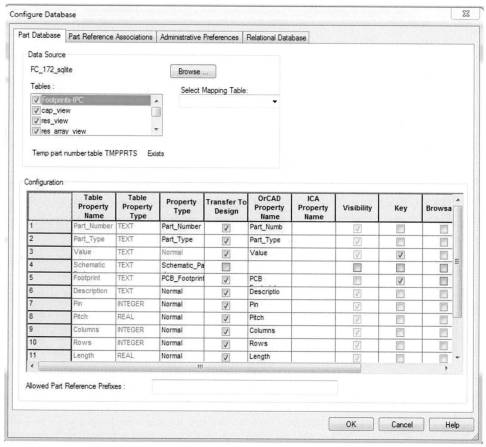

Figure 11.8 Configure Database dialog box.

The configuration area defines all these properties:
- *Table property name*—shows the property name as defined in the part database.
- *Table property type*—the data type of the property.
- *Property type*—one of six types indicating how Capture CIS interprets the property. Except for "Normal," as you assign a certain type to one property, it is removed from the list of choices for other properties in the table. These assigned properties have special functions. The first four properties must be defined.
  - **Part_number**—used by the Part Manager to identify the part in the database. It is the unique key.
  - **Part_type**—contains the component categories, which we can see as a tree structure in CIS Explorer.
  - **Schematic_part**—name of OLB/Schematic part. It is transferred as a property.
  - **PCB_footprint**—footprint name for footprint preview and netlist generation.
  - **PSpice_model**—not mandatory. When you set this property type, CIS activates PSpice functionality for placed parts with a value for this database property.
  - **Activepart_ID**—no longer used.
  - **Normal**—use this for all other properties in the table.
- *Transfer to design*—check to instruct Capture CIS to copy this property to the placed part when you place it on schematic sheet, or when you link the part with database.
- *OrCAD property name*—the name of the property as it appears on the placed part. This characteristic is only available when you check "Transfer to design". Select a property name from the drop-down list or type in the user property name. If you do not enter a value, the name of the property as it appears in the database is transferred to the placed part. Use this column to change property names.
- *Internet Component Assistant (ICA) property name*—no longer used.
- *Visibility*—sets the visibility of the property on the schematic design. You can change the visibility settings only for properties, which are transferred to the design.
  - ☐—property is always invisible.
  - ☑—property is always visible.
  - ☑—schematic part (from the OLB library) defines the visibility.
- *Key*—check to use the property as key during the initial part search when linking a database part. If you don't have a value property, don't set any key. You can specify any number of keys. Typically, you only set the key for the *value* property. When you link a part, Capture CIS searches for parts with that specific value in the database.

- *Browsable*—check to indicate that this property in the database contains references to datasheets, drawings, or other documents. When you browse your local part database, click on hyperlink (blue, underlined) values to view the referenced documents online.
- *Update part property*—check to instruct CIS to check the value of the property for placed parts against the database part's same property value, when you update the part status of your designs.

Database configuration file—Administrative Preferences tab

This tab in the DBC file configuration allows you to customize some CIS features for your work environment (Fig. 11.9).

Here you can select

- *Allow duplicate part numbers*—check to allow more than one part to use the same number in the database. This is most useful when you use parts with different associated mechanical parts. If you don't use associated mechanical parts, disable this.

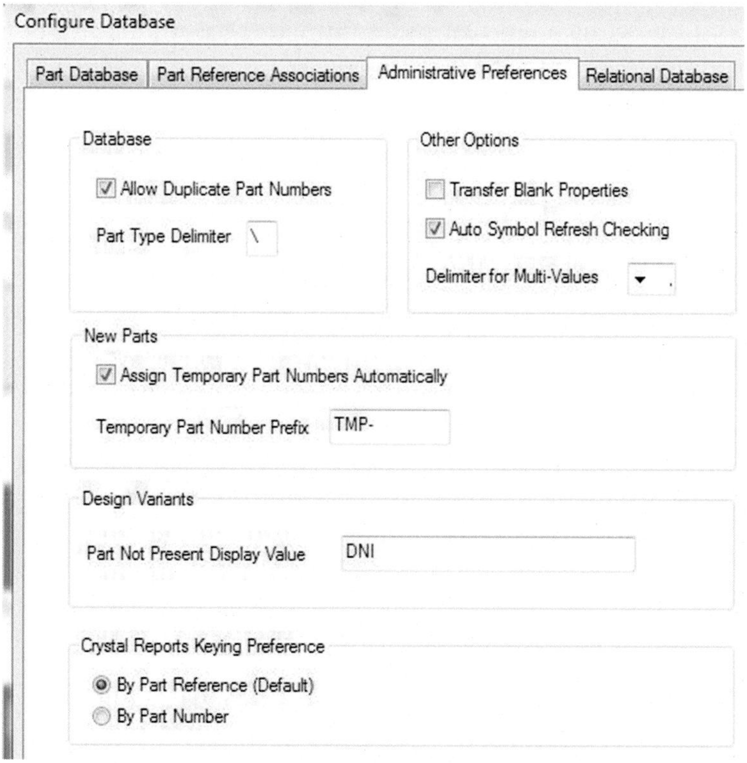

Figure 11.9 Administrative Preferences dialog box.

- *Part type delimiter*—indicates the folder hierarchy delimiter used in the database. The default delimiter is the backslash. For example, if you use *capacitor\ceramic* in the database, Capture CIS displays *ceramic* as a folder in the *capacitor* folder.
- *Transfer blank properties*—check to create a property on the placed part even if there isn't a value for the property in the database.
- *Auto symbol refresh checking*—check to enable CIS to automatically detect if symbols or footprints were updated in the configured libraries. If any changes are detected, the *refresh symbols from lib* command in the update menu of CIS Explorer and its corresponding icon on the toolbar are enabled. This indicates that you have to refresh the symbol or footprint information in CIS Explorer. If you do not select this check box, the *refresh symbols from lib* command in the update menu of CIS Explorer and its corresponding icon on the toolbar will always be enabled.
- *Delimiter for multivalues*—select the character you want to use to separate multiple values in your part database.
- *Assign temporary part numbers automatically*—check to have Capture CIS to create and track temporary part numbers. Temporary parts are tracked in the TMPPRTS table.
- *Temporary part number prefix*—type the prefix you want to use for temporary parts, TMP is the default. Capture CIS appends the temporary part number to this prefix. You can set up a different prefix (such as the user's initials) for each user's temporary parts.
- *Part not present display value*—the text description that you want CIS to use for variant parts set to not present (typical is DNI—Do Not Install). The property is displayed in the following locations:
  - Part Number and Value fields in the **Part Manager**.
  - Design variant columns in variant reports.
  - Variant parts on schematic page previews and printouts.
  - The part not present display value is not displayed in capture's schematic core design. This property also cannot be repositioned or edited in the schematic page editor. A long value is more likely to overlap a display; therefore, try to use a fairly short text equivalent for the default Not Present value.
- *By part reference (default)*—select this option to group and sort parts in BOM reports created using Crystal Reports grouped by *part reference*. Each part will have a unique item number. You can access part reference associations by clicking the Part Reference Associations tab in the Link Database Part command operation to improve the speed and accuracy of the search. When you choose Link Database Part, Capture CIS only searches for the associated database table and displays the parts of the appropriate type.
- *By part number*—select this option to sort parts in BOM reports created using Crystal Reports by part number. Since part numbers are unique, the resulting report assigns one item number to all parts with the same part number.

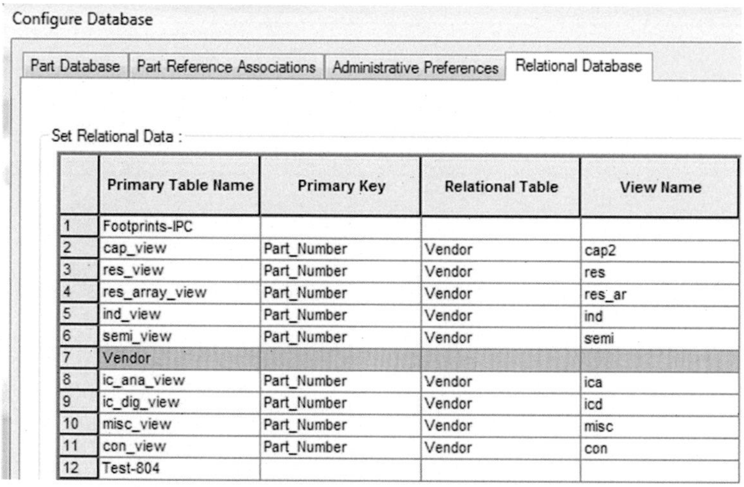

Configure Database

| Part Database | Part Reference Associations | Administrative Preferences | Relational Database |

Set Relational Data :

	Primary Table Name	Primary Key	Relational Table	View Name
1	Footprints-IPC			
2	cap_view	Part_Number	Vendor	cap2
3	res_view	Part_Number	Vendor	res
4	res_array_view	Part_Number	Vendor	res_ar
5	ind_view	Part_Number	Vendor	ind
6	semi_view	Part_Number	Vendor	semi
7	Vendor			
8	ic_ana_view	Part_Number	Vendor	ica
9	ic_dig_view	Part_Number	Vendor	icd
10	misc_view	Part_Number	Vendor	misc
11	con_view	Part_Number	Vendor	con
12	Test-804			

Figure 11.10 Relational Database dialog box.

Database configuration file—Relational Database tab

In the Relational Database tab (Fig. 11.10), you define the relationship between the part (primary) and relational tables in the database. In the part table the part number must be unique. In the relational table, you can have several datasheets with the same part number to collect several vendors, part documents, and so on. So you can easily see the possible replacements for your part. You can connect one part table with one relational table not with several. The "Set Relational Data" grid contains the following fields:

- *Primary table name*—use this list to define the part (primary) tables in your relational database. This is the only read-only field in the grid.
- *Primary key*—use this list to define the primary key that you want to use to form the relationship with the relational table.
- *Relational table*—use this list to specify a relational table that has a primary foreign key relationship with the selected primary table.
- *View name*—use this text field to define a friendly name for the view that will display when a user selects the primary table to create a relational query. For example, you may have the separate table called "Vendor" where you have additional entries for different vendors of some capacitors. Say, you have eight replacements for a 100n capacitor. The original capacitor has the unique entry with internal Part Number = "FC-CAP-0099" in your database. Then in the "Vendor" table you can have eight entries of possible replacements with the same internal Part Number, and you can choose which one to use as a replacement (see Fig. 11.11).

Once the database structure is defined, the ODBC driver is created, the DBC file is generated, and the capture.ini contains the proper directories, and it is possible to use CIS to design our schematics. To place the part from CIS database, run OrCAD

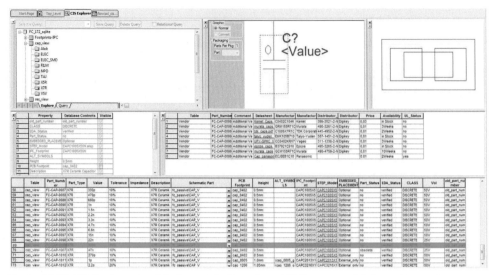

Figure 11.11 CIS Explorer graphical user interface view. *CIS*, Component information system.

Capture CIS, open the schematic page, and select `Place → Database part` from menu. The `CIS Explorer` window will appear, as shown in Fig. 11.11. Select the needed folder with parts in the database hierarchy, then choose the required part from the parts list in the bottom, and double click on it to place the selected component to the schematic sheet.

Component information system and PSpice

Simulation in PSpice is introduced in Chapter 9, Printed circuit board design examples. In order to perform the simulation, you need to assign proper PSpice models to your parts in Capture schematic design or Capture library. The OrCAD Component Information System helps to simplify this process and makes it more useful and convenient. This option is not very known by many users. It exists in addition to our database simulation properties as well. This section shows different situations that mostly cover the different scenarios connecting schematic parts and PSpice models. In this way, it is possible to keep parts not only for schematic design or PCB Editor, but for simulation as well. It makes the work much easier because everything is integrated in the same flow and the same component can be used for simulation, schematic design, and layout.

In the previous section, it was said that there are three mandatory PSpice attributes that must be considered and defined in a schematic part for simulation: `Implementation`, `Implementation Type`, and `PSpiceTemplate`. These attributes can be added as `Properties` on the schematic symbol (Fig. 11.12) or in the CIS database. If the property exists in both the symbol and the database, the database property

Figure 11.12 Schematic Part Properties.

overwrites the symbol property if placed through CIS. However, the recommendation is to define them in the schematic part, because using the option "Associate PSpice Model" such properties will be added automatically. Other properties such as Value or Tolerance should be defined in the database.

Note: Ideal resistors, capacitors, and inductors just need PSpiceTemplate, because they are not developed using model library (.lib). Real inductors and capacitors with parasitics use PSpice model (.lib), and they will have to be associated with the schematic part.

Ideal R, L, and C

Ideal resistors, inductors, and capacitors use only PSpiceTemplate property for simulation. This property should be defined in the schematic part. The procedure for that could be as follows:

- Use directly the ideal R, L, and C schematic parts from the analog.olb library, which is available in the default OrCAD library.
- If you want to use your own schematic part, then open the ideal R, L, or C from the analog.olb library in Capture, copy the value of PSpiceTemplate, and paste it in your schematic part. For that you will have to check that the pin names are the same ones as in the schematic part.

The value for such property depending on R, L, or C is the next one:

Resistor:

```
R^@REFDES %1 %2 ?TOLERANCE|R^@REFDES| @VALUE TC=@TC1,@TC2 ?TOLERANCE|\n.model
R^@REFDES RES R=1 DEV=@TOLERANCE% TC1=@TC1 TC2=@TC2|
```

As you can see, the resistor PSpiceTemplate value is referencing to different other property values. These values could be defined in the database or in the schematic part:

- *Value:* base resistance
- *Tolerance:* device tolerance
- *TC1:* linear temperature coefficient

- *TC2:* quadratic temperature coefficient

Please consider that the pin names of your schematic part must be the same as the pin names in the `PSpiceTemplate`. In this case the pin names are 1 and 2 (as you can see in the `PSpiceTemplate` definition for the resistor). So if you have a schematic part with the pin names A and B, you should change the `PSpiceTemplate` accordingly.

Inductor:
```
L^@REFDES %1 %2 ?TOLERANCE|L^@REFDES| @VALUE ?IC/IC=@IC/ ?TOLERANCE|\n.model
l^@REFDES IND L=1 DEV=@TOLERANCE% TC1=@TC1 TC2=@TC2 IL1=@IL1 IL2=@IL2|
```

As you can see, the inductor PSpiceTemplate value is referencing to different other property values:
- *Value:* inductance
- *Tolerance:* device tolerance
- *IC:* initial current through the inductor during bias point calculation
- *TC1:* linear temperature coefficient
- *TC2:* quadratic temperature coefficient
- *IL1:* linear current coefficient
- *IL2:* quadratic current coefficient

Capacitor:
```
C^@REFDES %1 %2 ?TOLERANCE|C^@REFDES| @VALUE ?IC/IC=@IC/ TC=@TC1,@TC2 ?
TOLERANCE|\n.model C^@REFDES CAP C=1 DEV=@TOLERANCE% TC1=@TC1 TC2=@TC2
VC1=@VC1 VC2=@VC2|
```

As you can see, the capacitor `PSpiceTemplate` value is referencing to different other property values:
- *Value:* capacitance
- *Tolerance:* device tolerance
- *IC:* initial voltage across the capacitor during bias point calculation
- *TC1:* linear temperature coefficient
- *TC2:* quadratic temperature coefficient
- *VC1:* linear voltage coefficient
- *VC2:* quadratic voltage coefficient

The way you could define these components in the database is quite simple. One proposal is to define the `PSpiceTemplate` property in the schematic part and the other properties in the database.

C and L with parasitics

In some precise circuits, it may be important to define the passive components with their real parasitic parameters, to make the simulation more accurate. In such case, C and L are defined using a PSpice model. It means that you could have a subcircuit as a PSpice library (*name*.lib) text file containing the preliminary created or downloaded PSpice model, as shown in Fig. 11.13 (where ESL, CAP, ESR, and LEAK are the parasitic

```
.SUBCKT REALCAP 1 2 PARAMS: ESL=1P CAP=1N ESR=1M LEAK=1G
R_R1        N00127 N00131  {ESR}  TC=0,0
L_L1        N00131 2  {ESL}
C_C1        1 N00127  {CAP}  TC=0,0
R_R2        1 2  {LEAK} TC=0,0
.ENDS
```

Figure 11.13 Subcircuit of capacitor model with parasitic parameters.

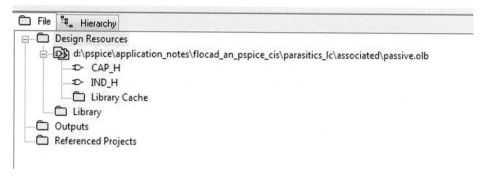

Figure 11.14 Schematic part library on OrCAD Capture.

parameters; C_C1 is the base capacitance; and R_R1, L_L1, and R_R2 are the parts of the subcircuit which serve in this model as the parasitic components of real capacitor). The creation of PSpice models as subcircuits directly in OrCAD Capture is described in Chapter 7, Making and editing Capture parts.

Consider that you have your schematic part for C or L without the PSpice properties defined. It means that such C and L are not yet defined for simulation. The procedure *to associate a PSpice Model to such schematic part* is

1. Open the parts library (.olb) where the schematic parts are included. For example, our parts are CAP_H and IND_H as shown in Fig. 11.14.
2. Double click on CAP_H. You will see that this schematic part is not prepared for simulation, because Implementation Type, Implementation, and PSpiceTemplate properties are blank (Fig. 11.15).
3. Close this window, highlight the capacitor, and right click on Associate PSpice Model.
4. Select the library (.lib) where the PSpice model is located and choose the appropriate model from the list of available models (Fig. 11.16).
5. Click on next and associate each model terminal with schematic part pin (Fig. 11.17).
6. Click OK. Your schematic part is now available for simulation.

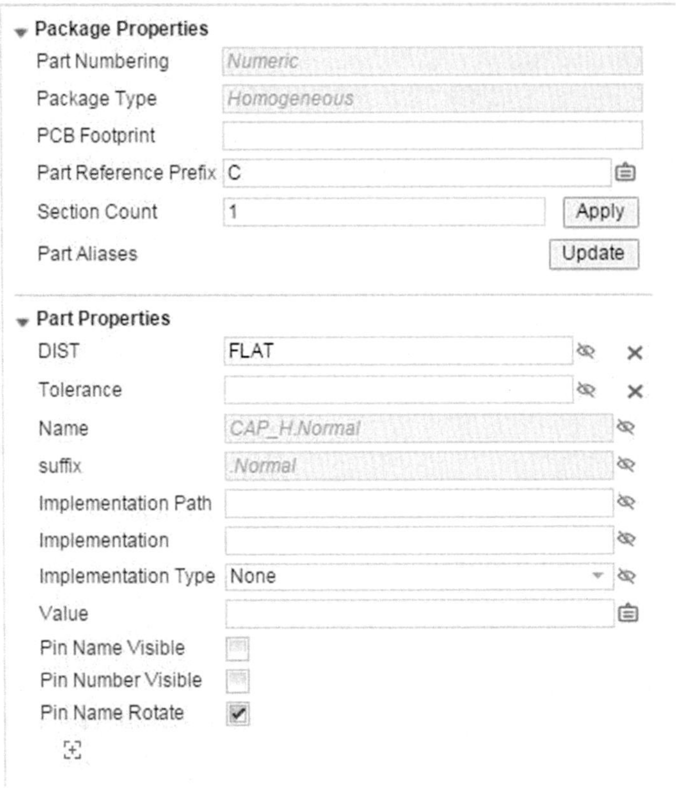

Package Properties

Part Numbering   Numeric

Package Type   Homogeneous

PCB Footprint

Part Reference Prefix   C

Section Count   1   Apply

Part Aliases   Update

Part Properties

DIST   FLAT

Tolerance

Name   CAP_H.Normal

suffix   .Normal

Implementation Path

Implementation

Implementation Type   None

Value

Pin Name Visible

Pin Number Visible

Pin Name Rotate   ✔

Figure 11.15 Properties of the schematic part CAP_H.

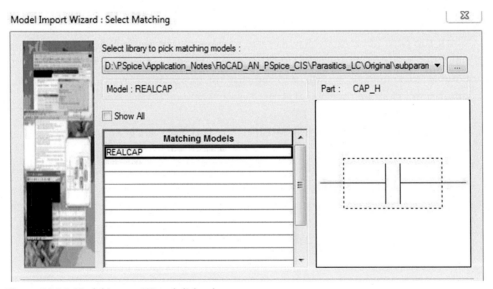

Figure 11.16 Model Import Wizard dialog box.

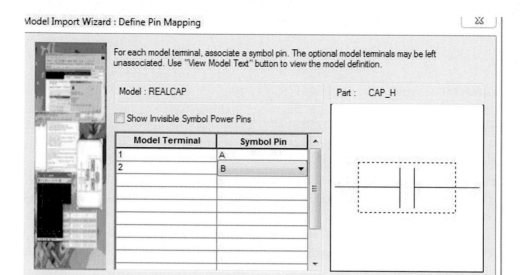

Model Import Wizard : Define Pin Mapping

For each model terminal, associate a symbol pin. The optional model terminals may be left unassociated. Use "View Model Text" button to view the model definition.

Model : REALCAP

Part : CAP_H

☐ Show Invisible Symbol Power Pins

Model Terminal	Symbol Pin
1	A
2	B

Figure 11.17 Model Import Wizard dialog box.

▼ Package Properties

Part Numbering	Numeric
Package Type	Homogeneous
PCB Footprint	
Part Reference Prefix	C
Section Count	1	Apply
Part Aliases		Update

▼ Part Properties

CAP	1N
DIST	FLAT
ESL	1P
ESR	1M
LEAK	1G
PSpiceTemplate	X^@REFDES %A %B @MODEL PARAMS: ?ESL
Tolerance	
Name	CAP_H.Normal
suffix	.Normal
Implementation Path	
Implementation	REALCAP
Implementation Type	PSpice Model
Value	
Pin Name Visible	
Pin Number Visible	
Pin Name Rotate	✔

Figure 11.18 Schematic Part Properties after the association of PSpice model.

7. Double click on the capacitor and observe its properties. `PSpiceTemplate`, `Implementation Type`, `Implementation`, and the parasitics parameters `ESL`, `ESR`, `CAP`, and `LEAK` are automatically defined (Fig. 11.18).

Now you can manually include this new schematic part into the database. In the database, you should define new columns, where the parasitics parameters `ESL`, `ESR`, `CAP`, and `LEAK` with particular values are considered. When you place such component, values of database are transferred to the design. Thus, you may have many different capacitors in the database, with different values of capacitance and parasitic parameters, but all of them can be based on the same library part (symbol), named `CAP_H` in this case and stored in the `PASSIVE.OLB` library (Fig. 11.14).

Association of schematic part with PSpice model

In the previous paragraph, we described the steps to associate a PSpice model to a schematic part. When you have a schematic part without a PSpice model (e.g., a transistor, an opamp, IC), all you have to do is to associate PSpice model to this part, so that the pin mapping and the PSpice properties are automatically defined.

Homogeneous parts and PSpice

Suppose you have an IC package with four identical opamps inside. All you have to do to use homogeneous parts with CIS and PSpice is to associate the schematic part with the PSpice model. Then each opamp of the package will be automatically assigned to the PSpice model, and it will be possible to run a simulation.

Heterogeneous parts and PSpice

The heterogeneous parts are not supported by PSpice.

Schematic parts with more pins than the PSpice model

Consider you have a schematic part, which has more physical pins than needed for simulation. In order to allow simulation, you have to define such extra pins as Unmodeled.

In order to understand this better, consider the operational amplifier ua741 as an example. This package has eight pins, but for simulation just five pins are sufficient (`in+`, `in−`, `V+`, `V−`, and `OUT`). In the schematic part, click on each pin that is not used for simulation (Fig. 11.19).

Then click on the + to add a new Pin Property, select `FLOAT` from the dropdown list of properties, enter `Unmodeled` and left-click the "tick" to set the property (Fig. 11.20).

Doing this, it is possible to simulate schematic parts that have more pins than the pins that are used for simulation.

Figure 11.19 Pin Properties dialog box.

Figure 11.20 FLOAT property definition.

How to simulate a project designed for both printed circuit board design and simulation?

In this type of projects, there are some blocks of schematic that are not necessary for simulation. For example, you could have a DDR2, DDR3, FPGA, power supply, etc. but only the power supply must be simulated. The questions would be *"Do I have to create a new project to simulate just the power supply?"*. The answer is no. If you are using the schematic—simulation—layout flow with CIS, the components that you placed for designing the power supply block will be already suited for simulation. All you have to do is to create a Test Bench, which will contain a part of your schematic design for simulation. *To create the Test Bench*, run OrCAD Capture CIS, and follow the next steps:

1. Open your project, select design icon or schematic folder, and click on `Tools →` `Test Bench → Create Test Bench` (Fig. 11.21).
2. Automatically a window pops up. Write the name for the Test Bench.
3. The Test Bench appears in the project manager tree (Fig. 11.22).

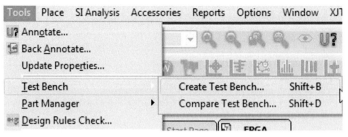

Figure 11.21 Creation of a Test Bench.

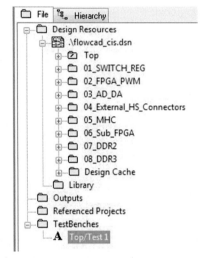

Figure 11.22 Test Bench in the OrCAD Capture structure.

4. Double click on it. Automatically a new design with the name *name*.dsn pops up. This design is identical to the original one.

5. Make root the schematic you want to simulate. In this case, make the root 01_SWITCH_REG using right click and selecting Make Root (Fig. 11.23).

6. Open the circuit on this schematic and you will see that all components are gray. It means that they actually are "invisible" for the simulation. So if you want to consider them in the simulation you have to select each one you need and right click on TestBench → Add Part(s) to Self (Fig. 11.24).

7. Now these components are colored again. It means that they are no longer invisible and they will be used for simulation.

8. Now you can place sources or whatever you need for simulation.

9. Create a simulation profile, simulate, and analyze the results.

10. Follow the same steps if you want to simulate another block (schematic) of your design.

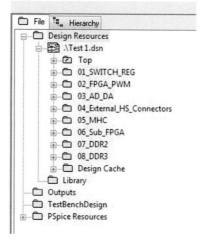

Figure 11.23 Test Bench structure.

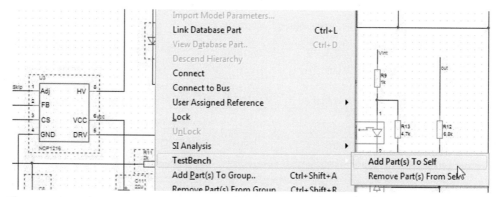

Figure 11.24 Selection of the components to be simulated in the Test Bench.

Working with component information system

This section explains the working of CIS for the daily work of Capture users. It is actually the common use of this option.

Find and place parts

To find and place the CIS database part to the schematic sheet, select `Place →` `Database Part` from the Capture CIS menu while you are in the schematic page view, then the CIS Explorer window will open (Fig. 11.11). The CIS Explorer window allows you to search for and retrieve a variety of part information.

Figure 11.25 Explore tab in CIS. *CIS*, Component information system.

	Property	Database Contents	Visible
1	old_part_number	old_part_number	☑
2	CLASS	DISCRETE	☑
3	EDA_Status	verified	☑
4	Part_Status	no	☑
5	EMBEDDED_PLACEME	External_only	☑
6	STEP_Model	CAPC3216X105N.step	☑
7	IPC_Footprint	CAPC3216X105N	☑
8	ALT_SYMBOLS	(cap_1206_gd)	☑
9	Height	1.05mm	☑
10	PCB Footprint	cap_1206	☐
11	Description	X7R Ceramik Capacitor	☑
12	Impedance		☑
13	Vol	50V	☑
14	Tolerance	10%	☑
15	Value	100n	☑
16	Part_Type	X7R	☑
17	Part_Number	FC-CAP-0003	☑
18	Schematic Part	fc_passive\CAP_V	☑

Figure 11.26 Visibility window to check the properties visible on the schematic.

The CIS Explorer window consists of several panes:
- Explorer window (Fig. 11.25) allows you to search for parts using local data from your preferred part database. The window contains two tabbed sections, Explore and Query.
  - In the Explore tab, you can search for parts using a hierarchical tree organized by part type.
  - The Query tab allows you to further filter your selection based on parametric or field data and to save the created queries for later use. You can combine several properties with different compare actions.
- The visibility window (Fig. 11.26) displays the default settings for which part properties are visible on your schematic page. You can use the visibility window to

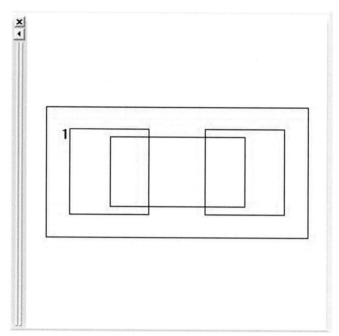

Figure 11.27 PCB footprint view in CIS. *CIS*, Component information system; *PCB*, printed circuit board.

override these default settings for single parts. You can also use the visibility window to display a compact summary of the part properties and their contents for the part you have selected in the database parts window. You can see more of the properties and contents in this view because the visibility window displays them in rows rather than columns.

- The footprint window (Fig. 11.27) displays the PCB Editor generated PCB footprint associated with the currently selected database part.
- The part window (Fig. 11.28) displays the schematic symbol and packaging data of the capture library part associated with the currently selected database part. For a multiple-part package, you can select the specific part in the package.
- The database part window (Fig. 11.29) displays the results of your part browsing and database queries. Double click on the selected part row will open the schematic window and will allow you to place the selected symbol on the schematic capture canvas. The placed part is still "linked" to the database, so it will not only inherit the properties from the database record, but also will remember from which record it was placed, and will allow to see that record later if you need.

When you select a part in the database parts window, the relational table view window (Fig. 11.30) displays all relational information that are connected to the current electrical part.

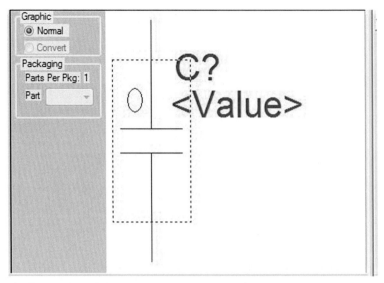

Figure 11.28 Schematic Part view in CIS. *CIS*, Component information system.

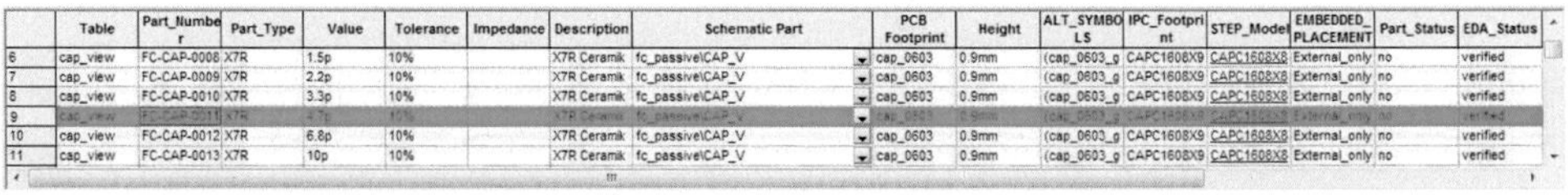

	Table	Part_Number	Part_Type	Value	Tolerance	Impedance	Description	Schematic Part		PCB Footprint	Height	ALT_SYMBOLS	IPC_Footprint	STEP_Model	EMBEDDED_PLACEMENT	Part_Status	EDA_Status
6	cap_view	FC-CAP-0008	X7R	1.5p	10%		X7R Ceramic	fc_passive\CAP_V	▼	cap_0603	0.9mm	(cap_0603_g	CAPC1608X9	CAPC1608X9	External_only	no	verified
7	cap_view	FC-CAP-0009	X7R	2.2p	10%		X7R Ceramic	fc_passive\CAP_V	▼	cap_0603	0.9mm	(cap_0603_g	CAPC1608X9	CAPC1608X9	External_only	no	verified
8	cap_view	FC-CAP-0010	X7R	3.3p	10%		X7R Ceramic	fc_passive\CAP_V	▼	cap_0603	0.9mm	(cap_0603_g	CAPC1608X9	CAPC1608X9	External_only	no	verified
9	cap_view	FC-CAP-0011	X7R	4.7p	10%		X7R Ceramic	fc_passive\CAP_V	▼	cap_0603	0.9mm	(cap_0603_g	CAPC1608X9	CAPC1608X9	External_only	no	verified
10	cap_view	FC-CAP-0012	X7R	6.8p	10%		X7R Ceramic	fc_passive\CAP_V	▼	cap_0603	0.9mm	(cap_0603_g	CAPC1608X9	CAPC1608X9	External_only	no	verified
11	cap_view	FC-CAP-0013	X7R	10p	10%		X7R Ceramic	fc_passive\CAP_V	▼	cap_0603	0.9mm	(cap_0603_g	CAPC1608X9	CAPC1608X9	External_only	no	verified

Figure 11.29 Database part window.

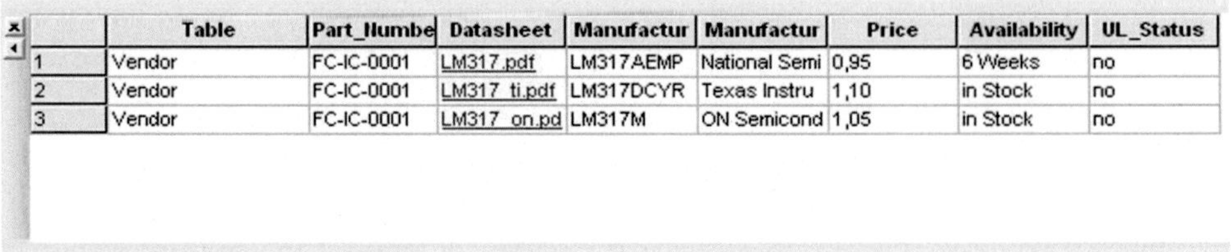

	Table	Part_Number	Datasheet	Manufactur	Manufactur	Price	Availability	UL_Status
1	Vendor	FC-IC-0001	LM317.pdf	LM317AEMP	National Semi	0,95	6 Weeks	no
2	Vendor	FC-IC-0001	LM317_ti.pdf	LM317DCYR	Texas Instru	1,10	in Stock	no
3	Vendor	FC-IC-0001	LM317_on.pd	LM317M	ON Semicond	1,05	in Stock	no

Figure 11.30 Relational information of a part.

Part Manager

The Part Manager window summarizes the status of all the parts in your design and provides a graphical interface for creating BOM variants. You can open the Part Manager window by selecting Tools → Part Manager → Open menu while you are in the Project Manager window (Fig. 11.31).

The left side pane displays a tree view similar to Windows Explorer. The tree view is a hierarchy of groups, subgroups, and BOM variants found in a design.

The right side pane of the Part Manager displays a list view with information of each part in your design. You can configure the Part Manager window to show other part properties in addition to the standard part information.

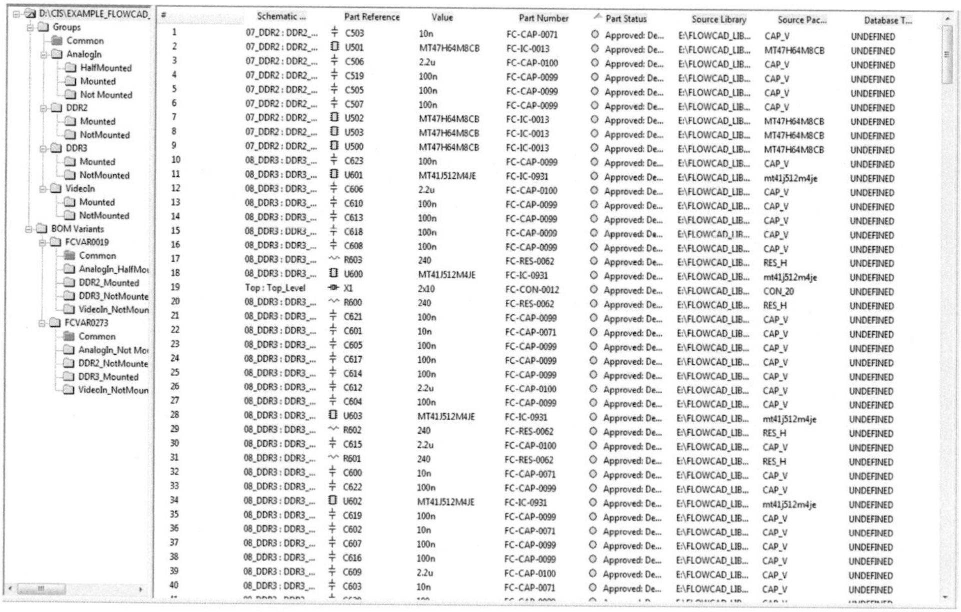

Figure 11.31 Part Manager view.

In this sense the green color in each part means that the part matches the definition in CIS database. Yellow means that the part is defined but not checked. Finally, red means that the part has no `Part Number` property value, or has a `Part Number` property value that has not been found in the current database.

Customize Part Manager

To configure the part property display, just click on `View → Configure Part Property Display`. After that a window pops up (Fig. 11.32). All `Properties` in `Selected part properties` column are shown in `Part Manager`. Use `Add/Remove` button or double click to move properties to the other column.

Never remove the three properties marked with an asterisk (`#`, `Schematic Page`, and `Part Status`).

Part Manager features

With `Part Manager` opened, you have many options to do with each component. By right clicking on a single part or group of similar parts you can start these actions:

- *Link database part*—replace the part with a different library part
- *View database part*—opens CIS Explorer and shows part
- *Update selected part status*—synchronize a single part with central database

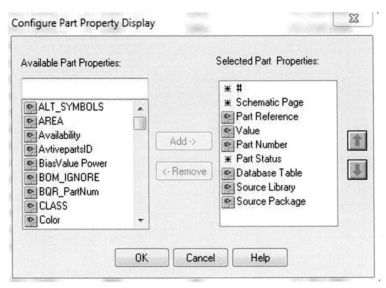

Figure 11.32 Part Property Display Configuration.

- *Update all part status*—synchronize all listed parts with central database
- *Goto Part on Schematics*—opens Capture page and shows part

Note: Under `Options` → `Update Part status`, you have the option to enable or disable `Verify Parts against .OLB libraries`. If it is enabled, the command `Update Part Status` also checks the cache against central library part.

Bill of materials

In Capture, it is possible to generate a BOM using the capture BOM command (`Tools` → `Bill of Materials`). CIS provides its own BOM command that lets you generate BOM reports based on the data in the parts database, as well as properties added to part instances after they have been placed in schematics. CIS `Bill of Materials` offers additional options for generating part reports.

You can generate a CIS `Bill of Materials` in two formats:
- Standard CIS BOM
- Crystal Report BOM

When generating a standard CIS BOM (`Reports` → `CIS Bill of Materials` → `Standard`), you use the standard `Bill of Materials` dialog shown in Fig. 11.33 to create a BOM template specifying which properties from the parts database to include in the BOM and how to sequence them. You can create multiple BOM templates specifying different combinations or arrangements of part data to report. When you need to generate a BOM, you choose the appropriate template for your purpose.

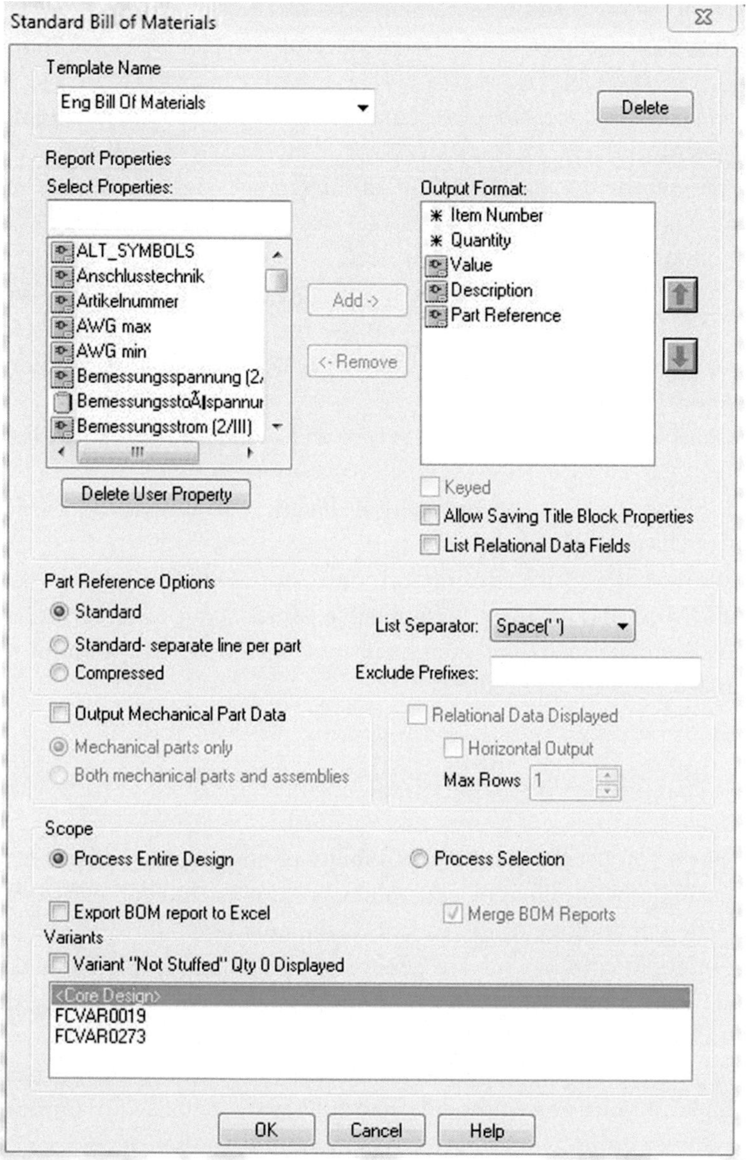

Figure 11.33 Standard Bill of Materials dialog box.

The Select Properties box lists all available properties you can include in the BOM report. The Output Format box lists the properties you want to include and the sequence in which their columns will appear in the BOM.

If you select a property in the Output Format list and check the Keyed check box under the list, parts having that property are grouped together on the BOM.

`Part Reference Options` can be

- *Standard*—parts are groups using the keyed property. All reference designators are listed (C1, C2, and C3).
- *Standard*—*separate line per part*—every part is in a separate row, ordered by a reference designator.
- *Compressed*—similar to standard but the reference designators are compressed (C1−C3).

You can choose the list separator you need.

The `Exclude Prefixes` box provides a way to filter out data for certain types of parts from the BOM. For example, suppose you use the prefix "TP" in part references for all test points. To exclude those parts from the BOM, you would enter `TP` in the `Exclude Prefixes` box.

You can enable `Relational Data Displayed` and define the format to include them to BOM.

You can define the `Scope`. In hierarchical designs, you can define BOM on entire design or hierarchical block level.

When you have the variant description with "not assembled" component you can choose the BOM format for these parts. Enable `Variant "Not Stuffed" Qty 0 Displayed` to get all not assembled parts with the quantity 0 in the BOM. Without this option, all not assembled parts are ignored.

Variants

With CIS option you get the powerful possibility of creating the variants of your schematics, for example, with different part values or "not assembled" parts depending on application. The terminology of the variants is the next one:

- *Groups*—multiple components are generally used to support a particular function or module (e.g., a power or memory module). These components are defined as a group and have varying version numbers.
- *Subgroups*—each subgroup represents a version or assembly of the parent group. For example, if your power module has different assemblies for Europe and Asia, then the Power group would have two subgroups. The set of components in each subgroup is the same as the parent group.
- *Core design*—the core design is the base schematic and PCB from which design variants can be created.
- *Common*—all components that are not part of a group but are still part of the core design. These modules or functions remain unchanged in all assemblies.
- *BOM variants*—the versions of the design that are manufactured from a BOM that is specific to each assembly.

For example, in the PCB project shown in Fig. 11.34, the hatched areas are functions with different variant information:

- The large red one, on the left side, is the power supply. It has three assembly variants.
- The green one in the top is the channel block with two assembly variants.
- We have marked eight different blocks and we need to define eight groups. The subgroups contain the assembly variants, three for power supply [220 V, 110 V, and Not included in Design (DNI)], and two for the channels (one channel or two channels).

All components in nonmarked areas have no variant and stay in common group.

The steps *to create the variants* are as follows:

1. Open `Part Manager` and create several new groups for the components that have variant assemblies, by right clicking and selecting `New Group`. In the case of the example of Fig. 11.34, there are eight different groups. As example, we show the definition for two of them: `Channel` and `Power` (Fig. 11.35).
2. Create subgroups for every variant of the group. For the `Channel` block, there are two variants: `1Chan` and `2Chan`. For the `Power`, there are three variants: `110`, `220`, and `DNI`, as shown in Fig. 11.36.

Finally, close the `Part Manager`.

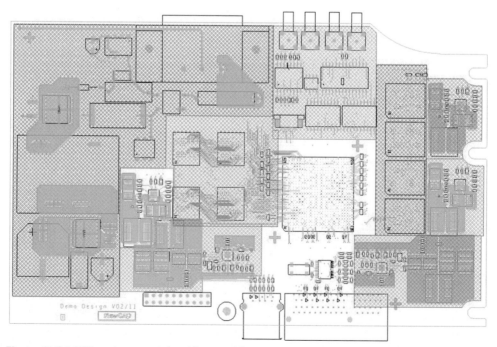

Figure 11.34 PCB project example with assembly variants. *PCB*, Printed circuit board.

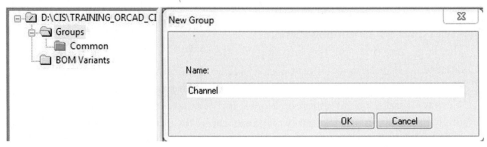

Figure 11.35 New Group window.

Figure 11.36 Created groups and subgroups.

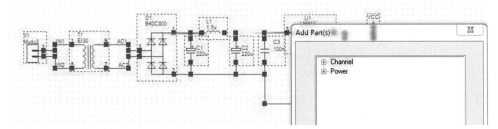

Figure 11.37 Adding parts to group.

3. Go to the schematic pages and sort the components into groups by selecting the needed components and right clicking on the menu Add Part(s) to Group, as shown in Fig. 11.37.
4. Open Part Manager and define the alternate and not assembled components in the groups (see Fig. 11.38).
5. Create BOM variants. Right click on BOM variants folder and create a new BOM variant (Fig. 11.39). For example, we define four variants Berlin, Miami, NewYork, and Paris.

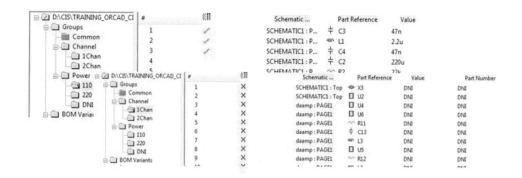

Figure 11.38 Mark alternate and not assembled components.

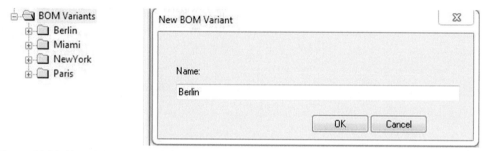

Figure 11.39 New BOM Variant definition.

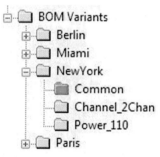

Figure 11.40 BOM Variant classification.

6. Move the subgroups to the BOM variants. For that drag the folder of the subgroup to the variant you want. For example, move the subgroups 110, 2Chan, and Common to NewYork (see Fig. 11.40) .

7. Move the Common group with the rest of the components to every BOM variant.

To see the variants in the schematic, click on View → Variant View Mode and select the desired variant. In default color settings, all alternate values are shown in green and all not present parts are shown in gray. You can change these colors in Options → Preferences.

Note: You can also generate variant BOM reports under Reports → Variant Report.

CHAPTER 12

Signal integrity simulation with OrCAD

Contents

What is signal integrity?

Signal integrity or "SI" is a measure of the quality of electrical signals sent from driver to receiver. Deep explanation of SI terms and issues is provided in Chapter 6, Printed circuit board design for signal integrity. Here we will explain briefly how the SI issues in your printed circuit board (PCB) design can be solved with the help of OrCAD® Signal Explorer (SigXP) tool (which is included to the OrCAD Professional bundle), or with more powerful tools, such as OrCAD PCB SI, Sigrity ERC, or Sigrity Advanced SI/Advanced PI.

When digital information shall be transmitted from sender to receiver, it is coded as binary information with logic values "1" and "0." This information is sent as a stream of binary values represented by a voltage (or current) waveform. For a logic "1" the amplitude is high, and for "0" the amplitude is low (Fig. 12.1).

Over short distances, and at low bit rates, the receiver sees the signal as it was sent. The information is read with the same clarity as it was written. But at high bit rates, and over longer distances, the physical parameters of the electrical connection between sender and receiver have an influence on the transmitted signals. This influence on the signals is coming from the physical interconnect between driver and receiver. This interconnect can be the copper traces and ground planes on a PCB, a connector or a cable, and the connectivity inside the integrated circuit (IC) package.

The physical implementation of the interconnect has an ohm resistance (R), an admittance (G), an inductance (L), and a capacitance (C), as shown in Fig. 12.2.

Complete PCB Design Using OrCAD® Capture and PCB Editor
DOI: https://doi.org/10.1016/B978-0-12-817684-9.00012-6

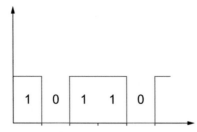

Figure 12.1 Digital data transmission with binary voltage states.

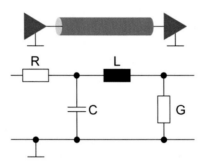

Figure 12.2 Transmission line between sender and receiver.

To describe the electrical behavior of this transmission line, a circuit of four components, RCLG, is used. The accuracy of this described model increases when the interconnect is divided into multiple segments and all circuits for each segment are connected as lumped elements on a long row of RCLG elements in one model.

This behavior of connecting copper wires is known since digital information was transmitted on telegraph lines using the Morse alphabet. The Morse speed limited the maximum distance between two operators to guarantee the readability of the information. After a certain combination of length (miles) and frequency (Hertz) the telegraph line became a transmission line and physical effects influenced and distorted the signal quality. The same effect can distort signals on small PCBs for MHz and GHz signals. When the influence of physics of the interconnect can no longer be ignored, we call the interconnect a *transmission line* or a channel, and our PCB design becomes a *high-speed design*.

Frequency is often used as a parameter to specify when a design is a high-speed one, but this is not really true. What is also important is the *rise time* and *fall time* of a signal. The rise time is specified as the time a signal takes when changing its state from low to high and is measured from 10% to 90% of the amplitude (Fig. 12.3). As a rule of thumb, a rise time faster than 1 ns on a postcard size (3″ × 4″) PCB is a high-speed design.

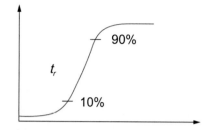

Figure 12.3 Rise time of a digital signal.

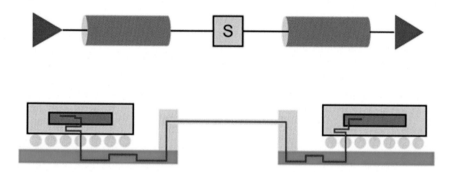

Sender – Silicon – IC-Package – PCB – Connector – Cable – Connector – PCB – IC-Package – Silicon – Receiver

Figure 12.4 Interconnect from sender to receiver through IC-packages, PCBs, connectors, and cable. *IC*, Integrated circuit; *PCB*, printed circuit board.

How to simulate in OrCAD Signal Explorer

In a simulation the whole interconnect (Fig. 12.4) needs to be modeled. The electrical behavior of the driver and receiver with its silicon dies and integrated circuit packages can be described in IBIS models. Even components, which adopt their behavior due to the data transmitted by special protocols (using *preemphasis*), can be modeled with *behavioral IBIS models*. For connectors and cables the models can be measured, and we can use the *s-parameter models*. In general, these elements of the transmission line cannot be changed by the engineer.

The Cadence SI tools, such as OrCAD SigXP, OrCAD PCB SI, and Sigrity Advanced SI, offer a possibility to simulate the effects the PCB has to the transmission line. Depending on when a simulation is performed we call it either prelayout or postlayout simulation.

Prelayout simulation

A prelayout simulation can be performed just with an OrCAD Capture schematic. If models are assigned to the components, then the wires are the connection to be made

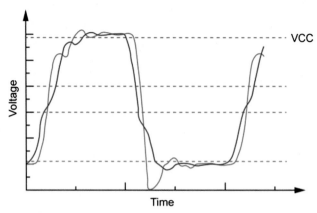

Figure 12.5 Waveform at sender and receiver.

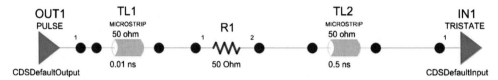

Figure 12.6 Prelayout simulation with transmission lines with delay as time parameter (ns).

inside the PCB. In OrCAD SigXP, which can be started from the OrCAD Capture SI Analysis menu, the parameters for the physical impact on SI can be simulated. After entering a few parameters such as the value for dielectric constant (typical ε_r for FR4 material is about 4.0—4.5) and setting the layers stackup inside the PCB, a simulation can be performed. A stimulus (a rectangular change of the signal from low to high) will be applied, and the waveform can be measured at the driver and the receiver (Fig. 12.5).

The simulation can consider how the waveforms are affected by changes in the length and impedance of the connections. The delay of the transmission line segments can be set as a time parameter or as a PCB trace length, and the impedance can be set directly in ohms, or as a parameter of trace width and dielectric thickness (Fig. 12.6). The effect of vias can be simulated with one or several vias inserted in the connection. The reflections caused by transmission line irregularity will lead to the signal waveform distortion, which can be analyzed by the engineer, to check the quality of the signal rising and falling edge, the overshoot, the setup and hold time, and the eye diagram correctness. As a result of running the simulations, the engineer can define the design rules and limitations, which can then be entered in the Constraint Manager. The maximum length of an interconnect can then be determined to maintain the specified setup time of a signal. The hold time is controlled by the minimum length of the connection. A via inside a PCB is a change of impedance and causes reflections. Default models for vias can be virtually added to the topology. The impact of these reflections will be seen at the waveforms, and the engineer can limit the number of vias to be used for a signal or specify that all signals of a bus must have the same number of vias. These constraints could then be used during PCB layout to maintain the signal integrity. OrCAD PCB Editor has a design rule check, which will verify that these constraints are not violated.

Postlayout simulation (verification)

A postlayout simulation is performed to verify that critical nets are still within the SI specification. A postlayout simulation is done when the net is already routed in PCB Editor and the exact dimensions are known. PCB Editor will extract the topology to OrCAD SigXP, which represents in its canvas the exact dimensions of the traces as well as the vias and termination components (Fig. 12.7).

Simulation not only helps to develop and verify design rules but also offers a virtual measurement of voltages and currents before a prototype is built. The engineer can measure his design by running a simulation sweep through a full range of parameters, that is, termination values for a resistor vary from 10 to 50 Ω in steps of 2 Ω. The result will show which tolerances are acceptable for these termination resistors (like 22 $\Omega \pm 10\%$), and based on these tolerances, a real existing resistor with its tolerance can be selected.

More comprehensive simulations can be done with higher tiers of the SI simulation tools. With Cadence Sigrity tools, issues such as ground bounce, electromagnetic interference, IR–drop, and thermal self-heating effects can be simulated. Simulation tools are built to simulate all aspects of today's designs in a fast and accurate way. To design and build the leading edge designs, a simulation is required. When someone already made a simulation and has derived the constraints, they can be reused by others too. SI is affected by reflections, cross talk, ground bounce, and power supply noise, and most of these issues can be controlled by PCB designer with a help of simulation tools and constraint rules.

How to use OrCAD PCB SI tool

OrCAD PCB SI can be used to create, modify, simulate, and save virtual prototypes of net topologies. Topologies are explored by modifying circuit parameters,

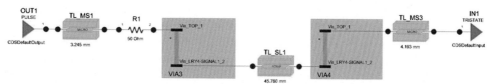

Figure 12.7 Postlayout simulation with transmission line in length (mm) and including vias.

simulating, and examining reports and waveforms. Then repeat this process to modify circuit parameters for optimum results. The tool graphically displays simulation results as spreadsheet data and waveforms.

Getting started

Invoke OrCAD PCB SI with a clean canvas by the following steps:

- Choose Start → Run → SigXP (Windows 7)
- Program shortcut is SigXplorer in the Start menu (Windows 10)

Or invoke OrCAD PCB SI by extracting a topology directly from OrCAD PCB Editor:

- In Constraint Manager:
  1. Select one of the Net Worksheets from the Electrical worksheet
  2. Select the net of the topology you want to extract, and then right-click to display the context-sensitive menu
  3. Select OrCAD PCB SI or SigXplorer as the license

The tool opens with an empty canvas or with the extracted topology. On the left side of the tool, there are icons to place the objects onto the canvas:

- Driver (OUT) and receiver (IN) or tristate elements
- Discrete elements such as a resistor, and a capacitor
- PCB traces, strip or microstrip lines
- Vias
- And so on

The set of icons and available functions depends on the license available: OrCAD Professional, Allegro PCB Designer, or most powerful—OrCAD PCB SI or Allegro High Speed Option.

Click on an icon and place the element onto the canvas. By clicking on the connecting points of each element, the elements can be connected with a connection line, which is recommended for the simulation (Fig. 12.8). Each element can be specified with its values. Right-click on the element and choose Parameters from the pop-up menu, then set the correct values of its parameters in the Parameters pane.

When the topology is entered you can start simulation using Analyze → Simulate, but before this a stimulus needs to be selected. Left-click on the driver element in the canvas to select it, then right-click and select Stimulus from the pop-up menu. In many simulations, you can use a pulse from low to high and then from high to low, as shown in Fig. 12.9.

In case there is no IBIS model available for a device, you can use one of the Cadence default models and change the parameters based on the information from the datasheet. You can create a custom stimulus using Stimulus → Custom from the pop-up menu.

The result is a waveform, which can be formatted, zoomed in, and otherwise analyzed (Fig. 12.10).

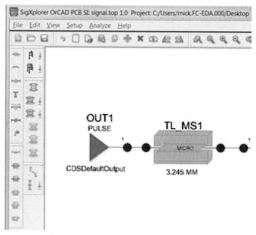

Figure 12.8 Icons to be placed as a topology on the canvas.

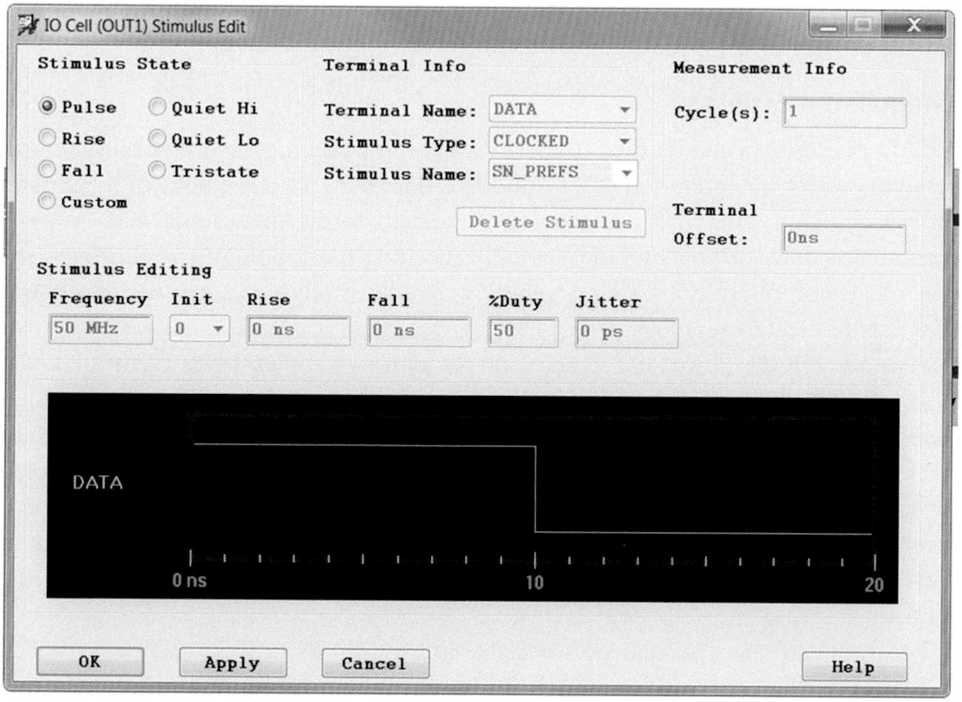

Figure 12.9 Stimulus setup in OrCAD SigXplorer.

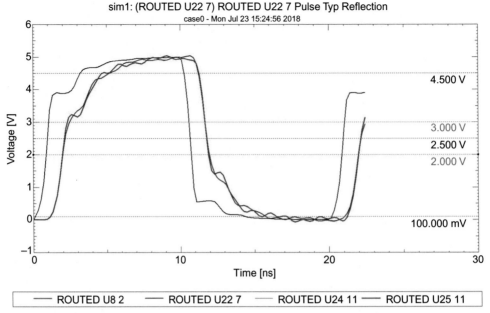

Figure 12.10 Simulation result with a critical glitch in one receiver (*red curve*).

Scheduling a topology

OrCAD PCB SI can also be used to specify a complex topology and schedule the way the components are connected to one another. There are two methods of scheduling a topology. You can wire the topology interactively in the canvas and create a template schedule, or you can automatically schedule the topology by selecting from a set of generic templates. Using a generic schedule is advantageous, because it is a faster way to create a topology.

Place the IOCells (driver and receiver) on the canvas, and then select a schedule using `Setup → Constraints...` menu choice, `Wiring` tab. All of the necessary TLines (transmission line elements) are immediately added and connected to the IOCell pins. If you decide not to select a generic schedule, add and connect each TLine manually to form the desired net schedule. When you extract a topology template with a generic schedule into Constraint Manager using `File → Update Constraint Manager` menu, the created Electrical Constraint Set stores the specified type of schedule as a *ratsnest schedule constraint* in the constraint set.

Following are the schedule types available in OrCAD PCB SI:

- Min Spanning Tree
- Daisy Chain
- Source Load Daisy Chain
- Star
- Far End Cluster

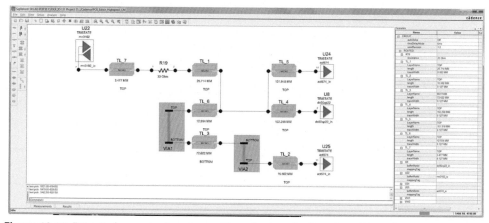

Figure 12.11 Extracted complex topology with real values for length and impedance.

If your net has a branch, a virtual *T-point* (tree-point) can be introduced in the design. SigXplorer automatically places a T-point element at the junction of two or more transmission lines (TLines) with no other pins on the node. This T-point is used to specify a net schedule or length properties in individual branches. The T-point can also be placed in PCB Editor and routed from and to it. It is also possible to create the complex topology for one net in the PCB Editor design, and then extract this topology to the SigXP (Fig. 12.11), make small corrections if needed, and export it back to a PCB Editor design as a new Electrical Constraint Set. This will allow you to apply this topology as a rule for other nets, for example, for all nets of a DDR address bus, which should have a Fly-By topology.

In the installation, there is a user guide for OrCAD PCB SI, which is called sigxpug.pdf that can be downloaded in the help environment.

Electrical rule checks in OrCAD Sigrity ERC

In OrCAD Sigrity ERC tool the electrical rule checks can be performed without any simulation models. To invoke Sigrity ERC, select `Analyze` → `ERC-SRC` from PCB Editor menu. From the layout of the PCB the tool can extract the parasitic parameters of the PCB. The extraction is made directly from the traces, vias, plane shapes, and stackup. Then the tool will sweep across all specified nets and will show, as result, the areas where SI might be compromised. OrCAD Sigrity ERC will calculate the impedance for each segment of an interconnect as well as the coupling of the signal to other traces next to it.

The results are available in interactive reports for each segment of a trace. By clicking on critical areas in the report the tool will jump to that location in the layout and zoom in on the trace segment. This interactive methodology will enable the user to quickly inspect the layout for SI violations such as routing across a split plane or too close to traces or vias of other signals (Fig. 12.12).

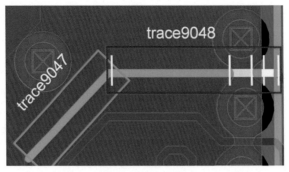

Figure 12.12 Trace segments of one net might have different impedances.

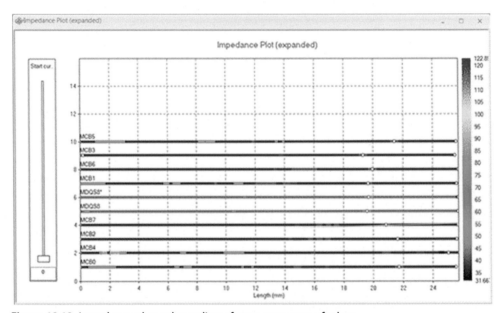

Figure 12.13 Impedance plot: color coding of trace segments of a bus.

Setup

To set up an electrical rule check analysis in OrCAD Sigrity ERC, you will need about 2 minutes. In Layout Setup folder of the workflow pane, check the PCB stack-up and prepare the power/ground nets. Choose Run ERC Sim & check violations. In Setup folder set up ERC Sim Options as you need. Click on Start ERC Sim in Simulation folder and browse the simulation results in Results and Report folder. No simulation models are required. The simulation time varies between 10 and 20 minutes for typical PCB with high-speed constraints.

Graphical reports

Results will be displayed as interactive graphical reports (Fig. 12.13). These reports allow the user to easily detect critical nets where SI violations probably occur. Each

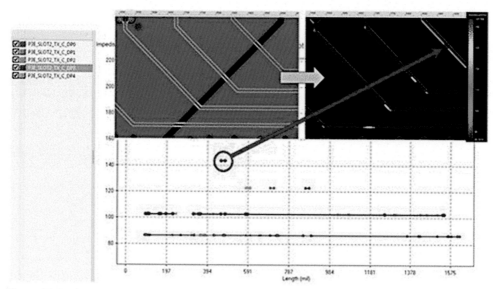

Figure 12.14 Interactive reports zoom in to critical location in OrCAD PCB Editor.

segment of a net is color coded, and the critical location where the violation is detected is easy to locate. When clicking on the red hotspot in the graphic report, the PCB Editor will zoom in to this location of the net (Fig. 12.14).

In accordance with the type of violation, clear instructions for the PCB designer will appear:

- impedance discontinuities
- coupling to other traces
- different number of vias in a bus
- differential signals out of phase

While looking at the critical location, the PCB designer normally knows how to modify the routing to avoid the coupling or impedance mismatch. This methodology detects critical routing, which might be introduced to the design at later routing stages and could be missed by the designer due to PCB complexity and tight schedule.

APPENDICES

Appendix A: List of design standards

ANSI standards

ANSI B94.11-197 (see also ASME B94.1 IM-1993) Twist drills
ANSI Y32.2-1975 (see also IEEE Std 315-1975) Graphic symbols for electrical and electronics diagrams

ASME standards

B18.2.8-1999, Clearance holes for bolts screws and studs
B18.6.3-2003, Machine screws and machine screw nuts
ASA 618.11-1961, Miniature screws
Y 14.5 M, Dimensioning and tolerancing
ASME B94.1 IM-1993, Twist drills

IEEE standards

IEEE Std 315-1975 (ANSI Y32.2-1975), Graphic symbols for electrical and electronics diagrams

IPC standards

IPC-1902/IEC 60097, Grid systems for printed circuits
IPC-2141A, Design Guide for High-Speed Controlled Impedance Circuit Boards
IPC-2221B, Generic standard on printed board design
IPC-2222A, Sectional design standard for rigid organic printed boards
IPC-2223D, Sectional Design Standard for Flexible/Rigid-Flexible Printed Boards
IPC-2224, Sectional standard for design of PWBs for PC cards
IPC-2225, Sectional design standard for organic multichip modules (MCM-L) and MCM-L assemblies
IPC-2226A, Sectional design standard for high density interconnect (HDI) printed boards
IPC-2251, Design guide for the packaging of high speed electronic circuits
IPC-2515A, Sectional requirements for implementation of bare board product electrical testing data description (BDTST)
IPC-2615, Printed board dimensions and tolerances
IPC-4101E, Specification for base materials for rigid and multilayer printed boards

IPC-7351B, Generic requirements for surface mount design and land pattern standard

IPC-9252B, Requirements for Electrical Testing of Unpopulated Printed Boards

IPC-A-600J, Acceptability of printed boards

IPC-A-610G, Acceptability of electronic assemblies

IPC-HDBK-610, Handbook and guide to IPC-A-610

IPC-CM-770E, Component Mounting Guidelines for Printed Boards

IPC-D-322, Guidelines for selecting printed wiring board sizes using standard panel sizes

IPC-D-350D, Printed board description in digital form

IPC-D-356B, Bare substrate electrical test data format

IPC/JPCA-2315, Design guide for high density interconnects (HDI) and microvias

IPC/JPCA-6801, Terms and definitions, test methods, and design examples for build-up/high density interconnect (HDI) printed wiring boards

IPC-SM-780, Component packaging and interconnecting with emphasis on surface mounting

IPC/WHMA-A-620C, Requirements and acceptance for cable and wire harness assemblies

JEDEC standards

Master Index for JEDEC Publication No. 95 (see also Appendix B for specific standards)

Military standards

MIL-HDBK-198A, Capacitors, selection and use of

MIL-HDBK-199A, Resistors, selection and use of

MIL-HDBK-5961A, List of standard semiconductor devices

MIL-STD-1276G, Leads for electronic component parts

MIL-STD-275E, Printed wiring for electronic equipment (superseded by IPC-2221B)

Appendix B: Partial list of packages and footprints and some of the footprints included in OrCAD PCB Editor

Table B.1 A list of package abbreviations.

Abbreviation	Full name	Footprint examples in library share\pcb\pcb_lib\symbols
BDIP (SDIP)	Butt-mounted dual inline package (surface DIP, std pitch)	—
BGA	Ball grid array	bga6m10020b60w700... bga39m1271521b76w5000 named as bgaNmSPbDwW where N = number of rows S = pitch in mm P = number of pads D = pad diameter W = package width
BQFP	Bumper quad flat package	quadb02544wg535... quadb025244wg1780ctx
CBGA	Ceramic column ball grid array	—
CFP	Ceramic flat packages	flat14...flat64
CGA	Column grid array	cpga68...cpga262
CQFP	Ceramic quad flat packages	quadflat24, mquad28 cqfp44_amd... cqfp208_alt
DIMM	Dual inline memory module	dimm050fingers8byte... dimm0508bvstm168
DIP	Dual inline package	dipswitch dip4_3...dip68_6 dip100b8w300l450... dip10064w900l3250
DO	Diode outline	smdo213aa... smdo214ac21
DPAK	Discrete packaging (type 1, TO-252)	Smdpak
D2PAK	Discrete packaging (type 2, TO-263)	—
D3PAK	Discrete packaging (type 3, TO-268)	—

(Continued)

Table B.1 (Continued)

Abbreviation	Full name	Footprint examples in library share\pcb\pcb_lib\symbols
LCC/LCCS	Leadless chip carrier/ leadless ceramic chip carrier	clcc28...clcc344
LGA	Land grid array	—
MELF	Metal electrode face	smdmll34...smdmll4112
MSOP	Micro (mini) small outline package	—
MLP	Micro leadframe package (no lead)	—
PGA	Pin grid array	pga68...pga179 pga805044d034w900... pga35050325d028w1900
PLCC	Plastic leaded chip carriers	plcc18... plcc28310x510 lcc20...lcc68
PLCCR	Plastic leaded chip carriers rectangular	lccs18...lccs32
PLCCS	Plastic leaded chip carriers square	—
PQFP	Plastic quad flat package	pqfp44... pqfp208_cyp quad10m36wg1200... quad050100wg1485ctx
QBCC	Quad bottom chip carrier	—
QFP	Quad flat packages	quad10m36wg1200...
QFN (QFPNL)	Quad flat no lead package	—
QLCCC	Quad leadless ceramic chip carrier (see LCC/LCCS)	—
POT	Adjustable resistor, potentiometer	resadj, pot
SIMM	Single inline memory module	simm050fingerslp38... simm100vstm45
SDIP	Shrink dual inline package	—
SIP	Single inline pins (through hole)	sip6...sip30 siptml2002...siptml120012
SIP SM dual	Single inline pins (dual, surface mount)	sipsmdill47516...sipsmdill67524
SIP SM line	Single inline pins (line, surface mount)	sipsml4758...sipsml67512

(Continued)

Table B.1 (Continued)

Abbreviation	Full name	Footprint examples in library share\pcb\pcb_lib\symbols
SOD	Small outline discrete (or diode)	sod57...sod87, smdsod87, smdsod8712, smdsod8721, dax2sod57, dax2sod64 dax1sod57, dax1sod64
SOIC (SOJ)	Small outline integrated circuit, J-lead	soj20...soj050266wb300l700
SOIC (SOG)	Small outline integrated circuit, gull wing	soic8...soic56w sog30m36wg2020l620... sog05044wg64211150
SON	Small outline nonleaded	—
SOP	Small outline package	sop24, sop44
SOT	Small outline transistor	sot23...sot143 smsot23bce... smsot2231234 mpak, mpak4
SSOT	Shrink small outline transistor	smsc59bec... smsc70123
SQFP	Shrink quad flat package	—
SSOP	Shrink small outline package	ssop14... ssop56
SWITCH	Switch or button	switch, switch6
SM, SMD	Surface mounted (parts or connectors)	telesm310face6x2... telesm420face8x8
TM, PTH	Through hole mounted (parts or connectors)	teletm310face6x2... teletm420face8x8
TO	Transistor outline	to3...to922
TP	Test point	tp...tpprobesocket
TQFP	Thin quad flat package	tqfp32, tqfp100
TSOP	Thin small outline package	tsop26_3...tsop50_4
TSSOP	Thin shrink small outline package	tssop14...tssop24
ZIP	Connector with diagonal pins	zip16...zip64 zigzag100l85016... zigzag100l205040
JUMPER	Jumpers, pins	jumper1...jumper1200
SOCKET	Generic name of connector used to insert daughter PCB or IC	xtsocket quad025132wg1085tms1 plcc68tms1225 pci32bit5vsocket isasocket

(Continued)

Table B.1 (Continued)

Abbreviation	Full name	Footprint examples in library share\pcb\pcb_lib\symbols
FINGERS		xtfingers pci32bit5vfingers isafingers
CONN	Connector, connector with latch	sbusvptm96, sbusvstm96 polcon100rhtm1sqsw4402... polcon156vhtm1sqsw42524 pc104j1p1header pc104j2p2header latcon100rhtm2oesw1100help... latcon100vhtm2oew35064 econ100vstm2anw40024... econ100vstm2sqw400120 dsubhdrptm15... dsubvstm50 conn6, conn9, conn10, conn20, conn26, conn50 blkcon100rhtm1sqsw3502... blkcon156 vhtm1sqsw31224
CONN SM	Connector surface mount	dincminsm3...dincminsm8
CONN TM	Connector through hole mount	db9, db15, db25 dincmintm3...dincstdtm8 din32abr...din96abcv dcon050rsalttm20... dcon085x169vstm50
PRESSFIT	Press-fit connector	fbusrs24...fbusvp264
RFCONN	RF connector, BNC connector	rftncv, rftncr1350 rfssmbv, rfssmbr325 rfsmbv, rfsmbr475 rfsmav, rfsmar725 rfsmar425, rfbncr1350
RELAY	Relays	relay1... relay11
OSC	Oscillators, quartz	osc84p
CONT	Contact field	multicon30, multicon43, ibmcon
MTHOLE	Mounting hole	mthole1, mthole2, mthole3 mtg125, mtg156, mtg250

IC, Integrated circuits; *PCB*, printed circuit board; *PTH*, plated through hole.

Table B.2 List of common discrete component packages.

Name or type	Name, case, or size	Standards	Footprints in library share\pcb \pcb_lib\symbols
Resistor, chip	0402, 0805, 1206, etc.	IEC 60115-B, JIS C 5201-B	smdcap (generic capacitor template) smdres (generic resistor template) smr0402. smc0402 smr0603, smc0603 0603rf_wv_12d (with vias) x2y_0603 (4 pins) smr0805, smc0805 smd0805 (diode—pins C and A) smd080521 (pin 2 marked) smd080512 (pin 1 marked) 0805rf_wv_12d (with vias) smr1206, smc1206 smd1206 (diode—pins C and A) 1206rf_wv_12d (with vias) smd120621 (pin 2 marked) smd120612 (pin 1 marked) smd1406 (diode —pins C and A) smd140621 (pin 2 marked) smd140612 (pin 1 marked) smd2309 (diode—pins C and A) smd230921 (pin 2 marked) smd230912 (pin 1 marked) smr2010, smr2512, sml2220, sml3312, sml1806, sml1614, sml1212, sml1210, sml1206, sml1110, sml0805
Capacitor, tantalum (molded)	A B C D E R T V X Y	EIAJ RC-2134B EIA 3216-18 EIA 3528-21 EIA 6032-28 EIA 7343-31 EIA 7260-38 EIA 2012-12 EIA 3528-12 EIA 7343-20 EIA 7343 EIA 7340	smct3216, smct321612 smct3528, smct352812 smct6032, smct603212 smct7343, smct734312 — — — — — —
Capacitor, polar	Axial Cylinder		cpax575x150031. . . cpax11775x675050 cpcyl1d150ls100031. . . cpcyld725ls325040

(Continued)

Table B.2 (Continued)

Name or type	Name, case, or size	Standards	Footprints in library share\pcb \pcb_lib\symbols
MELF (DL-41, LL-34)	Metal electrode face	EIC 10H01, LL-34	smdmll34 (diode—pins C and A) smdmll3421 (pin 2 marked) smdmll3412 (pin 1 marked) smdmll41 (diode—pins C and A) smdmll4121 (pin 2 marked) smdmll4112 (pin 1 marked)
SOD, SC-76 (molded)	Small outline diode	JEDEC DO215-D, EIAJ SC-76, EIAJ SC-76	daxsod57, daxsod64
SMA (molded)	SMT diode outline	JEDEC DO214-D (variation AC)	—
SMB (molded)	SMT diode outline	JEDEC DO214-D (variation AA)	smsmb
SMC (molded)	SMT diode outline	JEDEC DO214-D (variation AB)	—
DO-213	Diode outline (~MELF)	JEDEC DO213-D	—
DO-214 (see SMA, etc.) (See also SOT, DPAK, D2PAK packages)	Diode outline (molded)	JEDEC DO214-D	—
Inductor, chip	0805, 1206, etc.	See resistor, chip	sml0805, sml1206
Inductor, molded		IMC-2220 (see Vishay Web site)	
Inductor, wire wound	Power SMD inductor	MSS5131 (see Coilcraft website)	—

SMD, Surface-mounted devices; *SMT*, surface-mount technology.

Table B.3 Discrete package (DPAK).

Name	Package outline	JEDEC DWG no.	Variation—pitch
DPAK (TO-252)		TO252-E	AA—0.090 in. (3-lead) AB—0.090 in. (3-lead) AC—0.090 in. (3-lead) AD—0.045 in. (5-lead)

(*Continued*)

Table B.3 (Continued)

Name	Package outline	JEDEC DWG no.	Variation—pitch
D2PAK (TO-263)		TO263-D	AA—0.100 in. (4-lead) AB—0.100 in. (3-lead) BA—0.067 in. (6-lead) BB—0.067 in. (5-lead) CA—0.050 in. (8-lead) CB—0.050 in. (7-lead)
D3PAK (TO-268)		TO268-A	AA—5.45 mm (4-lead)

Table B.4 Some of the small outline transistor (SOT/SSOT/SC) packages.

Name	Package outline	JEDEC DWG No.	Variation—pitch
SOT23-3 (3-lead) SC-59 (3-lead) SSOT-3 (1.9 mm) (3-lead)	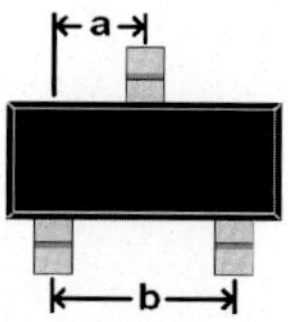	TO-236	AA—0.95 mm (a) AB—0.95 mm (a) —1.90 mm (b)
SOT23-5 (5-lead) EIAJ SC-74A (5-lead) SOT-26 (6-lead) SOT23-8 (8-lead)	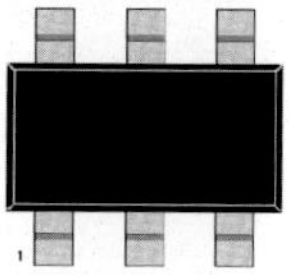	MO178-C MO193-C	AA—0.95 mm (5-lead) AB—0.95 mm (6-lead) BA—0.65 mm (8-lead)
SOT223-3 (4-lead) SOT223-4 (5-lead)	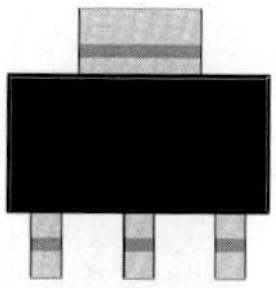	TO261-C	AA—2.30 mm (4-lead) AB—1.50 mm (5-lead)
SOT-89 (2- and 3-lead)		TO243-C	AB—3.0 mm (2-lead) AA—1.5 mm (3-lead)

(*Continued*)

Table B.4 (Continued)

Name	Package outline	JEDEC DWG No.	Variation—pitch
SOT-143 (4-lead) SOT-343 (4-lead)	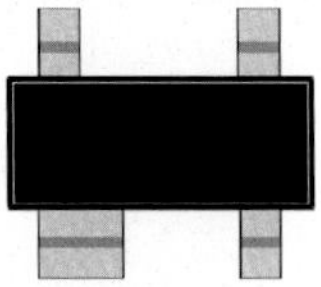	TO253-D (EIAJ SC-61B)	AA—1.92 mm —1.30 mm
SOT-353 (5-lead) SC-88 (5-lead) SC70 (6-lead) SC-74 (8-lead) SSOT-n (5, 6, 8-leads)	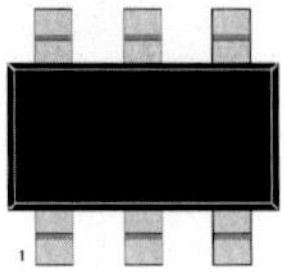	MO059-B MO203-B	AA—0.65 mm (5-lead) AB—0.65 mm (6-lead) BA—0.50 mm (8-lead)

Table B.5 Some of the small outline integrated circuit (SOIC/SOP/SO) packages.

Name	Package outline (no. leads)	JEDEC DWG no.	Variation—pitch
MSOP Body width 2.3 mm, 2.8 mm, 3.0 mm	 (8, 10)	MO187-E	AA—0.65 mm (8-lead) AA−T—0.65 mm (8-lead) DA—0.65 mm (8-lead) CA— 0.50 mm (8-lead) BA—0.50 mm (10-lead) BA−T—0.50 mm (10-lead)
SOIC Narrow body (0.150 in.) (3.8 mm)	 (8, 14, 16)	MS012-E	Width at lead—0.236 in. (6.0 mm) Pitch—0.050 in. (1.27 mm) Lead width—0.016 in. (0.40 mm) Lead spacing—0.034 in. (0.87 mm)
SOIC wide body (0.300 in.) (7.5 mm)	 (14, 16, 18, 20, 24, 28)	MS013-E	Width at lead—0.403 in. (10.3 mm) Pitch—0.050 in. (1.27 mm) Lead width—0.016 in.(0.4 mm) Lead spacing—0.034 in. (0.87 mm)

(Continued)

Table B.5 (Continued)

Name	Package outline (no. leads)	JEDEC DWG no.	Variation—pitch
SSOP narrow body (0.150 in.) (3.8 mm)	 (14, 16, 18, 20, 24, 28)	MO137-C	Width at lead—0.236 in. (6.0 mm) Pitch—0.025 in. (0.635 mm) Lead width—0.010 in. (0.25 mm) Lead spacing—0.015 in. (0.385 mm)
SSOP wide body (0.300 in.) (7.5 mm)	 (28, 48, 56, 64)	MO118-B	Width at lead—0.410 in. (10.4 mm) Pitch—0.025 in. (0.635 mm) Lead width—0.010 in. (0.25 mm) Lead spacing—0.015 in. (0.4 mm)
SOP	(44 - 90...)	MO174-A MO175-A MO180-B	Various package configurations

MSOP, Micro (mini) small outline package; *SOIC*, small outline integrated circuit; *SOP*, small outline package; *SSOP*, shrink small outline package.

Table B.6 Some of the common through-hole packages.

Package	Name	JEDEC standard(s)	Footprints in library share\pcb \pcb_lib\symbols
Axial leads	Resistor outline, diode outline, capacitor outline	TO-5...TO-18, TOP-205	res400...res1000, dio400, dio500 daxdo7...daxdo206ab dax11n746a759a...dax2950x150062 dax2do7...dax2do206ab dax1do7...dax1do206ab cb417, case17-02 ck12-10pf...ck17-10pf cap196...cap1500 capck05...capck62
Radial leads	Capacitor outline		radck05, radck06 rad100x050ls100031... rad800x375ls600034 disc200x100ls200x125034... disc1000x250ls375x275042 dipcap cy10, cy15, cy20

(Continued)

Table B.6 (Continued)

Package	Name	JEDEC standard(s)	Footprints in library share\pcb \pcb_lib\symbols
Cylinder	Capacitor cylinder		cyld150ls100031... cyld725ls325040
Crystal	Quartz/generator		crys11mhz, crys14
DIP	Dual inline package	MS001-D, TO250-A	dipswitch dip4_3...dip68_6 dip100b8w300l450... dip10064w900l3250
	(0.100 in. pitch)	MO043-A	
DO-15 (DO-41 glass/plastic)	Diode outline	DO204B-D	do15, do41
DO-35	Diode outline	DO-204-AH	do35, daxdo35, daxdo204aa... daxdo206ab
IPAK (TO-251)	Transistor outline, flange mount	TO251-D	
I2PAK (TO-262)	Transistor outline, flange mount	TO262-A	
TO-205AF/ TO-39	Transistor outline	TO205-E	
TO-3P/TO-41/ TO-247AD	Transistor outline	TO204-C	
TO-92	Transistor outline	TO226-G	
TO-92; TO-18, lead form STD	Transistor outline	TO226-G, TO206-B	
TO-126	Transistor outline	(Similar to TO-220)	
TO-218AC	Transistor outline	TO218-E	
TO-220 (and variations)	Transistor outline	TO220-K, TO262-A	
TO-226AE	Transistor outline	TO226-G	
TO-247, 2L, 3L	Transistor outline	TO247-E_01	
TO-264	Transistor outline	TO264-B	
	Diode outline, flange mount		r70
	Some more diode outlines		ns_h02a, mot17...mot361a, gi_axial7b gi_axial9, gi_br1w do4...do204ar dax1n746a759a...dax1n58235825 nec_sp
	Some more transistor outlines		

JEDEC, Solid State Technology Association.

Table B.7 List of ball grid array standards.

JEDEC doc.	JEDEC title
MO-151	Elevated to MS-034A. MS-034A was revised and became MS-034B, 2/20/03
MO-156	Square ceramic BGA family, 1.00, 1.27, and 1.50 mm pitch, CBGA
MO-157	Rectangular ceramic BGA family, CBGA
MO158-D	The addition of 47.5, 50.0, 52.5, and 55.0 mm body variations with 1.27 and 1.00 mm ball pitch to column grid array registration
MO163-B	Replaced by MS-028-A
MO-192	Low profile square ball grid array family
MO-195	Thin, fine pitch ball grid array family, 0.5 mm pitch
MO-205	Low profile, fine pitch BGA family, 0.80 mm pitch (rectangular)
MO-207	Square and rectangular die-size, ball grid array family
MO-210	Thin, fine pitch, rectangular ball grid array family, 0.80 mm pitch
MO-211	Die size ball grid array, fine pitch, thin/very thin/extremely thin profile
MO-216	Thin profile, square, and rectangular BGA family, for 1.00 and 0.80 mm pitch
MO-219	Low profile, FBGA registration, 0.80 mm pitch (square and rectangle)
MO-221	Extremely thin, two row cavity down, 0.50 mm pitch BGA family
MO-222	Ceramic BGA rectangular
MO-225	Addition of variations AB and BC to VFBGA
MO-228	Square, dual pitch and FBGA family
MO-233	Mixed pitch (0.80 and 1.00 mm), rectangular die size, fine DSBGA family
MO-234B	Low profile rectangular ball grid array family
MO-237E	DDR2 SDRAM DIMM package with 1.00 mm contact centers
MO-242B	Rectangular die-size, stacked ball grid array family, 0.80 mm pitch
MO-246C	Rectangular, fine pitch, thin ball grid array, 0.65 mm pitch
MO-261A	Thick and very thick, fine pitch, rectangular ball grid array family, 0.80 mm pitch
MO-264A	Rectangular die-size, stacked ball grid array family, dual pitch
MO-266A	Very thin, fine pitch, stackable BGA family, 0.50 mm pitch
MO-273A	Upper pop package, square, fine pitch BGA, 0.65 and 0.50 mm pitch
MO-275-A	Low profile, fine pitch ball grid array family, square
MO-280A	Ultrathin and very, very thin profile, fine pitch BGA family
MO-028-C	Addition of rectangular BGA variations to BGA family
MS-034-D	Elevation of registration MO-151, plastic BGA, 1.0, 1.27, 1.5 mm pitch, with increased ax dimensions to allow for thicker packages

Note: Because there are so many types and variations, only some of the JEDEC standards and titles are listed here. *BGA*, Ball grid array; *FBGA*, fine pitch ball grid array; *DSBGA*, dual pitch ball grid array; *JEDEC*, Solid State Technology Association.

Table B.8 List of quad flat packs standards.

JEDEC doc.	JEDEC title
MO-134-A	CQFP, 0.50 mm lead pitch with ceramic nonconductive tie bar
MO-143-C	Replaced—see MS-029-A
MO-148-A	MCM ceramic quad flatpack family, S-CQFP
MO-188-B	Power PQFP with heat slug

(Continued)

Table B.8 (Continued)

JEDEC doc.	JEDEC title
MO-189-A	Plastic QFP/heat slug (H-LQP/G) 2.00 mm thick/2.00 mm footprint
MO-198-A	3-tier family, PQFP-B
MO-204-B	PQFP outline with exposed heat sink, thermally enhanced PQFPs
MO-220K	Thermally enhanced plastic very thin and very, very thin fine pitch quad flat no lead package
MO-239-B	Thermally enhanced plastic very thin dual row fine pitch quad flat no lead package
MO-241-B	Dual in-line compatible, thermally enhanced, plastic very thin fine pitch, quad flat no lead package family
MO-243-A	Thermally enhanced plastic very thin and very very thin fine pitch bumped quad flat no lead package
MO-247C	Plastic quad no lead staggered multirow packages
MO-248E	Thermally enhanced plastic ultrathin and extremely thin fine pitch quad flat no lead package
MO-250-A	New family of thermally enhanced plastic very thin and very very thin pitch bumped quad flat no lead packages
MO-251-A	Thermally enhanced plastic very thick, quad flat no lead package
MO-254-A	Thermally enhanced plastic low and thin profile fine pitch quad flat no lead package
MS026-D	Standard—low/thin profile plastic quad flat package, 2.00 mm footprint, optional heat
TO271-A	4 lead quad flatpack

CQFP, Ceramic quad flatpack family; *JEDEC*, Solid State Technology Association; *MCM*, multichip module; *PQFP*, plastic quad flatpack.

Table B.9 List of some quad flatpacks—no lead (QFN) standards.

JEDEC doc.	JEDEC title
MO-220K	Thermally enhanced plastic very thin and very, very thin fine pitch quad flat no lead package
MO-241-B	Dual in-line compatible, thermally enhanced, plastic very thin fine pitch, QFN package family, includes addition of very thin profile variations
MO-247C	Plastic quad no lead staggered multirow packages
MO-255-B	Plastic very, very thin, ultrathin, and extremely thin, fine pitch quad flat small outline, nonleaded package family
MO-257-B	Plastic fine pitch quad no lead staggered two row thermally enhanced package family
MO-262A	0.50 mm pitch very thin and very, very thin flange-molded thermally enhanced (topside) QFNs
MO-263A	0.50 and 0.40 mm pitch very thin and very, very thin flange-molded QFNs

(Continued)

Table B.9 (Continued)

JEDEC doc.	JEDEC title
MO–265A	Thermally enhanced plastic very thin fine pitch quad flat no lead package, including corner terminals
MO–267B	Punch-singulated, fine pitch, square, very thin, lead-frame-based quad no lead staggered dual-row QFN package family

JEDEC, Solid State Technology Association.

Table B.10 List of some leadless chip carrier standards.

JEDEC doc.	JEDEC title
MO041-C	0.050 in. center leadless rectangular chip carrier type E, variations AA−AF
MO042-A	0.050 in. center leadless rectangular chip carrier type F
MO044-A	Leaded ceramic chip carrier 0.050 in. center, 68 and 84 terminals, item 11.11−138
MO047-B	PCC family, 0.050 in. lead spacing, square, item 11.11−242
MO052-A	Replaced—see MS-016−A. PCC family 0.050 in. lead spacing, rectangular
MO056-A	Ceramic 0.025 in. center chip carrier
MO057-A	Ceramic 0.020 in. center chip carrier
MO062-A	148 pin leadless ceramic chip carrier, 0.025 in. pitch
MO075-A	0.050 in. center nonhermetic leadless chip carrier quad series, square
MO076-A	0.050 in. center nonhermetic leadless chip carrier SO series, rectangular
MO107-A	Ceramic multilayer leaded chip carrier, 0.050 in. pitch, J−bend leadform, 20 mil min
MO110-A	Round lead, J−form square body, 0.050 in. pitch center ceramic chip carrier
MO111-A	Family of round leads, 0.050 in. pitch, gull−wing lead-form, center ceramic chip carrier
MO126-B	Leadless small outline ceramic chip carrier, 0.400 in. body, 0.050 in. pitch, 28, 32, 26 leads, variations AA−AC
MO129-A	Top brazed ceramic leaded chip carrier (0.020 in. lead pitch) with plastic nonconductive tie bar
MO130-A	Top brazed ceramic leaded chip carrier (0.015 in. lead pitch) with plastic nonconductive tie bar
MO131-A	Top brazed ceramic leaded chip carrier (0.025 in. lead pitch) with plastic nonconductive tie bar
MO144-A	Leadless small outline ceramic chip carrier, 0.350 in. body, 0.050 in. pitch, R-CDCC-N
MO147-A	Small outline J-lead ceramic chip carrier, 0.415 in. body, 0.050 in. lead spacing
MO217-B	Very very thin quad bottom terminal chip carrier family with addition of variations AE, AF, AG, BE, BF, and BG
MS002-A	0.050 in. leadless chip carrier, type A, variations AA−AH
MS003-A	0.050 in. leadless chip carrier, type B, variations BA−BH
MS004-B	0.050 in. center leadless chip carrier, type C, variations CA−CH

(Continued)

Table B.10 (Continued)

JEDEC doc.	JEDEC title
MS005-A	0.050 in. center leadless chip carrier, type D, variations DA–DH
MS006-A	0.050 in. center leaded chip carrier, 24 terminal leaded, type A
MS007-A	0.050 in. center lead chip carrier, type A, variations AA–AH
MS008-A	0.050 in. center lead chip carrier, type B, variations BA–BH
MS009-A	0.040 in. center leadless chip carrier packages, variations AA–AJ
MS014-A	Single-layer chip carrier family, 0.040 in. terminal spacing, ceramic, variations AA–AJ
MS016-A	PCC family, 1.27 mm/0.050 in. lead spacing, rectangular
MS018-A	Square plastic chip carrier family, 1.27 mm/0.050 in. pitch

JEDEC, Solid State Technology Association; *PCC*, plastic chip carrier.

Appendix C: Rise and fall times for various logic families

Logic family		Transition time (ns)		Rank (fastest)
		RT	FT	
BiCMOS	ABT	1.6	1.4	14
	BCT	0.7	0.7	12
	LVT	2.7	2.8	29
CMOS	AC	1.7	1.5	15
	ACT	1.7	1.5	25
	ACQ	2.4	2.4	16
	ACTQ	2.5	2.4	27
	AHCT	2.4	2.4	26
	C	35	25	39
	FCT	1.5	1.2	13
	HC	3.6	4.1	33
	HCT	4.6	3.9	34
	LCX	2.9	2.4	28
	LV	3.0	3.0	30
	LVQ	3.5	3.2	31
	LVX	4.8	3.7	35
	VCX	2.0	2.0	20
	VHC	4.1	3.2	32
ECL	10K	2.2	2.2	22
	10KH	1.7	1.7	17
	100K	0.6	0.6	11
	300K	0.5	0.5	10
	E	0.38	0.38	8
	EP	0.11	0.11	6
	LVEL	0.22	0.22	3
	EL	0.23	0.23	5
	SiGe-2.5V	0.02	0.02	1
	SiGe-3.3V	0.3	0.03	2
GaAs		0.3	0.1	4
LVDS		0.3	0.3	7

(Continued)

(Continued)

Logic family		Transition time (ns)		Rank (fastest)
		RT	FT	
SSTL		0.3	0.5	9
TTL	74nn	8.0	5.0	36
	ALS	2.3	2.3	24
	AS	2.1	1.5	18
	F	2.3	1.7	21
	FR	2.1	1.5	19
	H	7.0	7.0	37
	L	35	30	40
	LS	15	10	38
	S	2.5	2.0	23

FT, Fall time; *RT*, rise time.
Source: Partial data from IPC-2251, Design Guide for the Packaging of High Speed Electronic Circuit (2003. www.ipc.org) and Coombs (2001, Coombs' printed circuits handbook (5th ed.). New York: McGraw-Hill).

Appendix D: Drill and screw dimensions

Table D.1 English sizes.

Screw size no. or diam.	Threads (/in.)	Clearance hole (drill diameter) — Close fit Drill bit gauge	Close fit Min (in.)	Normal fit Drill bit gauge	Normal fit Min (in.)	Loose fit Drill bit gauge	Loose fit Min (in.)	Head diam. typ (in.)	Washer diam. typ (in.)	Nut size (in.)
0	80	No.51	0.067	No.48	0.079	3/32	0.104	0.116	0.188	5/32
1	72 or 64	No.46	0.081	No.43	0.092	No.37	0.114	0.141	0.219	5/32
2	64 or 56	3/32	0.094	No.38	0.105	No.32	0.126	0.167	0.250	3/16
3	56 or 48	No.36	0.106	No.32	0.119	No.30	0.140	0.193	0.312	3/16
4	48 or 40	No.31	0.120	No.30	0.130	No.27	0.156	0.219	0.375	1/4
5	44 or 40	9/64	0.141	5/32	0.160	11/64	0.184	0.245	0.406	1/4
6	40 or 32	No.23	0.154	No.18	0.174	No.13	0.197	0.270	0.438	5/16
8	36 or 32	No.15	0.180	No.9	0.200	No.3	0.225	0.322	0.445	11/32
10	32 or 24	No.5	0.206	No.2	0.225	B	0.250	0.373	0.500	3/8
1/4	28 or 20	17/64	0.266	9/32	0.286	19/64	0.311	0.492	0.625	7/16
5/16	24 or 18	21/64	0.328	11/32	0.349	23/64	0.373	0.615	0.688	9/16
3/8	24 or 16	25/64	0.391	13/32	0.411	27/64	0.438	0.740	0.813	5/8

Table D.2 Metric sizes (mm).

Screw size	Pitch (mm)	Clearance hole fit			Head, nut size	Washer diameter
		Close	Normal	Loose		
M 1.6	0.4	1.7	1.8	2.0	2.9	3.2
M 2.0	0.4	2.2	2.4	2.6	3.6	4.0
M 2.5	0.5	2.7	2.9	3.1	4.5	5.0
M 3.0	0.5	3.2	3.4	3.6	5.4	6.0
M 4.0	0.7	4.3	4.5	4.8	7.2	8.0
M 5.0	0.8	5.3	5.5	5.8	9.0	10.0
M 6.0	1.0	6.4	6.6	7.0	10.8	12.0
M 8.0	1.3	8.4	9.0	10.0	14.4	16.0
M 10	1.5	10.5	11.0	12.0	18.0	20.0

Appendix E: References by subject

Component package types and mounting (surface-mounted device)

IPC-7351B, Section 8.0-16.0, pp. 54–86.

Component placement, spacing, and orientation
General considerations

IPC-1902 (grid resolution considerations).
IPC-2221B, Section 7.2.4, p. 68; Section 8.1, p. 73–78.
IPC-7351B, Section 3.4, p. 31 (SMD).
IPC-AJ-820A. Assembly and Joining Handbook.
IPC-CM-770E, Section 8.1, p. 44.
Coombs, C. F., Jr. (2001). *Coombs' printed circuits handbook* (5th ed.). New York: McGraw-Hill. Section 49.4.

Capacitors/Bypassing/Placement

Avoiding passive-component pitfalls, Doug Grant and Scott Wurcer, Analog Devices Application Note AN-348.
Biasing and decoupling op amps in single supply applications, Charles Kitchin, Analog Devices Application Note AN-581.
Montrose, M. I. (1999). *EMC and the printed circuit board: Design, theory, and layout made simple. IEEE Press series on electronics technology*. New York: IEEE Press. p. 126.
Johnson, H., & Graham, M. (1993). *High-speed digital design: A handbook of black magic* (pp. 281–288). Englewood Cliffs, NJ: Prentice-Hall.
Ott, H. W. (1988). *Noise reduction techniques in electronic systems* (2nd ed.). New York: Wiley. p. 129.
Montrose, M. I. (2000). *Printed circuit board design techniques for EMC compliance* (2nd ed.). *IEEE Press series on electronics technology*. New York: IEEE Press. Chap. 3.

Fiducials

IPC-2221B, Section 5.4-5.7, p. 45–49.
IPC-CM-770E, Section 8.4, p. 46.

Wave soldering considerations

IPC-2221B, Section 8.1.3, p. 74.
IPC-7351B, Section 3.1, pp. 7–12 (SMD footprint design).

Coombs, C. F., Jr. (2001). *Coombs' printed circuits handbook* (5th ed.). New York: McGraw-Hill. Section 43.7.8.

Design rule checking

IPC-CM-770E, Section 7.2, p. 35.
Coombs, C. F., Jr. (2001). *Coombs' printed circuits handbook* (5th ed.). New York: McGraw-Hill. Section 19.5.1.

Land patterns (footprint design)
Surface-mounted device

IPC-7351B, Section 3.1, pp. 7–12 (SMD footprint design); Table 3-23, p. 24 (IPC SMD naming convention); Land Pattern Viewer (free demo version from PCB Libraries Inc. Web site).
IPC-SM-780, Section 6.7.2, p. 66.

Through-hole

(see the "Lead relationships" section)

Layer stack-ups

Advanced circuits web site (www.4pcb.com, search for "controlled impedance").
Coombs, C. F., Jr. (2001). *Coombs' printed circuits handbook* (5th ed.). New York: McGraw-Hill. Chap. 13, Fig. 13.9; Chap. 27.
DP83865 Gig PHYTER® V10/100/1000 Ethernet Physical Layer Design Guide, National Semiconductor application notes, August 2003.
Johnson, H., & Graham, M. (1993). *High-speed digital design: A handbook of black magic* (pp. 212, 217–221). Englewood Cliffs, NJ: Prentice-Hall. Chap. 5.
Ellwest Printed Circuit Boards Web site (http://www.ellwest-pcb.com, Ideas & Solutions).
Brooks, D. (2003). *Signal integrity issues and printed circuit board design. Prentice Hall modern semiconductor design series.* p. 308.

Lead relationships
Lead-to-hole ratio

IPC-AJ-820A, Section 2.5.2, p. 2-11.
IPC-SM-780, Section 8.6.1.5, p. 103.
Coombs, C. F., Jr. (2001). *Coombs' printed circuits handbook* (5th ed.). New York: McGraw-Hill. Section 42.2.1.

Lead bend radius and spacing

IPC-2221B, Fig. 8-10, p. 79.
IPC-AJ-820A, Section 2.5.1.4, p. 2-11; Section 2.5.3.1, p. 2-13.
IPC-CM-770E, Section 11.1.8, p. 67.
IPC-SM-780, Section 8.4.2, p. 94.

Materials

Copper

IPC-2221B, Section 4.4, p. 26; Table 4-2, p. 26.
IPC-4101E.

Prepreg/Laminate

IPC-4101E.
Coombs, C. F., Jr. (2001). *Coombs' printed circuits handbook* (5th ed.). New York: McGraw-Hill. Chaps. 8, 10, Section 10.3.

Plating

IPC-2221B, Tables 10-1, 10-2, p. 103.
Coombs, C. F., Jr. (2001). *Coombs' printed circuits handbook* (5th ed.). New York: McGraw-Hill. Sections 28.5, 29.4; Chaps. 30, 31.

Solder

(see the "Soldering and assembly" section)

Dielectric (relative permittivity) properties

IPC-2221B, Table 4-1, p. 23; Table 6-2, p. 60.
IPC-4101E.
Montrose, M. I. (1999). *EMC and the printed circuit board: Design, theory, and layout made simple. IEEE Press series on electronics technology*. New York: IEEE Press. Table 6-2.
Ott, H. W. (1988). *Noise reduction techniques in electronic systems* (2nd ed.). New York: Wiley. Table 4-1, p. 125.
Montrose, M. I. (2000). *Printed circuit board design techniques for EMC compliance* (2nd ed.). *IEEE Press series on electronics technology*. New York: IEEE Press. p. 111.
Bogatin, E. (2004). *Signal integrity—Simplified*. Upper Saddle River, NJ: Prentice Hall. p. 349.

Mounting hardware and specifications

ASME B1-10M Unified Miniature Screw Threads.
ASME B18.2.4.1M-2002 Metric Hex Nuts—Style 1.
ASME B18.2.4.2M-2005 Metric Hex Nuts—Style 2.
ASME B18.2.8-1999, Clearance Holes for Bolts, Screws, and Studs.
ASME B18.6.3-2003 Machine Screws and Machine Screw Nuts (inch series).
ASME B18.6.7M-2000 Machine Screws (metric series).
ASME B18.12-2001 Glossary of Terms for Mechanical Fasteners.
ASME B18.13-1996 Screw and Washer Assemblies (inch series).
ASME B18.13.1M-1999 Screw and Washer Assemblies (metric series).
ASME B18.21.1-1999 Lock Washers (inch series).
ASME B18-21-2M-1999 Lock Washers (metric series).
ASME B18.22.1-1981 Plain Washers (inch series).
ASME B18.22M-1981 Plain Washers (metric series).
ASME B94.11M-1993 Twist Drills (drill bit sizes—Inch and metric).

PCB

Fabrication

Coombs, C. F., Jr. (2001). *Coombs' printed circuits handbook* (5th ed.). New York: McGraw-Hill. Chaps. 13, 14.

Fabrication allowances

IPC-7351B, Section 3.1, p. 7—12 (SMD footprint design); Tables 3-25, 3-26; Fig. 3-20, p. 38.
IPC-CM-770E, Tables 8-1, 8-2, p. 50.
Coombs, C. F., Jr. (2001). *Coombs' printed circuits handbook* (5th ed.). New York: McGraw-Hill. Section 17.1.

Borders

IPC-2222A, Section 5.2.1.

General design information

ANSI/IPC-D-322 (standard panel sizes).
IPC-2615 (PCB dimensions and tolerances).
IPC-AJ-820A, Section 2.0, pp. 2-2—2-8.
Holden, H. T. *DFM in PWB Fab: A review of predictive engineering benefits.* Loveland, CO: Hewlett Packard Co.

PCB layout application notes, Maxim, Dallas, Web site: www.maxim-ic.com/ appnotes10.cfm/ac_pk/35/ln/en.

Holden, H. *Planning PCB design: For fun and profit.* Evergreen, CO: TechLead Corp.

Performance classes

IPC-2221B, Section 1.6.2, p. 2.
IPC-7351B, Section 1.3, p. 2.
IPC-CM-770E, Section 1.2.1, p. 1.

Assembly types

IPC-2222A, Section 1.6.
IPC-AJ-820A, Section 1.3.2, Fig. 1-1, p. 1-5.
IPC-CM-770E, Sections 1.2.2, 1.2.4, p. 1.

Density levels

IPC-7351B, Section 1.4, p. 2.

Producibility levels

IPC-2221B, Section 1.6.3, p. 2.
IPC-7351B, Section 1.3.1, p. 2.
IPC-CM-770E, Section 1.2.2, p. 1.

Power distribution concepts

IPC-2221B, Figs. 6-1, 6-2, 6-3, p. 54.

Sizes/Panels

IPC-2221B, Fig. 3-5, p. 18; Fig. 5-1, p. 41.

Testing

IPC-2221B, Sections 36, p. 10.

Plated through holes
Types: Through, blind, buried, mounting holes

IPC-2221B, Section 9.2, p. 98.
IPC-7351B, Section 1.5, p. 3.

Fabrication of through holes

Coombs, C. F., Jr. (2001). *Coombs' printed circuits handbook* (5th ed.). New York: McGraw–Hill. Chap. 11, Sections 21.2, 22.4.3.

Aspect ratio

IPC-2221B, Table 4–3, 4-4, 4-5, p. 27.
IPC-2222A, Table 9-6, p. 26.

Lead-to-hole ratio

Coombs, C. F., Jr. (2001). *Coombs' printed circuits handbook* (5th ed.). New York: McGraw–Hill. Section 42.2.1.
(see also the "Lead relationships" section)

Hole-to-land ratio

IPC-2221B, Table 4-2, p. 26.

Land requirements

IPC-2221B, Section 9.1.1, p. 95.

Annular ring requirements

IPC-2221B, Section 9.1.2, p. 96.
IPC-7351B, Section 3.4.7, p. 37.
Coombs, C. F., Jr. (2001). *Coombs' printed circuits handbook* (5th ed.). New York: McGraw–Hill. Sections 42.2.2, 48.8.1.

Clearance from planes

IPC-2221B, Section 6.3.1, Table 6-1 (column B1), p. 57.
IPC-2222A, Section 9.1.3, p. 22.

Nonfunctional lands

IPC-2222A, Section 9.1.4, p. 23.

Unsupported holes

IPC-2222A, Section 9.2.1, p. 25.

Fan-outs

IPC-SM-780, Section 6.7.1.1, p. 64.

Via (through-hole) guidelines and fabrication

IPC-7351B, Section 3.4.6, p. 35.
IPC-CM-770E, Section 8.6, p. 47.
Coombs, C. F., Jr. (2001). *Coombs' printed circuits handbook* (5th ed.). New York: McGraw-Hill. Section 23.8.

Filled/Plugged holes

IPC-2222A, Section 9.2.2.3, p. 25.

Rise and fall times of logic families

IPC-2251, Table 5-4, p. 29.
Coombs, C. F., Jr. (2001). *Coombs' printed circuits handbook* (5th ed.). New York: McGraw-Hill. Chap. 13, Table 13.2.
Montrose, M. I. (1999). *EMC and the printed circuit board: Design, theory, and layout made simple. IEEE Press series on electronics technology.* New York: IEEE Press. Table 3-1.
Ott, H. W. (1988). *Noise reduction techniques in electronic systems* (2nd ed.). New York: Wiley. p. 275.
(see also Appendix C)

Schematics
Symbols

ANSI 315-1975.

Diagrams

IPC-2221B, Section 3.3, p. 9.

Signal integrity

A practical guide to high-speed printed-circuit-board layout, John Ardizzoni, Analog Devices: Analog Dialogue, applications note.
Analysis on the effectiveness of high-speed printed circuit board edge radiated emissions based on stimulus source location, Mark I. Montrose, Jin Hong-Fang, Er-Ping Li.
Analysis on the effectiveness of high-speed printed circuit board edge termination using discrete components instead of implementing the 20-H rule, Mark I. Montrose, Jin Hong-Fang, Er-Ping Li.

Combating noise effects in adaptive cable equalizer designs, James Mears, National Semiconductor applications note.

Montrose, M. I. (1999). *EMC and the printed circuit board: Design, theory, and layout made simple. IEEE Press series on electronics technology.* New York: IEEE Press.

Johnson, H., & Graham, M. (1993). *High-speed digital design: A handbook of black magic.* Englewood Cliffs, NJ: Prentice-Hall.

Montrose, M. I. (2000). *Printed circuit board design techniques for EMC compliance (2nd ed.). IEEE Press series on electronics technology.* New York: IEEE Press.

QPI-AN1, PICOR general application note on QPI family.

Brooks, D. (2003). *Signal integrity issues and printed circuit board design. Prentice Hall modern semiconductor design series.* Upper Saddle River, NJ: Prentice Hall.

Cross talk

Montrose, M. I. (1999). *EMC and the printed circuit board: Design, theory, and layout made simple. IEEE Press series on electronics technology.* New York: IEEE Press. Chap. 7.

Johnson, H., & Graham, M. (1993). *High-speed digital design: A handbook of black magic.* Englewood Cliffs, NJ: Prentice-Hall. p. 191.

Montrose, M. I. (2000). *Printed circuit board design techniques for EMC compliance (2nd ed.). IEEE Press series on electronics technology.* New York: IEEE Press. p. 131.

Ground bounce/rail collapse

Bogatin, E. (2004). *Signal integrity—Simplified* (pp. 163, 249). Upper Saddle River, NJ: Prentice Hall.

Propagation speed/time delay

Johnson, H., & Graham, M. (1993). *High-speed digital design: A handbook of black magic.* Prentice-Hall. p. 178.

Bogatin, E. (2004). *Signal integrity—Simplified* (pp. 210–211, 214–216, 255). Upper Saddle River, NJ: Prentice Hall.

Reflections/Ringing

Montrose, M. I. (1999). *EMC and the printed circuit board: Design, theory, and layout made simple* (pp. 188–194). *IEEE Press series on electronics technology.* New York: IEEE Press.

Johnson, H., & Graham, M. (1993). *High-speed digital design: A handbook of black magic.* Englewood Cliffs, NJ: Prentice-Hall. p. 160.

Brooks, D. (2003). *Signal integrity issues and printed circuit board design. Prentice Hall modern semiconductor design series.* Upper Saddle River, NJ: Prentice Hall. Chap. 10.

Guard rings and traces

Montrose, M. I. (2000). *Printed circuit board design techniques for EMC compliance: A handbook for designers* (2nd ed.) (pp. 137–141). New York: IEEE Press.

Bogatin, E. (2004). *Signal integrity—Simplified* (pp. 452–459). Upper Saddle River, NJ: Prentice Hall.

Brooks, D. (2003). *Signal integrity issues and printed circuit board design.* Upper Saddle River, NJ: Pearson Education. p. 233.

Coombs, C. F., Jr. (2001). *Coombs' printed circuits handbook* (5th ed.). New York: McGraw-Hill. p. 15.6.

Johnson, H., & Graham, M. (1993). *High-speed digital design: A handbook of black magic* (pp. 201–204, 262). Englewood Cliffs, NJ: Prentice-Hall.

Ott, H. W. (1988). *Noise reduction techniques in electronic systems* (2nd ed.). New York: Wiley. p. 343.

Rich, A. Shielding and guarding, how to exclude interference-type noise, what to do and why to do it—A rational approach. Analog Devices Application Note AN-347: https://www.analog.com/en/search.html?q=AN-347.

Barnes, J. R. Designing electronic systems for ESD immunity. Conformity®: February 2003 (0302designing.pdf).

Terminations

Montrose, M. I. (1999). *EMC and the printed circuit board: Design, theory, and layout made simple. IEEE Press series on electronics technology.* New York: IEEE Press. p. 217.

Johnson, H., & Graham, M. (1993). *High-speed digital design: A handbook of black magic.* Englewood Cliffs, NJ: Prentice-Hall. Chap. 6.

Montrose, M. I. (2000). *Printed circuit board design techniques for EMC compliance* (2nd ed.). *IEEE Press series on electronics technology.* New York: IEEE Press. p. 143.

Ground areas (split planes, moated grounds)

Montrose, M. I. (1999). *EMC and the printed circuit board: Design, theory, and layout made simple* (pp. 107–111, 119). *IEEE Press series on electronics technology.* New York: IEEE Press.

Montrose, M. I. (2000). *Printed circuit board design techniques for EMC compliance* (2nd ed.) (pp. 152–170). *IEEE Press series on electronics technology.* New York: IEEE Press.

Brooks, D. (2003). *Signal integrity issues and printed circuit board design. Prentice Hall modern semiconductor design series.* Upper Saddle River, NJ: Prentice Hall. Chap. 16.

Guard shields

Coombs, C. F., Jr. (2001). *Coombs' printed circuits handbook* (5th ed.). New York: McGraw-Hill. Chap. 17, Section 17.2.3.2.

Johnson, H., & Graham, M. (1993). *High-speed digital design: A handbook of black magic.* Englewood Cliffs, NJ: Prentice-Hall. p. 201.

Ott, H. W. (1988). *Noise reduction techniques in electronic systems* (2nd ed.). New York: Wiley. p. 106.

Montrose, M. I. (2000). *Printed circuit board design techniques for EMC compliance* (2nd ed.) (pp. 137, 141). *IEEE Press series on electronics technology.* New York: IEEE Press.

Shielding and Guarding, Alan Rich, Analog Devices Application Note AN-347.

Noise reduction (general)

An applications guide for op amps, National Semiconductor Application Note AN-20, February 1969.

An IC amplifier user's guide to decoupling, grounding, and making things go right for a change, Paul Brokaw, Analog Devices Application Note AN-202.

Common mode ground currents, *High-Speed Digital Design, Online Newsletter* Vol. 7, Issue 02, Sigcon.

Montrose, M. I. (1999). *EMC and the printed circuit board: Design, theory, and layout made simple. IEEE Press series on electronics technology.* New York: IEEE Press. p. 41.

Grounding for low- and high-frequency circuits, Paul Brokaw and Jeff Barrow, Analog Devices Application Note AN-345.

Grounding rules for high-speed circuits, Don Brockman and Arnold Williams, Analog Devices Application Note AN-214.

Ott, H. W. (1988). *Noise reduction techniques in electronic systems* (2nd ed.). New York: Wiley. Chap. 6 (shielding).

Op amp selection guide for optimum noise performance, Glen Brisebois, Linear Technologies Design Ideas, *Linear Technology Magazine,* December 2005.

PCB design guidelines that maximize the performance of TVS diodes, Jim Lepkowske, ON Semiconductor application note AND8232/D.

Power supply effects on noise performance, Nicholas Gray, National Semiconductor application note AN-1261, September 2002.

Power supply ripple rejection and linear regulators: What's all the noise about? Gabriel A. Rincón-Mora and Vishal Gupta, Texas Instruments Planet Analog application note (https://pdfs.semanticscholar.org/95d9/d3c47f4a72247ede4788a02430f796df95a1.pdf).

Shielding and guarding: How to exclude interference-type noise, Alan Rich, Analog Devices Application Note AN-347.

EMI/EMC

Coombs, C. F., Jr. (2001). *Coombs' printed circuits handbook* (5th ed.). New York: McGraw-Hill. Section 17.2.3.2.

Montrose, M. I. (1999). *EMC and the printed circuit board: Design, theory, and layout made simple. IEEE Press series on electronics technology.* New York: IEEE Press.

Montrose, M. I. (2000). *Printed circuit board design techniques for EMC compliance* (2nd ed.). *IEEE Press series on electronics technology.* New York: IEEE Press.

Brooks, D. (2003). *Signal integrity issues and printed circuit board design. Prentice Hall modern semiconductor design series.* Upper Saddle River, NJ: Prentice Hall. Chaps. 8, 9.

Soldering and assembly

IPC J-STD-001G (soldering requirements).

IPC-SM-780, Section 8.1, p. 84.

Assembly issues and processes

IPC-2221B, Section 8, p. 72.

IPC-7351B, Section 7.4, Fig. 7-1, p. 50.

IPC-CM-770E, Sections 6.3, 7.1, p. 37.

IPC/WHMA-A-620C.

Coombs, C. F., Jr. (2001). *Coombs' printed circuits handbook* (5th ed.). New York: McGraw-Hill. Chap. 13, Section 13.3; Chaps. 41, 42, 43.

Reflow soldering

IPC-CM-770E, Sections 6.2.2.3, 6.2.2.4, p. 35.

Coombs, C. F., Jr. (2001). *Coombs' printed circuits handbook* (5th ed.). New York: McGraw-Hill. Section 43.6.

Wave soldering

IPC-CM-770E, Section 6.2.2.1, p. 35, and Section 8.1.2.

Coombs, C. F., Jr. (2001). *Coombs' printed circuits handbook* (5th ed.). New York: McGraw-Hill. Section 43.7.

Components, placing/spacing

IPC-2221B, Section 8.1.2, p. 73.

Components, orientation

IPC-2221B, Fig. 8-1, p. 75.

Components, mounting

IPC-2221B, Section 8.1.9, p. 76.

Reliability

Coombs, C. F., Jr. (2001). *Coombs' printed circuits handbook* (5th ed.). New York: McGraw-Hill. Section 53.3.

Thermal management

IPC-2221B, Section 7, p. 65; Fig. 7-1 (heat sink and spacing), p. 68.

ADN2530—11.3 Gbps, active back-termination, differential VCSEL driver, Analog Devices data sheet (333411103ADN2530.pdf).

Analytical modeling of thermal resistance in bolted joints, S. Lee, M. M. Yovanovich, S. Song, and K. P. Moran, HTD-Vol. 263, Enhanced cooling techniques for electronics applications ASME 1993.

Evaluation of power MOSFET thermal solutions for desktop and mobile processor power, Tim McDonald and John Ambrus, International Rectifier Corp.

How to select a heat sink, Seri Lee, Aavid Thermal Technologies, Inc., application note. JEDEC Standard 51.

LFPAK thermal design guide—LFPAK thermal performance junction to ambient, Philips application note, Rev. 01, June 16, 2003.

LT1806/LT1807 data sheet (18067f[1].pdf), Linear Technology Corp.

Monitor heat dissipation in electronic systems by measuring active component die temperature, Maxim, Dallas, Semiconductor application note 3500, March 18, 2005.

Package thermal characteristics, Actel Corp. application note, February 2005.

Thermal considerations of QFN and other exposed-paddle packages, Maxim Integrated Products application note HFAN-08.1, November 2001 (4hfan081.doc 11/20/2001).

Thermal design considerations, Philips Semiconductors IC Packages, Chap. 6 (IC26_CHAPTER_6_2000.pdf).

Thermal performance of national's LLP package, highlighting the LM2750 regulated voltage converter, Brian J. Conner, National Semiconductor application brief 111.

Thermal Resistance v PCB Area Comparison, Zetex application note.

Trace routing

Trace spacing (voltage withstanding)

IPC-2221B, Table 6-1 (trace to trace), p. 57; Fig. 8-4 (trace to mounting hardware), p. 76.

IPC-AJ-820A, Fig. 2-12 (trace to mounting hardware), p. 2-13.

Coombs, C. F., Jr. (2001). *Coombs' printed circuits handbook* (5th ed.). New York: McGraw-Hill. Section 49.3.2.

Trace spacing (cross talk/3-W rule)

Montrose, M. I. (1999). *EMC and the printed circuit board: Design, theory, and layout made simple. IEEE Press series on electronics technology.* New York: IEEE Press. p. 207.
Johnson, H., & Graham, M. (1993). *High-speed digital design: A handbook of black magic* (pp. 215, 219). Englewood Cliffs, NJ: Prentice-Hall.
Montrose, M. I. (2000). *Printed circuit board design techniques for EMC compliance* (2nd ed.) (pp. 131, 136). *IEEE Press series on electronics technology.* New York: IEEE Press.
Bogatin, E. (2004). *Signal integrity—Simplified* (pp. 421–425). Upper Saddle River, NJ: Prentice Hall.

Trace width—current handling

IPC-2221B, Section 6.2, p. 56.

Trace width—transmission lines

IPC-2221B, Fig. 6-4, p. 59.
IPC-2251, Section 5.5–5.5.5.4, pp. 31–36.
Johnson, H., & Graham, M. (1993). *High-speed digital design: A handbook of black magic.* Englewood Cliffs, NJ: Prentice-Hall. p. 214.

Microstrip

IPC-2221B, Section 6.4.1, p. 59.

Embedded microstrip

IPC-2221B, Section 6.4.2, p. 60.

Stripline

IPC-2221B, Section 6.4.3, p. 61.

Asymmetric stripline

IPC-2221B, Section 6.4.4, p. 61.
(see also the "Transmission lines" section)

Transmission lines
Characteristic impedance (microstrip, stripline topologies, etc.)

IPC-2141A, Sections 4.1–4.7, pp. 15–27.
IPC-2251, Sections 5.5–5.5.5.4, pp. 31–36.
IPC-2221B, Fig. 6-4, p. 59.

Coombs, C. F., Jr. (2001). *Coombs' printed circuits handbook* (5th ed.). New York: McGraw-Hill. Chap. 16.

Montrose, M. I. (1999). *EMC and the printed circuit board: Design, theory, and layout made simple* (pp. 171–179). *IEEE Press series on electronics technology*. New York: IEEE Press.

Johnson, H., & Graham, M. (1993). *High-speed digital design: A handbook of black magic* (pp. 186–188). Englewood Cliffs, NJ: Prentice-Hall.

Montrose, M. I. (2000). *Printed circuit board design techniques for EMC compliance* (2nd ed.) (pp. 101–190). *IEEE Press series on electronics technology*. New York: IEEE Press.

Brooks, D. (2003). *Signal integrity issues and printed circuit board design* (pp. 185, 203, 265). *Prentice Hall modern semiconductor design series*. Upper Saddle River, NJ: Prentice Hall.

Bogatin, E. (2004). *Signal integrity—Simplified* (pp. 220–233, 258–260, 274). Upper Saddle River, NJ: Prentice Hall.

(see also the "Trace routing" section)

Critical length

Montrose, M. I. (1999). *EMC and the printed circuit board: Design, theory, and layout made simple. IEEE Press series on electronics technology*. New York: IEEE Press. p. 195.

Johnson, H., & Graham, M. (1993). *High-speed digital design: A handbook of black magic*. Englewood Cliffs, NJ: Prentice-Hall. p. 166.

Ott, H. W. (1988). *Noise reduction techniques in electronic systems* (2nd ed.). New York: Wiley. p. 308.

Bogatin, E. (2004). *Signal integrity—Simplified*. Upper Saddle River, NJ: Prentice Hall. p. 297.

Propagation delay time

IPC-2141A, Section 3.4.7, p. 9.

IPC-2251, Section 5.4, p. 30.

Montrose, M. I. (2000). *Printed circuit board design techniques for EMC compliance* (2nd ed.). *IEEE Press series on electronics technology*. New York: IEEE Press. p. 111.

Bogatin, E. (2004). *Signal integrity—Simplified* (pp. 214–216). Upper Saddle River, NJ: Prentice Hall.

Reflections/Impedance mismatches

IPC-2251, Section 5.6.2, p. 36.

Coombs, C. F., Jr. (2001). *Coombs' printed circuits handbook* (5th ed.). New York: McGraw-Hill. Chap. 16.

Ott, H. W. (1988). *Noise reduction techniques in electronic systems* (2nd ed.). New York: Wiley. p. 169.

Bogatin, E. (2004). *Signal integrity—Simplified* (pp. 278–283, 302). Upper Saddle River, NJ: Prentice Hall.

Sharp corners

Johnson, H., & Graham, M. (1993). *High-speed digital design: A handbook of black magic.* Englewood Cliffs, NJ: Prentice-Hall. p. 174.

Bogatin, E. (2004). *Signal integrity—Simplified.* Upper Saddle River, NJ: Prentice Hall. p. 315.

Index